# Professional Locksmithing Techniques

Bill Phillips

**TAB BOOKS**
Blue Ridge Summit, PA

*To my sons, Daniel and Michael*

FIRST EDITION
FIRST PRINTING

© 1991 by **Bill Phillips**
Published by TAB BOOKS
TAB BOOKS is a division of McGraw-Hill, Inc.

**Library of Congress Cataloging-in-Publication Data**

Phillips, Bill, 1960—
    Professional locksmithing techniques / by Bill Phillips.
      p.   cm.
    Includes index.
    ISBN 0-8306-7523-X    ISBN 0-8306-3523-8 (p)
    1. Locksmithing.    I. Title.
TS520.P55  1990
683'.3—dc20                  90-47231
                                    CIP

TAB BOOKS offers software for sale. For information and a catalog, please contact TAB Software Department, Blue Ridge Summit, PA 17294-0850.

Questions regarding the content of this book should be addressed to:

**Reader Inquiry Branch**
**TAB BOOKS**
**Blue Ridge Summit, PA 17294-0214**

Acquisitions Editor: Kimberly Tabor
Book Editor: Cherie R. Blazer
Production: Katherine G. Brown
Book Design: Jaclyn J. Boone

# *Contents*

# Acknowledgments

$W$riting this book wasn't a solo effort. I received assistance from many companies and individuals. Much of it was in the form of technical information, photographs, drawings, advice, and encouragement. Without such help, this book could not have been written—at least not by me.

Although I am grateful to everyone who offered assistance, I'd like to give special thanks to: A-1 Security Manufacturing Corp.; Alarm Lock Systems, Inc.; Associated Locksmiths of America; Cherie R. Blazer; Dom Security Locks; ESP Corporation; Framon Manufacturing Co. Inc.; Gloria Glenn; Anthony "A. J." Hoffman, CML; Ilco Unican Corporation; Lori Corporation; Master Lock Company; Medeco Security Locks; Bert Michaels; Schlage Lock Company; Simplex Access Controls Corporation; Slide Lock Tool Company; and Merlynn J. Smith, and Kimberly Tabor.

# Introduction

*T*his book is written for the do-it-yourselfers who want to save money and for people who want to work as locksmiths. It is also for security professionals who want to learn about the latest developments in physical and electronic security. It is designed to be used by those who have had no prior knowledge of locksmithing.

The following pages contain practical information about nearly every type of locking device in use today. The locking devices reviewed in this book range from those commonly found on homes to those used in banks, U.S. Mint buildings, and the White House. Included are explanations on how the devices are constructed and step-by-step instructions for installing, servicing, buying, and selling them.

The do-it-yourselfer might be particularly interested in the sections on types of locks and keys, locksmithing tools (how to buy and make them), automobile opening techniques, automotive lock servicing, forced entry techniques, special door hardware, and home and office security.

For individuals planning to begin a career in locksmithing, every chapter of this book can be helpful. View the book as a course and follow a study schedule. Especially helpful is the information on how to locate prospective employers, how to get hired as a locksmith, how to start a successful locksmithing business, the pros and cons of joining a locksmith association, laws, ethical issues, locksmith licensing, and locksmith certifications.

The security professional will be particularly interested in the chapters on high-security cylinders; pushbutton combination locks, lock picking, key impressioning, master keying; special door hardware; emergency exit door devices; basic electricity for locksmiths, electromagnet locks, and electric strikes; and automotive lock servicing.

Extensive research was done to make this book current and complete. The information in it is based on the following sources: interviews with noted physical and electronic security experts, technical service manuals and trade journals, product catalogs from manufacturers and distributors of locking devices, and questionnaires sent to locksmithing schools and shops throughout the United States and Canada.

After you've read this book, I'd like to hear any questions or comments you have about it. Let me know how it has been helpful to you and your ideas about how it can be improved. Send your letter to: Bill Phillips, c/o TAB BOOKS, Blue Ridge Summit, PA 17294-0850.

*Chapter* **1**

# The Locksmithing trade

"*E*xperienced locksmith wanted. Relocate to beautiful southwest Florida." "Locksmiths wanted . . . Work in Las Vegas." "Experienced locksmiths wanted . . . in several major metropolitan areas . . . We offer our employees: health, life, and long term disability insurance, pension plan, paid vacation and holidays, tuition reimbursement. . . ."

These are excerpts from advertisements that appeared in a locksmithing trade journal two months before this book was published. Similar ads are regularly in such journals. They also appear in local newspapers every day. There have never been more opportunities for locksmiths than there are right now.

As stated in the seventh edition of the *Encyclopedia of Vocational Guidance* (J. G. Ferguson Publishing Company, 1987): "The locksmith employment outlook is excellent for the next decade. . . . The occupation itself has remained a fairly stable one, and locksmiths with an extensive knowledge of their trade need rarely be unemployed."

## RECENT CHANGES TO THE TRADE

In the past, a locksmith was basically a hybrid of carpenter and mechanic. The locksmith's work was limited mainly to installing mechanical locksets, opening locked car doors, rekeying mechanical locks, and duplicating keys. As more hardware and department stores began doing those jobs, locksmiths responded by offering more services.

In addition to the usual services, locksmiths now sell and install a wide array of sophisticated security devices (FIG. 1-1). They also offer a wide range of security advice. Locksmithing has become a specialty trade that relies heavily on education.

**I-I**  In addition to locks, todays locksmith installs sophisticated security devices.

A person planning to become a locksmith should be able to use most common hand and power tools, be mechanically inclined, be a good reader, and have no serious criminal record. Because many locksmithing jobs require the locksmith to drive, a good driving record can also be helpful.

## EMPLOYMENT OPTIONS

Some of the work opportunities for locksmiths include owning or working in a locksmithing shop, working as an in-house locksmith for a private or public organization, working for a manufacturer or distributor, teaching locksmithing, and designing security-related products.

### Owning a shop

Successful owners of locksmithing shops usually have good business acumen and a broad base of locksmithing skills. Many small shops are one-man or family operations, with the owner wearing all the hats. Other locksmithing shops have 30 or more full-time employees, and the owners never have to go out on calls. The number of employees a shop has isn't as important to its success as the range and quality of work the shop can perform.

There are two basic types of locksmithing shops: store front and mobile. A store front shop is one that is operated from a building customers can walk into. A mobile shop is operated from a vehicle, usually a van or truck. Mobile shops always go to the customer to perform services. Most store front shops use vehicles to allow them to offer both in-store and mobile services (FIG. 1-2).

This flexibility gives the store front more money-making opportunities than the mobile shop. A building gives a business a more professional and stable image, which helps in obtaining work. A building also provides a place to display a variety of merchandise. However, a store front shop is more expensive to start and operate than a mobile shop. Rent, utility bills,

Roy's Lock Shop

**1-2** Using both a storefront and a service vehicle allows a locksmith to sell merchandise and offer out-of-shop services.

merchandise, and additional equipment are a few of the extra expenses a store front shop has.

Low start-up and operating costs are two reasons many locksmiths prefer operating a mobile shop. Another is that a mobile shop owner doesn't need a broad range of locksmithing skills, because mobile shops generally offer only a few services.

### Working in a shop

Although there are a few notable exceptions, most mobile shops don't hire employees. When mobile shop owners need help they usually sub-contract work to other locksmiths. This allows mobile shop owners to avoid the cost of employee compensation insurance, employee benefits, extra record keeping, etc.

A locksmith who works for a small to mid-size shop has varied tasks daily. On any given day he or she may go out to install several locksets, open a car door, and change a safe combination, then go back to the shop to rekey some locks and wait on customers.

Some large shops assign specific duties to their locksmiths. For example, one may be assigned to only wait on customers, another to just service locking devices in the shop, and another to go out on calls. Sometimes the jobs are even more specific—such as only servicing safes.

Locksmiths who work in a shop (excluding owners) usually earn an hourly wage plus extra pay for night and weekend work. Some hard-working locksmiths earn more money working overtime than they make as their base pay. A few also earn commissions for selling merchandise.

### In-house locksmiths

Universities, school systems, hotels, and cities are major employers of in-house locksmiths. Competition is fierce for in-house jobs, because they usually offer good pay, job security, and a controlled work environment.

Most locksmiths who work in-house once worked for a locksmithing shop. In-house locksmiths often have a broad range of locksmithing skills, but rarely have to use most of those skills. Their work is usually limited to installing and servicing a few types and brands of security devices and hardware. In-house locksmiths don't ordinarily have to sell merchandise.

## Working for a manufacturer or distributor

Manufacturers and distributors of security products and supplies are good sources of employment for many locksmiths (FIG. 1-3). They hire locksmiths to stock products, sell merchandise, help develop new products, and conduct seminars. Sometimes apprentice locksmiths are hired to sell merchandise or stock products.

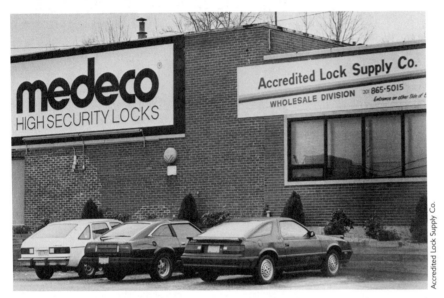

**1-3**   Manufacturers and distributors of locksmithing products are good sources of employment for locksmiths.

## Instructors

Locksmithing instructors usually have at least 5 years of locksmithing experience. Some schools require their instructors to also have at least 120 hours of teacher training.

Manufacturers of locking devices frequently use instructors to conduct seminars. However, few locksmiths make their living solely by teaching; most also work in a shop.

## Designing security-related products

Some locksmiths have made money designing new locking devices and products to make locksmithing easier. Many of them are employed by manufacturers. But it isn't unusual for locksmiths to invent new tools or to creatively modify locking devices while working for a locksmithing shop. Some locksmiths have formed their own companies to manufacture their inventions. Others have sold their ideas.

# Chapter 2

# Basic types of locks and keys

*T*erms such as mortise bit-key lock and Medeco key-in-knob lock are meaningless to most people. But the terms provide useful information to locksmiths. Like other trades, locksmithing has its own vocabulary to meet special needs.

## TERMINOLOGY

Laymen frequently use a generic name like padlock, automobile lock, or cabinet lock when referring to a lock. Such a name has limited value to locksmiths, because it is very general. It simply refers to a broad category of locks that are used for a similar purpose, share a similar feature, or look similar to one another.

Locksmiths identify a lock in ways that convey information needed to purchase, install, and service it. The name they use is based not only on the purpose and appearance of the lock, but also on the lock's manufacturer, key type, method of installation, type of internal construction, and function.

The names used by a locksmith consist of several words. Each word in the name provides important information about a lock. The number of words a locksmith will use for a name depends on how much information he or she needs to convey.

When ordering a lock, for instance, the locksmith needs to use a name that identifies the lock's purpose, manufacturer, key type, appearance, etc. But a name that simply identifies the lock's internal construction may be adequate for describing a servicing technique to another locksmith.

## Generic names

Some of the most commonly used generic lock names include: automobile lock, bike lock, ski lock, cabinet lock, deadbolt lock, gun lock, key-in-knob lock, luggage lock, lever lock, padlock, combination lock, and patio door lock. Sometimes generic terms have overlapping meanings. A padlock, for instance, can also be a combination lock. Figure 2-1 shows a variety of padlocks.

The *key-in-knob* lock refers to a style of lock that is operated by inserting a key into its knob (FIG. 2-2). A *lever* lock has a lever as a handle (FIG. 2-3). A *deadbolt* lock contains a deadbolt for maximum security (FIG. 2-4). As the names imply, the automobile lock, bike lock (FIG. 2-5), ski lock (FIG. 2-6), patio door lock, etc., are based on the purposes for which the locks are used. Sometimes locks that share a common purpose look very different from one another. Figure 2-7 shows several styles of patio door locks.

## Manufacturer's names

Locksmiths often refer to a lock by the name of its manufacturer, especially when all or most of the company's locks share a common characteristic. Locks manufactured by Medeco Security Locks, Inc., for example, all have similar internal constructions. Simply by knowing a lock is a Medeco lock, a locksmith can consider the options for servicing it.

Several lock manufacturers are so popular in the locksmithing industry that every locksmith is expected to be familiar with their names and the common characteristics among each manufacturer's locks. Those manufacturers include: Arrow, Best, Corbin, Dexter, Ilco/Unican, Kwikset, Master, Medeco, Russwin, Sargent, Schlage, Weiser, and Yale.

Warded padlocks

Brass, 4-pin
tumblers

Solid brass case
padlocks

Combination
locks

**2-1**   Padlocks come in assorted shapes and styles. Belwith International

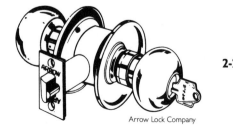
Arrow Lock Company

**2-2**   A popular key-in-knob lock.

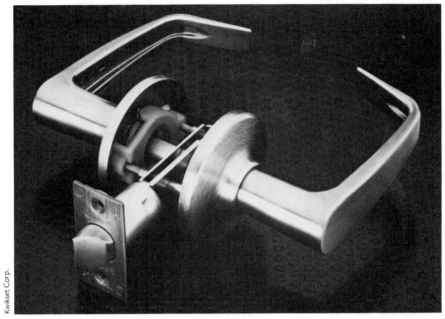

Kwikset Corp.

**2-3**   Lever locks are frequently used in facilities for the physically impaired.

**2-4**   A deadbolt lock can provide good security.

Master Lock Company

**2-5**  A bike lock secures bikes to racks, posts, and other anchor points.

## Key type

Many times a lock is identified by the type of key used to operate it. Bit key lock and tubular key lock are two common examples. *Tubular key locks* (sometimes called *Ace locks*) are primarily used on vending machines and coin-operated washing machines (FIG. 2-8). *Bit key locks* are on many closet and bedroom doors. When speaking about a bit key lock, locksmiths usually use a name that reveals how it is installed.

## Installation method

*Rim lock* and *mortise lock* identify locks based on installation method. A rim lock (also called a *surface mounted lock*) is one whose body is designed to be installed on the surface (or rim) of a door (FIGS. 2-9, 2-10, and 2-11).

A mortise lock is designed to be installed in a mortise (or recess) in a door. Figure 2-12 shows a mortise bit key lock installed. Not all mortise locks are operated with a bit key; the lock in FIG. 2-13 uses a cylinder key.

Master Lock Company

**2-6** A ski lock safeguards both skis and poles by securing them to a rack, tree, or post.

Patio
door
pin lock

Patio
door
key lock

Patio door &
window lock

**2-7** Patio locks can look very different from one another. Belwith International

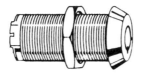

**2-8** A typical tubular key lock.

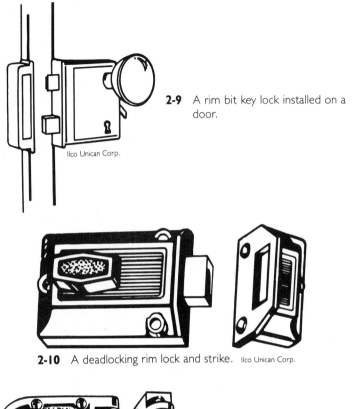

**2-9**   A rim bit key lock installed on a door.

Ilco Unican Corp.

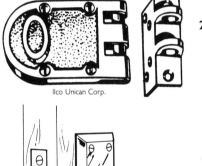

**2-10**   A deadlocking rim lock and strike.   Ilco Unican Corp.

**2-11**   A jimmy-proof rim lock can provide excellent security if its strike is properly installed.

Ilco Unican Corp.

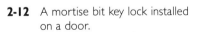

**2-12**   A mortise bit key lock installed on a door.

Ilco Unican Corp.

**2-13** Most modern mortise locks are operated with a cylinder key.

Adams Rite Mfg. Co.

### Internal construction

For servicing locks, names based on their internal constructions are most helpful to a locksmith. Examples include: warded lock, pin tumbler lock, disc tumbler lock, wafer tumbler lock, lever tumbler lock, and side bar lock.

Lock names based solely on internal construction seldom indicate the lock's purpose, installation method, function, or appearance. Most such names refer to types of cylinders. A lock that uses a pin tumbler cylinder, for example, is called a *pin tumbler lock* or a *pin tumbler cylinder lock*.

(Note: Some people use the terms lever lock and lever tumbler lock synonymously. However, the latter refers to a type of internal construction, whereas the former refers to a type of handle used [refer to FIG. 2-3]).

Most types of cylinders can be used with a wide variety of locks. A key-in-knob lock, for example, can use a wafer tumbler cylinder or a pin tumbler cylinder. Both cylinder types can also be used with many other types of locks. (Later chapters provide more information about the internal constructions of locks.)

### Lock functions

Entrance lock, classroom lock, and vestibule lock are names based on how a lock functions. A classroom lock, for example, is one whose inside knob is always in the unlocked position for easy exiting, and whose outside knob can be locked or unlocked with a key. An institution lock, however, has both knobs always in the locked position to prevent easy exiting; a key must be used on either knob to operate the lock. (Lock functions are listed in appendix A.)

At this point, you should have a good idea of how locksmiths identify locks. They simply combine several applicable terms that provide the necessary specificity. Now when you hear a name like mortise bit key lock, you should better understand what it means. Don't worry if you don't remember all the names used for locks. The purpose of this chapter is to help you understand the logic behind some of the most commonly used names.

## TYPES OF KEYS

A key is the device that operates locks. It also refers to any instrument that resembles and functions like such a device. Keys come in a wide variety of shapes and sizes. Some look like rings, others like jackknives.

A locksmith doesn't need to know everything about all the different kinds of keys, but it is a good idea to become familiar with the basic types. There are eight basic types of keys locksmiths commonly sell and work with: bit key, barrel key, flat key, corrugated key, cylinder key, tubular key, angularly bitted key, and dimple key. Virtually all other keys are variations of these types.

Keys usually have all or most of the following features: a bow, a blade (or bit), ward cuts or throat cuts, a stop, and tumbler cuts.

The *bow* (rhymes with "toe") is the handle of the key; it's the part you hold when using the key to operate a lock. The *blade* is inserted into a lock's keyway.

*Ward cuts* and *throat cuts* on a key's blade permit it to bypass obstructions on or within a lock; these cuts are needed to allow a key to enter or be rotated in a lock.

The *stop* of a key (usually the key's shoulder or tip) is the part that makes the key stop within a lock at the position the key needs to be in to operate a lock. Without a stop, you would need to pull a key in and out of a lock until you aligned it into the right position.

The *tumbler cuts* (also called *bitting*) on a blade manipulate tumblers within a lock into position for the lock to operate. To operate a lock, each tumbler cut on a key must correspond in spacing (position) and depth to a tumbler within the lock.

### Bit key

A *bit* key is used for operating bit key locks. It is usually made of iron, brass, steel, or aluminum (FIG. 2-14). This key is sometimes erroneously called a skeleton key. (Chapter 5 explains the differences between the two.) The main parts of a bit key are: bow, shank, shoulder, throat, post, bit, tumbler cuts, and ward cuts.

### Barrel key

The *barrel* key comes in a variety of sizes and styles. Some barrel keys look similar to bit keys and have most of the same parts. The major difference between the two types is that the barrel key has a hollow shank. Another difference is that many barrel keys don't have a shoulder.

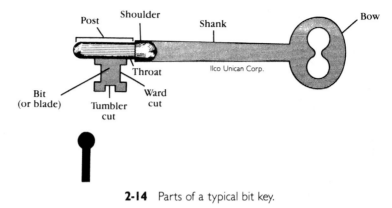

**2-14** Parts of a typical bit key.

### Flat keys

As the name implies, a *flat* key (sometimes called flat steel key) is flat on both sides. Most are made of steel or nickel silver (FIG. 2-15). Such keys are often used for operating a lever tumbler lock, a type of lock used on luggage and safe deposit boxes.

The parts of a flat key are: bow, blade, tip, stop, throat cut, and tumbler cut. The throat cut allows a key to bypass an obstruction found on most lever tumbler locks.

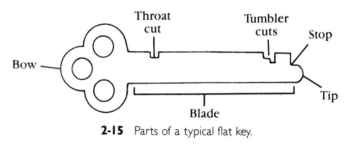

**2-15** Parts of a typical flat key.

### Corrugated key

Many corrugated keys look similar to flat keys. Both types usually have the same parts. *Corrugated* keys have corrugations or ripples along the length of their blades. They are designed to allow the key to fit into correspondingly shaped keyways. Unlike most flat keys, corrugated keys have cuts on both sides of their blades.

Corrugated keys are most frequently used with warded padlocks. Some corrugated keys are designed to operate other types of locks. For example, Schlage Lock Company manufactures a key-in-knob lock that uses special types of corrugated keys. Those keys look more like cylinder keys than they look like flat keys (FIG. 2-16).

### Cylinder key

The most popular today is the *cylinder* key. It is used to operate pin tumbler locks and disc tumbler locks. You probably have several cylinder

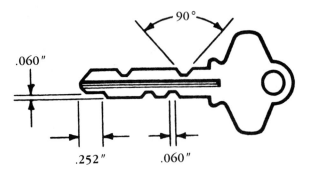

**2-16**   A corrugated key for a Schlage Wafer Lock. Schlage Lock Company

keys to unlock the front door of your home or the doors of your car (FIG. 2-17).

The parts of a cylinder key are: bow, shoulder, blade, tumbler cuts, keyway grooves, and tip. The shoulder acts as a stop; it determines how far the key will enter the keyway. Some cylinder keys don't have shoulders; those keys use the tip as a stop. The *keyway grooves* are millings along the length of a key blade that allow the key to enter a keyway. The other parts of a cylinder key are like the corresponding parts of a flat key.

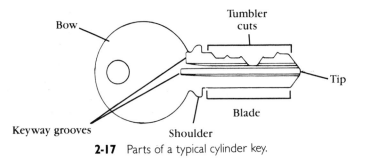

**2-17**   Parts of a typical cylinder key.

## Tubular key

The *tubular* key has a tubular blade with cuts (depressions) milled in a circle around the end of the blade (FIG. 2-18). The key is used to operate tubular key locks, which are often found on vending machines and coin-operated washing machines.

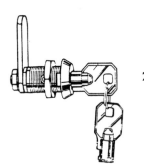

**2-18**   A tubular key lock and tubular keys are used for coin-operated machines.

Tubular keys are often improperly termed Ace keys. The term "Ace key" is a short form of a brand name, but it doesn't apply to all tubular keys. The first tubular key was patented by Chicago Lock Company to operate its Chicago Ace Lock brand tubular key lock. Today many companies manufacture tubular key locks and tubular keys.

Parts of a tubular key include: bow, blade, tumbler cuts, and nib. The *nib* shows which position the key must enter the lock to operate it. The purposes of the bow, blade, and tumbler cuts are similar to the purposes of corresponding parts of a cylinder key.

### Angularly bitted key

The *angularly bitted* key is used with some high-security locks. The key has cuts that angle perpendicularly from the blade.

The key is designed to cause pin tumblers within a cylinder to rotate to specific positions. Medeco Security Locks, Inc. popularized the angularly bitted key (FIG. 2-19). Chapter 9 provides more information about Medeco locks.

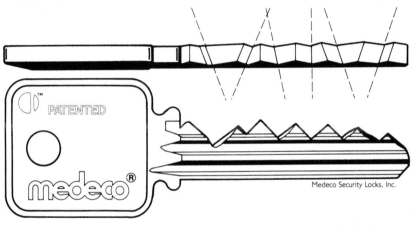

**2-19** An angularly bitted key is often used to operate high-security locks.

### Dimple key

The *dimple* key is used to operate some high-security pin tumbler locks. It has cuts that are drilled or milled into its blade surface; the cuts normally don't change the blade silhouette (FIG. 2-20). Lori Corporation's Kaba locks are popular locks operated with dimple keys.

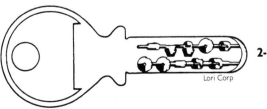

**2-20** A dimple key has drilled or milled cuts on its blade.

*Chapter* **3**

# Key blanks and key blank directories

*A* key *blank* (or blank, for short) is basically an uncut key. It looks similar to a key, but doesn't have cuts that allow it to operate a lock (FIG. 3-1).

Blanks come in the same types as do keys. The differences among blanks correspond to the differences among keys. By copying the cuts of a key onto a proper blank, a duplicate key is made.

An *original key blank* (or *genuine key blank*) is one that is supplied by a lock manufacturer to duplicate keys for that company's locks. Several companies make blanks that can be used in place of original key blanks.

**3-1** Blanks look like keys without cuts.

ESP Corp.

## CHOOSING THE RIGHT BLANK

Before duplicating a key, you have to choose the right blank. This is easy to do if you know which factors to consider. First decide which basic type of blank you need—it must be the same type as the key you're duplicating. Then select a blank that matches the key in the important areas for the particular key type.

### Bit keys and barrel keys

Important factors to consider when choosing a blank to duplicate a bit key or a barrel key are thickness of the bit and diameter of the post and shank. The key and blank should closely match in those areas (FIGS. 3-2 and 3-3). Because there is usually a lot of tolerance in locks that use these types of keys, the blank might not have to match the key perfectly.

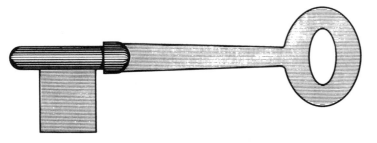

**3-2**   A bit key blank. · Ilco Unican Corp.

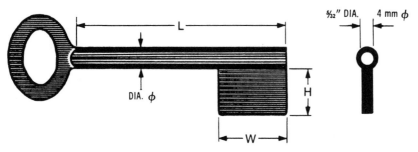

**3-3**   Before duplicating a barrel key, you need to find the right blank. Ilco Unican Corp.

### Flat keys and corrugated keys

Thickness, length, width, and shape of the blade are primary factors to consider when choosing a blank to duplicate a flat key. The key and blank should closely match in these areas. Figure 3-4 shows some flat key blanks. A blank for a corrugated key should have the same corrugated configuration as the key has (FIG. 3-5).

### Cylinder keys

Choosing a blank to duplicate a cylinder key can be a little tricky. There are more varieties of cylinder keys than there are of other keys, and the

**3-4** Flat key blanks come in various sizes.

**3-5** A typical corrugated key blank for warded padlocks.

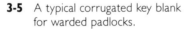

differences among cylinder keys are often slight. However, finding the right blank for a cylinder key can be simple if you approach the task methodically. First examine the bow for clues.

**The bow** When you know the manufacturer of the lock a key operates, your choices of a blank are greatly reduced. Sometimes the key bow provides that information.

Most major lock manufacturers use distinctive bow shapes for their factory original keys and blanks. And many aftermarket key blank manufacturers use similar bows when making corresponding key blanks. With a little practice, you'll be able to quickly identify distinctive bows (FIG. 3-6).

A company name is on the front of most key bows. The name on factory original keys is that of the manufacturer of the lock the key operates. But the name on aftermarket keys is usually of the manufacturer of the key blank that was used to make the key. (Occasionally it's the name of the locksmith shop that duplicated the key).

Most bows also have letters and numbers on them. Usually those numbers and letters are coded in a way that makes it easy for you to identify the manufacturer of the lock the key operates. For example, SC1, SC2, SC3, etc. are used to indicate the key or blank is for a lock manufactured by Schlage Lock Company. Another example: MA3, and MA7 indicate the

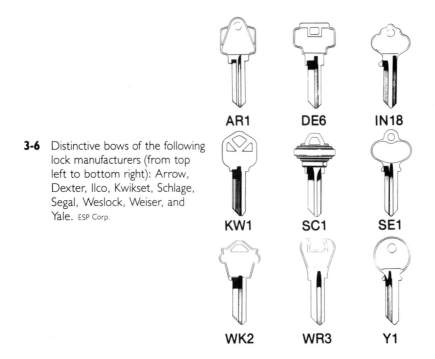

**3-6** Distinctive bows of the following lock manufacturers (from top left to bottom right): Arrow, Dexter, Ilco, Kwikset, Schlage, Segal, Weslock, Weiser, and Yale. ESP Corp.

blank or key is for a lock manufactured by Master Lock Company. Different key blank manufacturers use different codes.

It isn't necessary for you to remember all the codes used. However, you will be able to duplicate keys more quickly if you remember the codes most frequently found on keys. After determining the manufacturer of the key you're duplicating, you might still have to choose among several blanks designed for that manufacturer's products—but you've greatly narrowed down your choices.

**Comparing keyway grooves**   At this point you'll need to compare the keyway grooves of the key with those of a blank. Pick any blank you think might be the right one. Turn the key and blank on the same side, and make sure both tips point in the same direction.

See if the blank has the same number of keyway grooves on each side as the key has on each side. If not, find one that does. After finding one, look at the points where the keyway grooves of the key and blank touch, or nearly touch, the bows. Use those points to compare each keyway groove of the blank to its corresponding groove of the key. Check if each keyway groove has the same shape as does its corresponding groove. Make that comparison on both sides of the key and blank. The five standard shapes are: right angle, left angle, square, V, and round. All these shapes are illustrated in FIG. 3-7.

If any keyway groove of the blank doesn't match the shape of its corresponding groove, you don't have the proper blank. After finding a blank with keyway grooves that match those of the key, compare the widths of the grooves with the widths of their corresponding grooves on

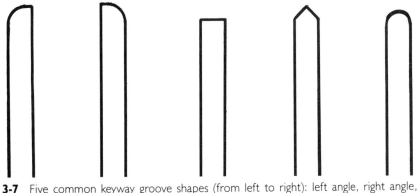

**3-7** Five common keyway groove shapes (from left to right): left angle, right angle, square, V, and round.

the key. Then compare the thickness, width, and length of the blade of the blank with those of the key. Those dimensions are measured by placing the blank and key together, and aligning them at the shoulders. If both blades match in all those respects, you have the right blank.

If the blank matches the key in every respect except blade thickness, width, or length, it can sometimes still be used to duplicate the key. A blank that is longer than the key can be cut down to the length of the key. A blank that is thinner than the key may have a loose fit in the keyway, but it can still be used.

The process of finding a blank to duplicate a cylinder key might seem long and tedious. With a little practice, however, you'll be able to complete the process within a few seconds. You will also notice that you're using certain blanks much more frequently than you're using others. In a short period of time, you'll remember what the keyway grooves of those commonly used blanks look like. Then when you see a key with grooves that match one of those blanks, you'll automatically know which blank you need.

### Tubular keys

Choosing a blank to duplicate a tubular key is very simple, because there are few significant differences among tubular keys. Figure 3-8 shows some tubular keys. The important areas of such keys are: the size of the nib, and the inside and outside diameters of the shank. Find a blank that matches a tubular key in those respects and you can duplicate the key.

### KEY BLANK DIRECTORIES

Several key blank manufacturers publish directories that have drawings of blanks, silhouettes of keyway grooves, and side-by-side listings of the key coding systems of various key blank manufacturers. Although such directories are inexpensive (and sometimes free), they're invaluable to locksmiths who regularly duplicate keys. Figure 3-9 shows a page from a key blank directory.

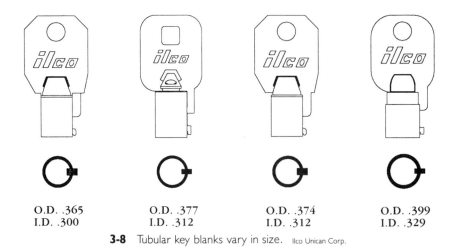

| O.D. .365 | O.D. .377 | O.D. .374 | O.D. .399 |
| I.D. .300 | I.D. .312 | I.D. .312 | I.D. .329 |

**3-8** Tubular key blanks vary in size.  Ilco Unican Corp.

Why would a key blank manufacturer provide information about blanks made by other companies? To let locksmiths know which of that company's blanks can be used in place of blanks made by other manufacturers. But the information in a key blank directory can helpful in other ways.

## Using the directories

Key blank directories are commonly used in two ways. One is to compare the numbers and letters on a key to those in the directory to determine which of several blanks a key can be duplicated with. Those key codes are listed not only near drawings of key blanks in a directory, but also in cross-reference charts in the directory (TABLE 3-1).

If you're duplicating a key that has 110 on its bow, for example, you could use the directory entries, such as those in FIGS. 3-9 and 3-10, to learn several things. The directory would let you know that 110 is a number from Illinois' key coding system, and that Star's blank number IS1 could be used to duplicate the key. You could then find out which other manufacturer's blanks are compatible with Star's IS1. That's done by looking at each Star cross-reference chart to find IS1, and then seeing what number is beside it.

Keep in mind that the publisher of a directory usually shows drawings of its company's blanks in its directory. Likewise, all the listed key code systems in a key blank directory are compared to the system of the publisher.

Another way a directory can be used is to compare a key to the drawings and silhouettes. Many times that comparison makes it unnecessary to handle a lot of blanks. By laying a key on top of the drawing of a blank, aligning both at the shoulders, you can usually determine if that blank's blade is the right size. By standing the key up on its tip directly above the drawing's corresponding keyway groove silhouette (with the key facing

**STAR**

# Table 3-1   Excert from Key Blank Directory.

| Hazelton | STAR |
|---|---|
| 4637 | 5HO1 |
| 4647 | 5WK1 |
| 4650 | 5KW1 |
| 4652 | 5SN1 |
| 4655 | 5AR2 |
| 4838 | HYA4 |
| 4840 | HYA5 |
| 4844 | HPL2 |
| 4846 | OPL2 |
| 4854 | HPL1 |
| 4856 | OPL1 |
| 4858 | HPL3 |
| 4864 | OPL4 |
| 4867 | OPL5 |
| 4870/72 | HPL68. |
|  | HPL73 |
| 4900 | HBR11 |
| 4902 | OBR11 |

Hollymade
(See Challenger)

| Huber | STAR |
|---|---|
| 700 | 5HU1 |

| Hurd | STAR |
|---|---|
| 4932 | HBR7 |
| 4933 | OBR7 |
| 4934 | HBR9E |
| 4935 | OBR9H |
| 4936 | HBR10J |
| 4937 | OBR10K |
| 4938 | HBR12A |
| 4939 | OBR12B |
| 9000 | 5HD1 |
| 9015 | OFD2 |
| 9016 | HFD7 |
| 9020 | OFD2 |
| 9025 | OFD2 |
| 9025RH | OFD2 |
| 9026 | HFD7 |
| 9026-2 | HFD7 |
| 9040 | HFD2 |
| 9041 | OFD2 |
| 9043 | OFD2 |
| 9044 | YJ3 |
| 9046 | SYJ1 |
| 9047 | OBR1DB |
| 9058 | HFD2 |
| 9059 | OFD2 |
| 9070 | HFD1 |
| 9071 | OFD1 |
| 9072 | HFD1 |
| 9073 | OFD1 |
| 9074 | HFD1 |
| 9082 | HFD3 |
| 9083 | OFD3 |
| 9084 | HFD3 |
| 9086 | HFD3 |
| 9087 | OFD5 |
| 9090 | HFD6 |
| 9091 | HFD6 |
| 9098 | HFD4 |
| 9099 | OFD3 |
| 9124 | HFD6 |
| 9125 | OFD6 |
| 9128 | OFD10 |
| 9129 | HFD10 |
| 9133B | YJ4 |
| 9147 | YJ1 |
| 9148 | YJ4 |
| 9174 | HPL6 |
| 9175 | OPL4 |
| 9299 | HPL6 |
| 9300 | OPL68 |
| 9301 | HPL68 |
| 9305 | OPL68 |

| Hurd | STAR |
|---|---|
| 9337 | HFD9 |
| 9338 | OFD9 |
| 9340 | HBR5 |
| 9341 | OBR5 |
| 9356 | HFD9 |
| 9357 | OFD9 |
| 9421 | OFD4 |
| 9422 | HFD9 |
| 9423 | HFD4 |
| 9424 | OFD9 |
| 9427 | HFD9 |
| 9428 | OFD9 |
| 9431 | HFD9 |
| 9432 | OFD9 |
| 9433 | HFD4 |
| 9434 | OFD4 |
| 9518 | HFD5 |
| 9520 | HFD5 |
| 9521 | OFD5 |
| 9522 | HFD4 |
| 9523 | OFD4 |
| 9524 | HFD4 |
| 9525 | OFD4 |
| 9526 | HFD8 |
| 9530 | HBR2 |
| 9531 | OBR2 |
| 9532 | HPL3 |
| 9533 | OPL1 |
| 9534 | OBR1 |
| 9535 | HBR1 |
| 9537 | OPL4 |
| 9539 | OPL5 |
| 9542 | HFD4 |
| 9543 | OFD4 |
| 9544 | HFD4 |
| 9545 | OFD4 |
| 9546 | HFD4 |
| 9547 | OFD4 |
| 9549 | OFD4 |
| 9556 | YJ1 |
| 9557 | YJ4 |
| 9571 | HPL6 |
| 9572 | OPL4 |
| 9576 | HBR3 |
| 9577 | OBR3 |

| Illinois | STAR |
|---|---|
| 110 | IS1 |
| 260 | IS3 |
| 360 | IS2 |

| Ilco | STAR |
|---|---|
| DC1 | LDC1 |
| MG1 | UN4 |
| YS1 | CP2 |
| AA | HN1 |
| FC. | CP1 |
| WS. | 5DA2 |
| YS2 | CP3 |
| MZ4 | MZ4 |
| MZ5 | MZ5 |
| PA6 | AD1 |
| VO6 | 6VL1 |
| MZ9 | MZ1 |
| MZ10 | MZ2 |
| MZ11 | MZ3 |
| MZ12 | 5DA2 |
| DA20 | 5DA1 |
| TR26 | TO2 |
| FT37 | FT3 |
| FT38 | FT2 |
| F44 | 5FT1 |
| HO44 | HN4 |
| P54F | 5DO4 |
| RE61F | 5RP2 |
| T61C | TO1 |

| Ilco | STAR |
|---|---|
| T61F | LU1 |
| RE61N | RP1 |
| 62DP | UN2 |
| 62DT | DA4 |
| 62DU | DA3 |
| 62FS | UN3 |
| 62VW | VW1 |
| VW67 | VW2 |
| 70S | SM1 |
| HD70,HD71 | HN2-3 |
| VW71 | VW3 |
| VW71A | VW5 |
| 73VB | VW4 |
| HD74 | HN5 |
| VR91 | 5VR1 |
| VR91AR | 5VR3 |
| VR91B | 5VR4 |
| 100AM | CG3 |
| 995M,996M | 5YA11 |
| 997D,997E | 5YA6 |
| 997X | 4YA6 |
| J997M | 6YA9 |
| 0997E | 6YA6 |
| 998 | 6YA3 |
| 998GA | 6YA12 |
| 998GST | 6YA8 |
| L998GST | 7YA8 |
| 998R | 6YA7 |
| 999 | 5YA1,5YA1M |
| 999A | 6YA1 |
| 999B | 4YA1 |
| 999N | 5YA1E |
| 999R | 5YA13 |
| C999 | 5YA2 |
| 1000 | 5CO3 |
| 1000F | 6CO6 |
| 1000G | 5CO6 |
| 1000T | LCO15 |
| 1000V | LC07 |
| S1000V | CO7 |
| X1000F | 6CO10 |
| X1000FR | 6CO14 |
| X1000KC | 5CO8 |
| X1000KR | 5RU3 |
| 1001 | 5CO4 |
| 1001ABM | 5CO12 |
| 1001E | 7CO2 |
| 1001EA | 6CO2 |
| 1001EB | 5CO2 |
| 1001EG | 6CO1 |
| 1001EH | 5CO11 |
| 1001EL | 7CO1 |
| 1001EN | 5CO1 |
| 1001GH | 5CO13 |
| A1001ABM | 6CO12 |
| A1001C1/ C2/D1/D2 | 6CO16 |
| A1001EH | 6CO11 |
| L1001ABM | 7CO12 |
| L1001C1/ C2/D1/D2 | 7CO16 |
| L1001EH | 7CO11 |
| R1001EF | 5CO9 |
| R1001EG | 6CO5 |
| R1001EL | 7CO5 |
| R1001EN | 5CO5 |
| 1003M, L1003M | 5AU1 |
| R1003M | 5BO1 |
| 1004 | 5LO1 |
| 1004A | 6LO1 |
| 1004AL | 7L01 |
| 1004KA | 6IL2 |
| 1004KL | 7IL2 |
| 1004M | 5IL6 |
| 1004N | 5IL11 |
| 1007 | 5SA4 |

| Ilco | STAR |
|---|---|
| 1007K | 5SA8 |
| 01007K | 5SA9 |
| 01007KC | 5SA10 |
| 1007KMA | 6SA3 |
| 1007KMB | 5SA3 |
| 1007RMA | 6SA6 |
| 1007RMB, N1007RMB | 5SA6 |
| L1007KMA | 7SA3 |
| 1009 | 5SA2 |
| 1010 | 5SA5 |
| 1010N | 5SA7 |
| L1010N, A1010N | 6SA7 |
| 01010 | 5SA1 |
| 1011 | 5RU1 |
| 1011D1 | 5RU7 |
| 1011D41 | 5RU8 |
| 1011GH | 5CO13 |
| 1011M | 5RU4 |
| 1011P | 5RU2 |
| 1011PB to PY | 5RU5 |
| 1011PZ | 5RU5 |
| A1011D1 | 6RU7 |
| A1011D41 | 6RU8 |
| A1011M/P/S/T | 6RU9 |
| A1011P | 6RU2 |
| A1011PB to PY | 6RU5 |
| A1011PZ | 6RU5 |
| L1011D1 | 7RU7 |
| L1011D41 | 7RU8 |
| L1011P | 7RU2 |
| L1011PZ | 7RU5 |
| N1011M/P/S/T | 5RU9 |
| 1012 | 5RU6 |
| 1014 | 5EA2 |
| 1014A | 6EA2 |
| 1014D.1014DX | 4EA4 |
| 1014F | 5EA1 |
| 1014J | EA3 |
| 1014JS | EA5 |
| 1014K | 4EA1 |
| L1014A | 7EA2 |
| 01014S | EA6 |
| X1014F | 5EA1 |
| 1015 | 5CH3 |
| 1015C | 5CH1 |
| 1015M | 5CH2 |
| A1015M | 6CH2 |
| A1015MR | 6CH4 |
| L1015M | 7CH2 |
| L1015MR | 7CH4 |
| 1016 | 5PE2 |
| 1016N | 5PE1 |
| 1017 | 5NW3 |
| 1017B | 5NW1 |
| 1017BA | 6NW1 |
| 01017ML | 6NW2 |
| 01017MX | 7NW1 |
| 1019 | 5RE1 |
| 1019A | 6RE1 |
| 1019D | 5RE2 |
| A1019M | 6AR3 |
| 1020 | 5HU1 |
| 1021BA | 5NW3 |
| 1022 | 5SE1. 5SE1M |
| 1022AB | 6SE1 |
| 01022 | 5SE2 |
| 01022AB | 6SE2 |
| 01022AR | 6SE4 |
| R1022AB | 6SE5 |
| 1023 | 5CL1 |
| 1025 | AR1 |
| 1033N | 5UN1 |
| 1034 | 5PO1 |

| Ilco | STAR |
|---|---|
| 1034H | 5PO2 |
| 1041C | JU1 |
| 1041E | 5CG5 |
| 1041G | CG1 |
| 1041GA | CG6 |
| 1041GR | CG2 |
| 1041H | IS1 |
| 1041T | 5CG7 |
| 1041Y | 5CG4 |
| 1043B | IS2 |
| 1043D | IS3 |
| 1046 | 5JU2 |
| 1047CR,1047M | 5YA2 |
| 1054 | 5IL3 |
| 1054F | 5IL1 |
| 1054FN, X1054FN | 5IL9 |
| 1054K | 5IL2 |
| 1054KD | 5DE1 |
| 1054MT | 5IL11 |
| 1054TW | 5IL2 |
| 1054UN | IL10UN |
| 1054WB | 5WR2 |
| A1054F | 5DO1 |
| A1054KD | 6DE1 |
| A1054WB | 6WR2 |
| D1054K | 5DE3 |
| D1054KA | 6DE3 |
| L1054B | IL5 |
| L1054K | 5DO3 |
| S1054F | 5DO2 |
| X1054F | 5IL7 |
| X1054JA | 5IL8 |
| X1054K | 5IL4 |
| X1054WA | 5WR1 |
| 1064,N1069G | 4RO2 |
| R1064D | 5RO4 |
| 1069 | RO1 |
| 1069FL | RO3 |
| 1069G | RO5 |
| 1069H,1069N | RO6 |
| 1069LA | 5AU2 |
| 1071 | 5WI1 |
| 1073H.1073K | 5FR1 |
| 1079 | 5KE2 |
| 1079B | 5KE1 |
| 1092 | MA1 |
| 1092B | 4MA2 |
| 1092D | 5MA7 |
| 1092DS | 4MA7 |
| 1092H | 5MA6 |
| 1092J | 6MA8 |
| 1092N | 5MA5 |
| 1092NR | 5MA3, 5MA4 |
| 1092V | 5MA3 |
| 1092VM | 5MA4 |
| 1096 | 5EL2 |
| 1096L | 5EL1 |
| E/C/S1096CN | 5EL4 |
| E/C/S1096LN | 5EL3 |
| L1096CN | 6EL4 |
| 1098DB | OBR1DB |
| 1098M | OBR1 |
| 1098NR | OBR3 |
| D1098X | DE2 |
| H1098A | HBR5 |
| H1098A/B | HBR6 |
| H1098A/C | HBR5M |
| H1098C | HBR7 |
| H1098LA | HBR2 |
| H1098M | HBR1 |
| H1098NR | HBR3 |
| H1098C/A | HBR8 |
| L1098LA | OBR4 |
| O1098B | OBR5 |
| O1098B A | OBR6 |

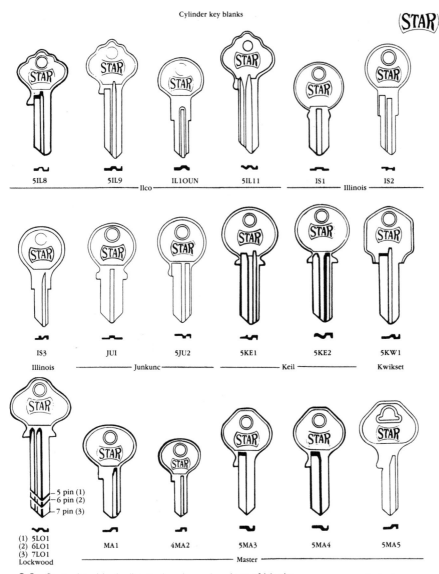

**3-9** Some key blank directories show drawings of blanks. Star Lock and Key Mfg. Co., Inc.

the same direction as the drawing is facing), you can see if that blank's keyway groove matches those of the key.

## CUTTING KEYS BY HAND

Today locksmiths rarely need to cut keys by hand. Sophisticated machines can be used to cut virtually any key. Nevertheless, every locksmith should be able to duplicate keys by hand. This skill can be useful in emergency situations when a proper key machine isn't available. Developing the skill can also prepare a locksmith for more technical skills such as key impressioning.

To duplicate a key by hand you will need a vise, small files (triangle, pippin, and warding), and key blanks. The vise holds the blank in place while you carefully file the proper cuts into the blank. Proper cuts are those with the same angles, depths, and positions (spacing) as those on the key.

## Cutting procedures

When cutting a bit key by hand, first tightly wrap a piece of soft metal around the blade of the bit to create a pattern of the key's cuts. Remove the pattern, and wrap it around the blade of the blank. Clamp the blank and pattern into a vise and carefully file down the blank to match the pattern (FIG. 3-10).

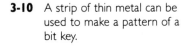
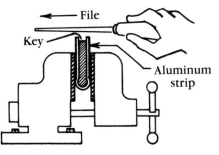

**3-10**  A strip of thin metal can be used to make a pattern of a bit key.

To duplicate a flat or corrugated key by hand, both the key and the blank need to be clamped into a vise. While holding both in the same direction, align the key and blank at the shoulders (key cuts on top of blade) and clamp them together in the vise. Using the key as a guide, carefully copy the cuts onto the blank with a warded file.

A cylinder key can be duplicated by hand by clamping the key and blank together (aligned at the shoulders) in a vise. Use a pippin file or a triangle file to carefully copy the cuts.

## Smoking a key

Some locksmiths find it easier to duplicate a flat, corrugated, or cylinder key by smoking the key before cutting the blank. The soot on a smoked key helps a person know when the blank has been filed down enough. When the file wipes the soot off the key, you know you've filed deep enough at that cut.

To smoke a key, use a pair of pliers to hold it over a candle flame, cuts facing down. Move the key over the flame until soot has built up in all the cuts.

After cutting a smoked key, be sure to clean the soot off because the soot can damage a lock.

*Chapter* **4**

# Warded, lever tumbler, disc tumbler, and side bar wafer locks

*T*he locks covered in this chapter—warded, lever tumbler, disc tumbler, and side bar wafer—are all named by their internal construction.

## WARDED LOCK

The Romans are credited with inventing the warded lock. It is the oldest and least secure type of lock still in use today.

It is difficult to describe the appearance of a warded lock because "warded" refers to a type of internal construction. A large key hole on a lock indicates that it is warded (FIG. 4-1).

The warded lock relies primarily on wards for its security. Other types of locks also use wards, but not as the primary means of security. *Wards* are fixed obstructions within a lock case; they are designed to prevent unauthorized keys from being rotated in the keyway. In theory, only a key with notches corresponding with the sizes and positions of the wards can operate a warded lock; the wards prevent other keys from being rotated into a position to open the lock (FIG. 4-2).

Warded locks aren't used for high-security applications because the typical warded lock provides fewer than 50 keying possibilities. That means a person with 50 different keys for such locks can open virtually all of them.

Changing ward sizes and configurations provides few keying possibilities, because the number of ward changes that can be made depend on

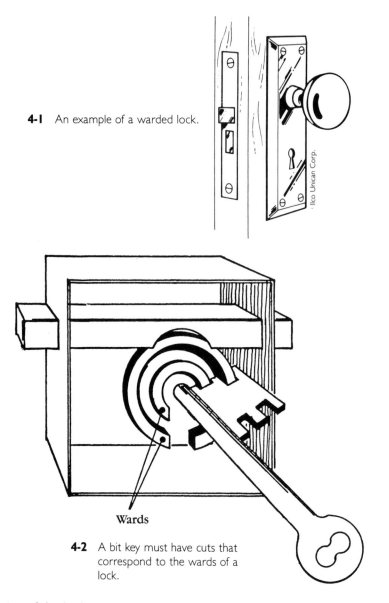

**4-1**   An example of a warded lock.

Ilco Unican Corp.

**Wards**

**4-2**   A bit key must have cuts that correspond to the wards of a lock.

the size of the lock. A warded keying system could never be large enough to match modern high-security locks.

A more significant problem with warded locks is that a properly notched key isn't needed to operate them. Wards can be easily bypassed by using a very thin key, known as a skeleton key.

## Warded padlock

Warded padlocks are the least expensive type of padlock available. They are also the least secure (FIG. 4-3). The locks are usually operated with cor-

**4-3** A warded padlock offers little security.

rugated keys; some of the more expensive models use flat keys. A group of five skeleton keys sold by locksmithing supply houses can be used to open most warded padlocks. A properly bent paper clip works nearly as well.

**Parts** All parts of a warded padlock are housed in the lock's case. The case of most warded padlocks consists of several steel plates or laminations. The parts of a typical warded padlock include: a shackle, pins, a shackle spring, a stop plate, a retainer spring, a shackle retainer, several dummy plates, one or more ward plates, and a keyway.

When in the locked position, the shackle spring presses the shackle retainer against the shackle, which holds the shackle in the locked position. The key that bypasses the wards in the lock can be turned to push the shackle retainer against the spring to release the shackle from the lock.

As the name implies, a ward plate is a plate that is used as a ward. The plate has a cutout in the center shaped to allow only certain keys to enter.

A dummy plate is used to build the size of a lock case. Such a plate has a large hole in the center. The stop plate is the one directly beneath the shackle and furthest from the keyway. That plate holds one side of the shackle in the lock case at all times. The pins are used to hold all the various types of plates together.

Because the locks are so inexpensive, locksmiths don't generally repair warded padlocks. If you'd like to disassemble one, just file down the pins. The plates can then easily be removed and you'll be able to see all the parts. However, don't expect to find replacement parts at a locksmith supply house.

### The warded bit key lock

The oldest type of warded lock is the warded bit key lock. It usually has a metal case and a large keyhole that accepts a bit key. Such locks come in two styles: mortise and rim.

The mortise bit key lock is designed to be mortised (or recessed) into a door. The rim bit key lock is designed to be surface mounted on a door. Figure 4-4 shows an outside view of a rim bit key lock.

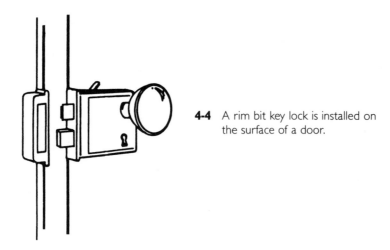

**4-4**   A rim bit key lock is installed on the surface of a door.

**Parts**   The construction of most warded bit key locks is very simple. The type and number of parts those locks have depend primarily on the functions of the locks. Those designed for the same function usually have few construction variations among them.

Basic parts found in warded bit key locks include: a latch bolt, a spindle hub, a latch bolt spring, an inside latch lock, an inside latch lock spring, a deadbolt, and a lever tumbler. All these parts are housed in a lock case (FIG. 4-5). The front of the lock case is the cover plate.

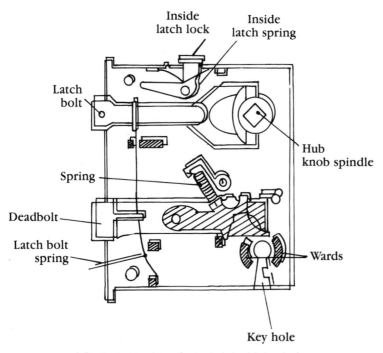

**4-5**   An inside view of a typical rim bit key lock.

Fixed obstructions on the lock case are positioned to prevent the deadbolt from moving freely when in the fully locked or fully unlocked position. When a key designed to bypass the wards is inserted into the keyhole and turned, it raises the deadbolt over the obstructions and slides it to the locked or unlocked position.

When the deadbolt is in the unlocked position, the latch bolt can be used to keep the door closed. Turning the door knob causes the hub spindle to pull the latch bolt back into the lock case and allow the door to be opened. When the door knob is released, the latch bolt spring pushes the latch bolt back out of the lock case.

The inside latch lock is used to deadlatch the latch bolt. When the inside latch lock is pushed down, it prevents the latch bolt from being pulled back by the hub spindle from outside the door.

The tumbler spring applies pressure to keep the deadbolt in the proper positions while it's being moved with a key. Some locks use additional springs, but all serve the purpose of applying pressure to parts. Both flat springs and coil springs are commonly used in warded bit key locks.

**Servicing**    Most problems with a warded bit key lock are caused by foreign material—usually dirt or paint—in the lock, and/or a weak or broken spring. Often the lock can be serviced simply by cleaning the lock case and parts, then using a little graphite to lubricate the parts. Oil shouldn't be used, because it can cause dust to collect on the parts. Occasionally servicing the lock involves replacing a spring.

## LEVER TUMBLER LOCK

The *lever tumbler lock* is so named because it relies primarily on levers within the lock for its security. It is frequently found on luggage, mail boxes, and school lockers.

Most lever tumbler locks sold in the United States, Mexico, and England use flat keys. Many sold in other countries are operated with bit or barrel keys. The type of key used, however, doesn't affect the basic operation of a lever tumbler lock.

Generally speaking, the lever tumbler lock offers slightly more security than does the warded lock. Some specially designed lever tumbler locks offer a high degree of security. An example is the type found on safe deposit boxes. Unlike a typical lever tumbler lock, the high-security version often has six or more tumblers, is made of high-quality materials, and has little tolerance between its parts.

### Parts

The parts of a lever tumbler lock are housed in a case consisting of a base (back cover) and a cover. The parts within the case include the trunnion (key plug), lever tumblers (usually three or more), and bolt. A post (or stop) is fixed on the bolt (FIG. 4-6).

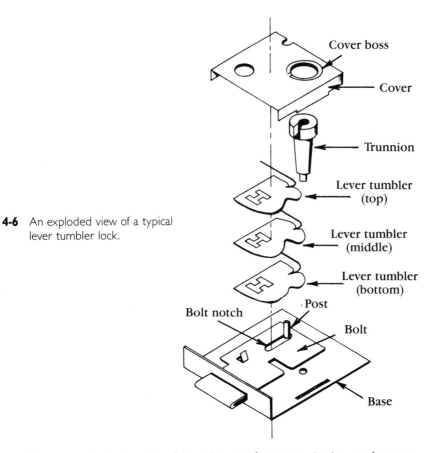

**4-6**  An exploded view of a typical
lever tumbler lock.

The cover boss is a raised opening on the cover; it sits on the trunnion and exposes the keyway. A properly cut key has tumbler cuts that
correspond in height and width to the lever tumblers. The key also has a
throat cut that allows it to ride over the raised ridge of the cover boss. The
cover boss of a lever tumbler lock acts like a ward of a warded lock.

To understand how lever tumbler locks operate, you have to know
the parts of a lever tumbler. They include: front trap, gate (or fence), rear
trap, pivot hole, spring, and saddle (FIG. 4-7). The lever tumblers sit on the
bolt and are held in place by spring tension. The traps and gate are open
areas on the tumbler that restrict the movement of the bolt post.

## Operation

When the bolt post moves from one trap to another, the lock bolt extends
or retracts. When a properly cut key is inserted, it slides over the saddles
of the tumblers and lifts each to the height necessary to allow the bolt
post to be moved from one trap to another.

For increased security, modern lever tumbler locks use tumblers with
staggered saddle heights or staggered trap heights. Such tumblers make it
difficult to pick the locks.

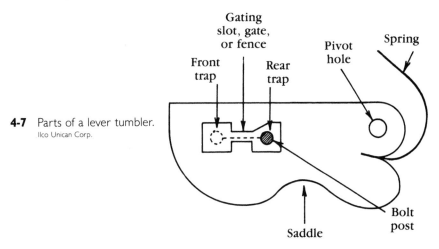

**4-7** Parts of a lever tumbler.
Ilco Unican Corp.

### Servicing

Some lever tumbler lock cases are riveted or spot welded. These locks usually don't cost enough to justify servicing them. To service them you would need to drill out the rivets or chisel through the case, then spend time putting the case back together.

Other lever tumbler locks can easily be disassembled by removing a screw holding the cover, or by prying up the metal tabs holding the cover. After removing the cover, look for a broken spring. If you find one, replace it.

### DISC TUMBLER LOCK

The *disc tumbler lock* is used on automobiles, desks, cabinets, and vending machines. As the name implies, this lock relies primarily on disc tumblers for its security. Sometimes disc tumbler locks are called wafer tumbler locks.

The lock provides better security and offers more keying possibilities than either the warded or lever lock offers. A typical disc tumbler lock allows over 3,000 possible key changes.

### Parts

Basic parts of a disc tumbler lock include: lock housing (or shell), bolt (or cam), retainer, plug, springs, and disc tumblers—also called wafer tumblers (FIGS. 4-8 and 4-9).

A disc tumbler is a flat piece of metal with a rectangular hole in its center. It also has a leg on one side for a spring to sit on. All disc tumblers within a lock are the same height and width, but their center holes vary in height. In order to operate a disc tumbler lock, a key must have cuts that correspond to the center holes of the tumblers. The smaller the center hole, the deeper the corresponding cut must be (FIG. 4-10).

The plug is the cylindrical part of the lock that contains the keyway. To move the bolt to the locked or unlocked position, the plug must be

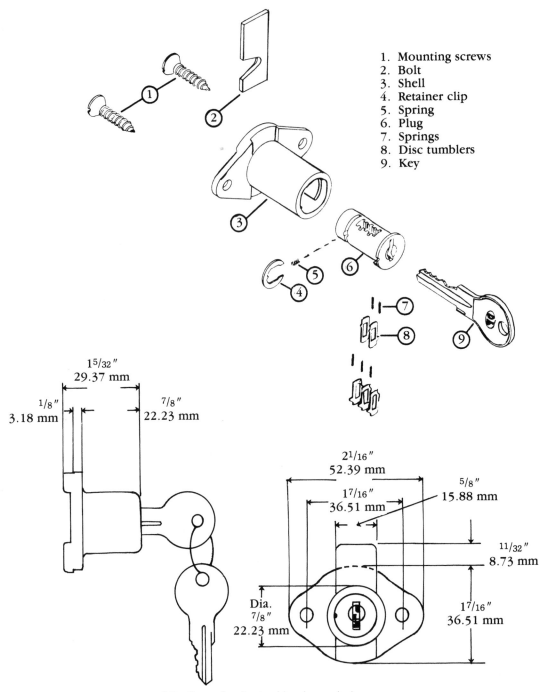

1. Mounting screws
2. Bolt
3. Shell
4. Retainer clip
5. Spring
6. Plug
7. Springs
8. Disc tumblers
9. Key

**4-8**  Parts of a disc tumbler drawer lock. Ilco Unican Corp.

Parts
desk lock

1. Nut
2. Bolt
3. Shell
4. Retainer clip
5. Spring
6. Plug
7. Springs
8. Disc tumblers
9. Key

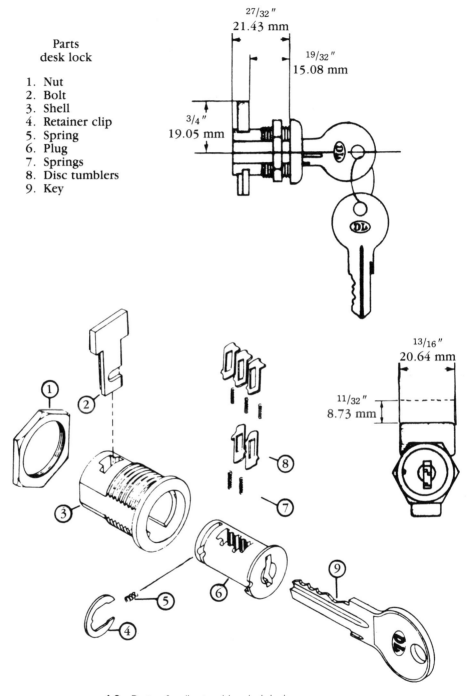

**4-9** Parts of a disc tumbler desk lock. · Ilco Unican Corp.

**4-10**  Disc tumblers have center holes of varying lengths that correspond to the depths of key cuts.

rotated. Whenever the proper key isn't in the keyway, the disc tumblers prevent the plug from rotating.

The plug has rectangular slots that each hold a spring and a disc tumbler. The springs maintain constant pressure on the tumblers, forcing them to protrude out of the plug and partially enter the shell. In that position, the tumblers connect the plug and shell together. A properly cut key is needed to pull all the tumblers fully into the plug, which allows the plug to rotate (FIG. 4-11).

## SIDE BAR WAFER LOCK

Many automobiles manufactured by General Motors Corporation use a type of disc tumbler lock called a *side bar wafer lock*. It has a V notch on one side of each disc, a V shaped side bar, and a special slot within the

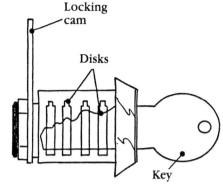

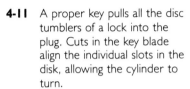

**4-11** A proper key pulls all the disc tumblers of a lock into the plug. Cuts in the key blade align the individual slots in the disk, allowing the cylinder to turn.

lock housing for a portion of the side bar to protrude into when in the locked position.

The side bar must be fully retracted from the slot in the housing before the plug can be rotated. Before the side bar can fully retract from the housing, all the V notches of the disc tumblers must be aligned to allow the V shaped portion of the side bar to fit into those notches.

Spring pressure forces the side bar to constantly press against the discs. When the proper key is inserted into the lock, all the disc tumblers are properly aligned and the side bar pushes into the V notches and clears the slot in the lock housing.

# Chapter 5

# Pin tumbler locks

*A* pin tumbler lock is any lock that has a pin tumbler cylinder. The pin tumbler cylinder is the most popular type of cylinder in use today. It is used in key-in-knob locks, deadbolt locks, rim locks, padlocks, and automobile locks.

The basic parts of a pin tumbler cylinder for rim and mortise locks include the following: cylinder case (or shell or housing or body), plug (or core), keyway, upper pin chambers, lower pin chambers, springs, top pins (or drivers), and bottom pins (FIG. 5-1). Pin tumbler cylinders for key-in-knob locks have a few extra parts (FIG. 5-2).

The cylinder case houses the other basic parts. The plug rotates when the proper key is inserted into it. The keyway is an opening in the plug that accepts the key. The drilled holes (usually five or six) across the length of the plug are lower pin chambers that each hold a bottom pin. The corresponding drilled holes in the cylinder case directly above the plug are upper pin chambers that each hold a spring and a top pin.

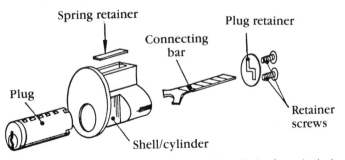

**5-1**  Partially exploded view of a pin tumbler cylinder for a rim lock.

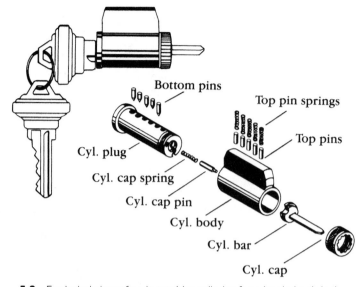

Bottom pins

Top pin springs

Top pins

Cyl. plug

Cyl. cap spring

Cyl. cap pin

Cyl. body

Cyl. bar

Cyl. cap

**5-2** Exploded view of a pin tumbler cylinder for a key-in-knob lock.
courtesy Schlage Lock Company

## OPERATION

The downward pressure of the springs drive the top pins down into the plug until they rest on the bottom pins. When a top pin is partially in an upper pin chamber and partially in a lower pin chamber, it prevents the plug from being rotated.

There's a small amount of space between the plug and the cylinder case. This space is called a *shear line*. Without a shear line, the plug could never be rotated, because it would fit too tightly in the cylinder case. When the right key is inserted into a pin tumbler lock, it lifts the bottom pins to the shear line, which causes the top pins to also be at the shear line. Figure 5-3 illustrates how the right key aligns the pins to the shear line.

When all the pins are at that position, the plug is free to rotate into the locked or unlocked position. While the plug is rotating, the lower pin chambers separate from the upper pin chambers. The top pins then rest on top of the plug (FIG. 5-4).

Usually all the top pins in a cylinder are the same height, but the bottom pins vary in length. In no case, however, is a bottom pin long enough to protrude into an upper pin chamber when no key is in the keyway. The key cuts can only lift the pins up, not pull them down. The depth of the cuts in a key must correspond to the depths of the bottom pins for the key to operate the lock.

## REMOVING A BROKEN KEY

Sometimes you might need to remove a piece of broken key from the plug of a pin tumbler lock. One way to accomplish this is to insert a thin

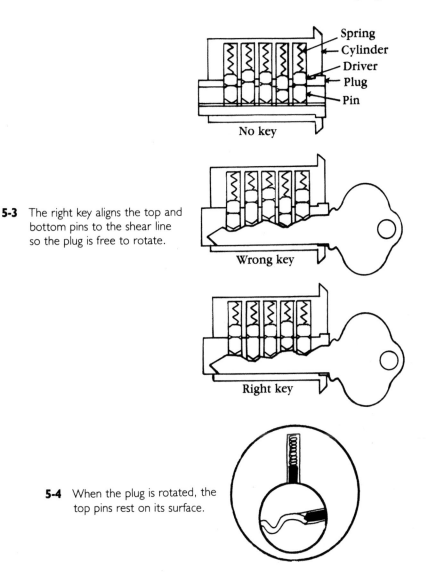

**5-3** The right key aligns the top and bottom pins to the shear line so the plug is free to rotate.

**5-4** When the plug is rotated, the top pins rest on its surface.

hooked piece of wire into the keyway, hook pointing down, and catch the end of the broken key (FIG. 5-5). Sometimes it's easier to use a thin saw blade to poke the broken key and manipulate it out of the plug (FIG. 5-6).

If part of the broken key is protruding out of the plug, you might be able to use a pair of pliers to pull it out. Locksmith supply houses sell a wide variety of broken key extractor tools.

## REKEYING

To rekey a pin tumbler cylinder, first remove the device retaining the plug—which will be a retainer clip, cam, or cylinder cap. The device varies among different pin tumbler cylinders.

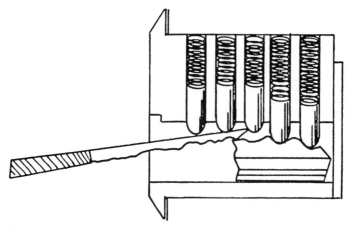

**5-5**   A thin hooked tool can be used to remove a piece of broken key from a cylinder.

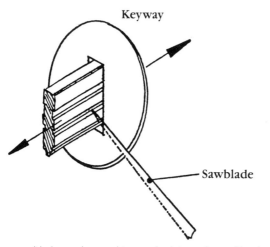

**5-6**   Sometimes a saw blade can be used to manipulate a piece of broken key out of a cylinder.

A retainer clip can be removed by prying it off with a small screwdriver. A cam can be removed by unscrewing the two small screws holding it in place.

To remove a cylinder cap (found on key-in-knob locks), depress the cap pin with an awl or edge of a small screwdriver, then rotate the cap counterclockwise until it comes off (FIG. 5-7). Remove both the cap pin and the small spring beneath it from the plug.

Rotate the plug about 15 degrees clockwise or counterclockwise. The easiest way to do this is by using a key. If no key is available, you'll need to pick or shim the lock. Be sure not to pull the plug forward while rotating it.

Now hold a plug follower or following bar firmly against the back of the plug (FIG. 5-8). Push the tool through the cylinder body. Make sure the plug follower is in constant contact with the back of the plug; if the plug and plug follower separate too soon, the top pins and springs will fall out of the upper pin chambers.

After the plug is pushed out of the cylinder body, set the cylinder body aside and set the plug into a plug holder, sometimes called a holding fixture (FIG. 5-9). It isn't essential to use a plug holder, but using one makes rekeying cylinders easier.

Remove the pins from the plug. Insert a new key. Then add appropriate size pins into the plug. Use bottom pins that reach the shear line of the plug with the new key in place. Figure 5-10 shows how the pins should be aligned.

There are three ways you can find the correct pins to use. One is by comparing the key bitting numbers written on the bows of some factory original keys to the cross-reference charts listed on most key pinning kits.

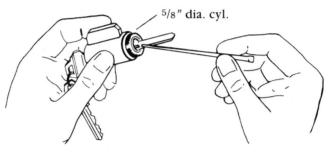

**5-7** Depress the cylinder cap pin and rotate the cylinder cap counterclockwise to remove the cap. Schlage Lock Company

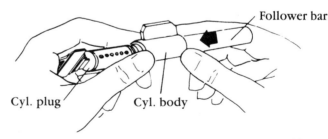

**5-8** Use a plug follower to hold the springs and top pins in place while removing the plug. Schlage Lock Company

**5-9** A plug holder can be used to hold the plug you're working on. Schlage Lock Company

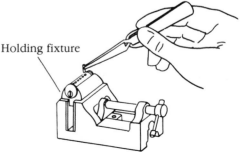

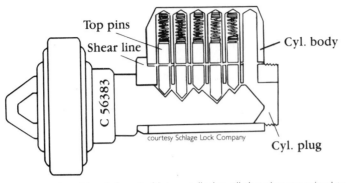

Top pins

Shear line

Cyl. body

C 56383

courtesy Schlage Lock Company

Cyl. plug

**5-10**   When a properly cut key is inserted into a cylinder, all the pins are raised to the shear line.

If the key doesn't have such a number you can use a key gauge or a caliper to measure the cuts of the key, then compare your readings with the information listed on your key pinning kit.

Another way to find the right pins is by trial and error. Look at the key and compare the depths of the cuts to one another. The deeper the cut in the key, the longer the pin for that cut must be. After finding the right pin for the first cut, use the size of that pin as a reference point for locating the other sizes you need.

Ideally, you should find pins that fit perfectly in the plug. In an emergency, you might want to use bottom pins that are too long and file them down with a fine mill file. Be sure to use the key to rotate the plug while filing, so you don't flatten the top of the plug (FIG. 5-11). Most locksmiths consider it unprofessional to file pins down.

After fitting the pins to the plug, insert the plug into the face of the cylinder body. Push the back of the plug against the plug follower until the tool comes out of the cylinder (FIG. 5-12). Do not allow the plug and the plug follower to separate from each other.

Hold the plug and cylinder body so the plug doesn't slip out of the body, and test the new key. Make sure the plug rotates easily and the key smoothly slides in and out of the plug (FIG. 5-13). If there is a problem at

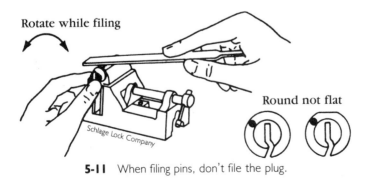

Rotate while filing

Round not flat

Schlage Lock Company

**5-11**   When filing pins, don't file the plug.

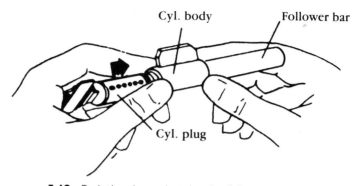

**5-12** Push the plug against the plug follower. Schlage Lock Company

this point, use the plug follower to remove the plug, and check the heights of the pins. Then put the plug back into the housing and test the key again.

After the key is working properly with the plug in the cylinder body, slowly remove the key. Don't pull hard enough for the plug to slip out of the cylinder body. Then reassemble the cylinder by reversing the procedure you used to disassemble it (FIG. 5-14).

## REPLACING TOP PINS

Many locksmiths purposely remove the springs and top pins every time they rekey locks. This allows the locksmith to clean the upper pin chambers and to replace any worn or broken top pins.

To reload the upper pin chambers, use a pair of tweezers or similar device to hold pins. With this device and a plug follower, you can easily replace the springs and top pins (FIG. 5-15). However, it takes a little practice. The first time you try loading upper pin chambers, you'll probably frequently drop pins and springs.

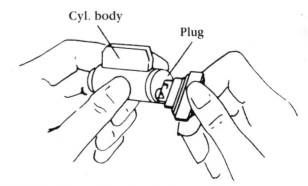

**5-13** Test the new key after rekeying a cylinder. Schlage Lock Company

## TUBULAR KEY LOCKS

The tubular key lock is basically a pin tumbler lock that has its tumblers arranged in a circle. Like any other pin tumbler lock, it has springs, top pins, and bottom pins. The tubular key has notches of varying sizes to correspond to the lengths of the pins. When the key is inserted into the keyway, it pushes the pins to the shear line so the plug can rotate.

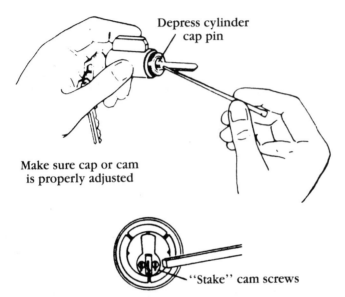

**5-14** Put the retaining device back in place after rekeying the cylinder. Schlage Lock Company

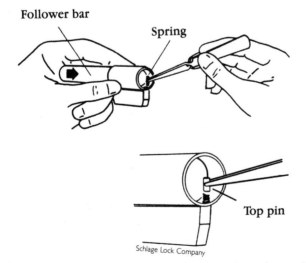

**5-15** It takes a little practice to feel comfortable replacing springs and top pins.

*Chapter* **6**

# Locksmithing tools

*I*n addition to the hand and power tools commonly found in hardware stores, locksmiths work with a variety of special tools. Some of the special tools are used only for locksmithing tasks. This chapter reviews many of those tools, and explains how to make some of them. This chapter also includes a list of all the tools needed to start a locksmithing business.

Any of the locksmithing tools you can't make yourself or find locally can be purchased from locksmith supply houses. Appendix E has a list of such suppliers. You can obtain catalogs from them by sending them a locksmith business card or letterhead, a locksmith license, a copy of your locksmith bond card, or similar evidence that you're a locksmith. Some suppliers only sell to locksmiths, but many also sell to other security professionals.

## ELECTRIC DRILL

An electric drill is a locksmith's most often used power tool (FIG. 6-1). It is used for installing locks, drilling locks, installing various exit devices, etc. A high-quality drill can cost several times more than a low-quality drill, but it is usually worth the extra money. It will provide many years of heavy-duty service, and save you a lot of time and sweat.

There are three basic drill sizes: 1/4 inch, 3/8 inch, and 1/2 inch. A drill's size is based on the largest diameter drill bit shank the drill's chuck can accept without using an adapter. For example, a drill whose chuck can hold a drill bit shank up to 1/2-inch in diameter is a 1/2-inch drill.

A drill's power is a combination of chuck speed and torque. Chuck speed is measured in RPM, or revolutions per minute when spinning freely in the air. Torque refers to the twisting force at the chuck when the drill is being used to drill a hole. RPM alone isn't a good measure of a drill's power, because a drill slows down when doing work. More important than free-spinning speed is the speed of a drill's chuck while it is drilling a hole.

Chuck speed and torque is largely determined by the type of reduction gears a drill has. Reduction gears work somewhat like car gears. One gear, for example, allows a car to move quickly on a flat road; another is used to give the car more power to climb a steep hill. The analogy isn't perfect, however, because a drill (unlike a car) comes with a fixed set of reduction gears—you can't shift the gears of a drill.

A drill with a single-stage reduction gear set has a chuck that spins very fast in the air (high RPM), but slows down considerably when drilling a hole. A drill with a three-stage reduction gear set has fewer RPM but more torque. Generally speaking, the higher the reduction gear set, the slower and more powerful the drill.

Most $1/4$-inch drills have single-stage reduction gear sets. Chucks commonly spin at 2500 RPM or more. Such drills are lightweight, and are used primarily for drilling plastic, thin softwood, and sheet metal. Using a $1/4$-inch drill to drill hardwood or steel would be time consuming and could damage the drill.

A $3/8$-inch drill is faster than the $1/2$-inch drill, and provides more torque than the $1/4$-inch drill. The chuck of a $3/8$-inch drill usually spins at about 1000 RPM. The drill can be useful for drilling metals up to $3/8$ inch thick and wood up to $3/4$ inch thick.

**6-1**    An electric drill is an important tool for locksmiths.

The 1/2-inch drill is by far the most popular size among locksmiths. A 1/2-inch drill will usually have either a two-stage reduction gear set or a three-stage reduction gear set, and its chuck spins at about 600 RPM. The 1/2-inch drill is useful for drilling thick hardwood and steel; it is also useful for drilling softwood, sheet metal, plastic, etc.

Not all 1/2-inch drills are alike, however; they often differ greatly in quality and price. Some manufacturers label their drills as "heavy-duty," "professional," or "commercial." Such labels have no standard meaning. When looking for a high-quality drill it's best to ignore such labels, and look for specific features. Important features to look for include: two-stage or three-stage reduction gear sets, at least 600 RPM, variable speed reversible control switch, double insulation, anti-friction (needle or ball) bearings, and at least 4.5 amps of power.

Drills come with one or two fixed speeds or variable speeds. Variable speed drills are the most flexible; they have a switch that allows you to use any speed from 0 RPM up to the drill's highest speed. This enables you to drill different materials at different speeds.

Many drills also have a switch that reverses the direction of the chuck. This is useful for backing out screws or a stuck drill bit. Drills that have both variable speed control and a reversible drive switch are called variable speed reversible (or VSR).

Some drills feature double insulation. The drill is housed in nonconductive material such as plastic, and nonconductive material is also used to isolate the motor from other metal parts of the drill. Double insulation is a safety feature that protects the user from getting shocked. Most high-quality drills are double insulated, so don't mistake plastic housing as a sign of low quality.

Anti-friction bearings help a drill run smoothly and help it last longer. Low-quality drills use all plain sleeve bearings. A few high-quality drills use a well-planned mixture of both types of bearings.

The amperage (measured in amps) that a drill uses is a good indicator of its power. Usually the more amps a drill uses, the more powerful it is. Look for a drill that uses at least 4.5 amps.

## CORDLESS DRILL

A cordless drill is usually lighter and more convenient than an electric drill (FIG. 6-2). The cordless drill is operated with battery power. It can be very useful when ac power isn't available, but a cordless drill isn't as powerful as an electric drill.

## BROKEN KEY EXTRACTOR

A broken key extractor is used to remove broken pieces of a key from a lock. You can make the tool with a 4-inch piece of a hacksaw blade. Starting at either end of the blade, grind off a 1 inch length of the non-toothed side of the blade until that 1 inch length is about 1/4 inch thick. Wrap electrical tape around the other 3 inches of the blade. The taped 3-inch part is

courtesy Porter Cable Corp.

**6-2** A cordless drill can be useful to a locksmith when electric power isn't available.

used as the handle, and the narrow 1-inch section is used for entering a keyway.

## PLUG FOLLOWER

A plug follower is used to hold top springs and pins in the cylinder when the plug is removed. It must be about the same size as the plug. Locksmithing supply houses sell plug followers in various sizes; most of them are made of bar stock.

You can make your own plug followers out of wooden dowels or copper tubing. They should each be about 4 inches long and about 1/2 inch in diameter; you might need smaller diameters for some locks.

## PLUG HOLDER

A plug holder holds a plug while you're servicing or rekeying it. You can make one by using a hacksaw to cut off the bottom quarter of an old cylinder that's the same size as the plug you're working on.

## LIST OF TOOLS NEEDED TO START A LOCKSMITHING BUSINESS

| | |
|---|---|
| Allen wrenches | Bench grinder with wire brush |
| Auger bits, 1-inch | Bezel nut wrench (FIG. 6-4) |
| Automobile opening tools (FIG. 6-3) | Bolt cutter, 16-inch |
| Aviation snips, 10-inch | Boring jigs |
| Awls (steel) | Broken key extractors (FIG. 6-5) |

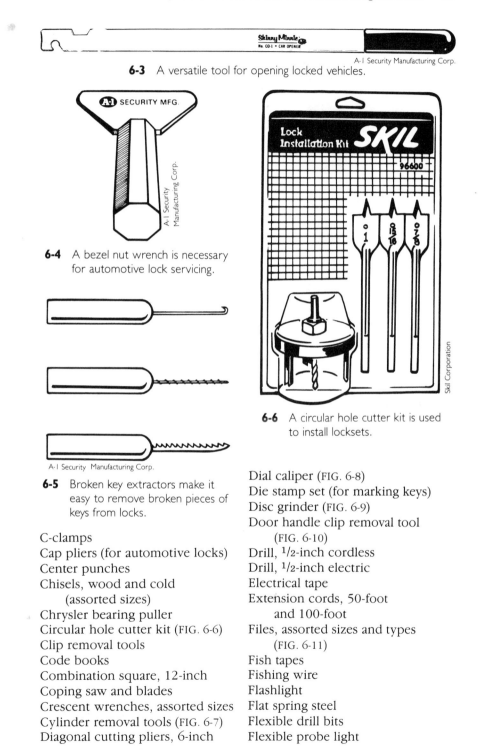

**6-3** A versatile tool for opening locked vehicles.

A-1 Security Manufacturing Corp.

**6-4** A bezel nut wrench is necessary for automotive lock servicing.

A-1 Security Manufacturing Corp.

**6-5** Broken key extractors make it easy to remove broken pieces of keys from locks.

**6-6** A circular hole cutter kit is used to install locksets.

C-clamps
Cap pliers (for automotive locks)
Center punches
Chisels, wood and cold
 (assorted sizes)
Chrysler bearing puller
Circular hole cutter kit (FIG. 6-6)
Clip removal tools
Code books
Combination square, 12-inch
Coping saw and blades
Crescent wrenches, assorted sizes
Cylinder removal tools (FIG. 6-7)
Diagonal cutting pliers, 6-inch

Dial caliper (FIG. 6-8)
Die stamp set (for marking keys)
Disc grinder (FIG. 6-9)
Door handle clip removal tool
 (FIG. 6-10)
Drill, 1/2-inch cordless
Drill, 1/2-inch electric
Electrical tape
Extension cords, 50-foot
 and 100-foot
Files, assorted sizes and types
 (FIG. 6-11)
Fish tapes
Fishing wire
Flashlight
Flat spring steel
Flexible drill bits
Flexible probe light

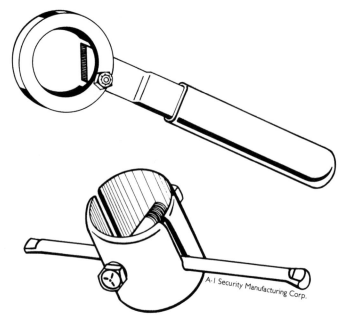

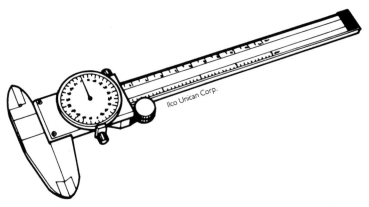

**6-7**  Cylinder removal tools are used to forcibly remove mortise cylinders.

**6-8**  A dial caliper is used to measure pin tumblers, keys, and blanks.

**6-9**  A disc grinder can be useful for removing high-security screws.

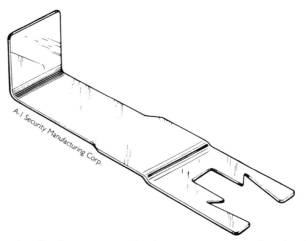

**6-10** A door handle clip remover makes it easy to remove a door handle from an automobile.

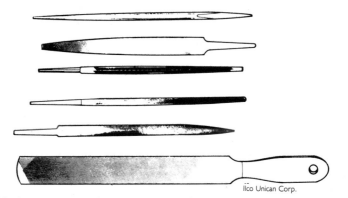

Ilco Unican Corp.

**6-11** Locksmiths use files to impression keys and to duplicate keys by hand.

GM lock decoder (FIG. 6-12)
GM puller
Grounding adapter
Hacksaw and blades
Hacksaw, mini
Hammers, claw and ball peen
Inspection scope
Key bitting punch (FIG. 6-13)
Key blank assortment
Key coding machine (FIG. 6-14)
Key duplicating machines
    (FIG. 6-15)
Key gauges
Level, pocket size
Lock parts assortment (FIG. 6-16)
Lock picking gun

Lock picks (FIG. 6-17)
Lock plate compressor tools
Lubricant (such as WD-40)
Mallet, rubber
Masking tape
Masonry bits, assorted sizes
Measuring tape, 50-foot
Moto-Tool
Multimeter
Nails and screws, assorted types
    and sizes
Needle-nose pliers, 6-inch
Pin kits, assorted types
Pin tray (FIG. 6-18)
Pin tweezers (FIG. 6-19)
Plastic glue

A-1 Security Manufacturing Corp.

**6-12**   A GM decoder is used to decode tumbler combinations of GM locks.

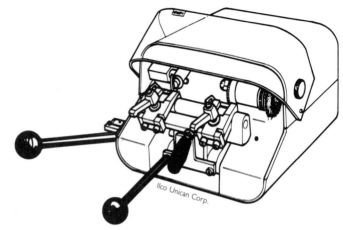

Ilco Unican Corp.

**6-15**   A key duplicating machine is used to copy keys.

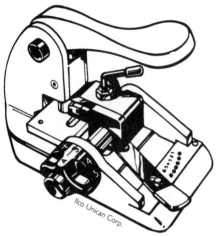

Ilco Unican Corp.

**6-13**   A key bitting punch stamps cuts into a key blade rather than milling them.

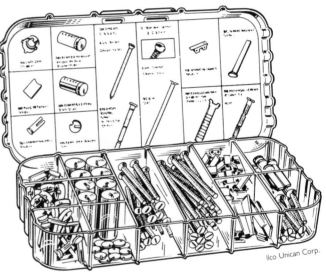

Ilco Unican Corp.

**6-16**   Lock parts are often needed when repairing locks.

Ilco Unican Corp.

**6-14**   A key coding machine is used to make a key when the original isn't available.

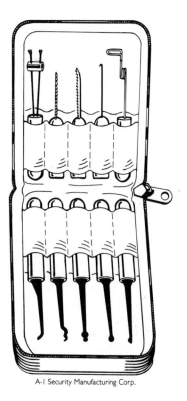

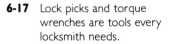
**6-17** Lock picks and torque wrenches are tools every locksmith needs.

A-1 Security Manufacturing Corp.

**6-18** A pin tray allows a locksmith to keep lock tumblers in order.

A-1 Security Manufacturing Corp.

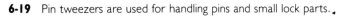
Ilco Unican Corp.

**6-19** Pin tweezers are used for handling pins and small lock parts.

Plug followers, assorted sizes
(FIG. 6-20)
Plug holders, assorted sizes
(FIG. 6-21)
Plug spinner (FIG. 6-22)
Pocketknife
Pry bar
Punch and die set in single- and
double-D configurations
Riveter and rivets
Sandpaper and emery cloth
Screwdrivers, Phillips and flat
(assorted sizes)
Shims (FIG. 6-23)
Slide hammer (dent puller)
Snap ring pliers, assorted sizes
Socket sets, standard and deep
Solder, rosin core
Soldering gun or iron
Spanner wrenches
Spindle assortment
Spring assortment
Spring steel assortment
Staple guns and staples
Steering wheel puller
Stramer remover
(for automotive work)
Strike locator (for installing locksets)
Strike mortising tools
(for installing locksets)
Stud locator
Terminal crimper
Tool boxes
Torque wrenches (FIG. 6-24)
Trim panel removal tool
Tubular key decoder (FIG. 6-25)
Tubular key lock picks (FIG. 6-26)
Tubular key lock saw (FIG. 6-27)
Vacuum cleaner, portable
VATS decoder
Vise
Vise grips, 5-inch and 10-inch
Warded padlock pick set
Wedges (FIG. 6-28)
Wire stripper
Wrenches, adjustable and pipe

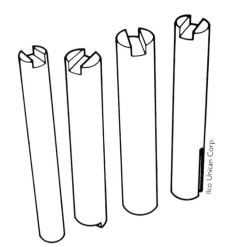

**6-20**   Plug followers are used for rekeying locks.

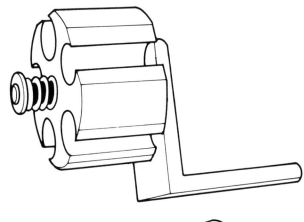

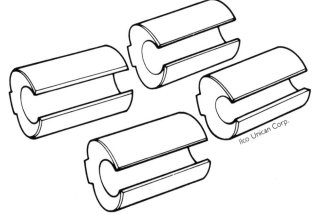

**6-21**   A plug holder holds plugs while they're being worked on.

**6-22** A plug spinner is used to spin a plug around when a lock has been picked in the wrong direction.

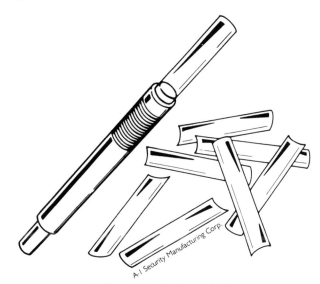

**6-23** A shim holder and shims.

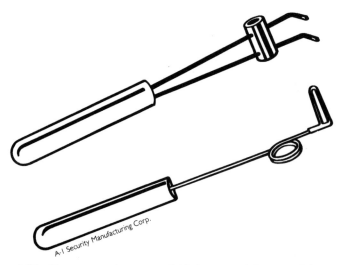

**6-24** Torque wrenches are available in assorted shapes and sizes.

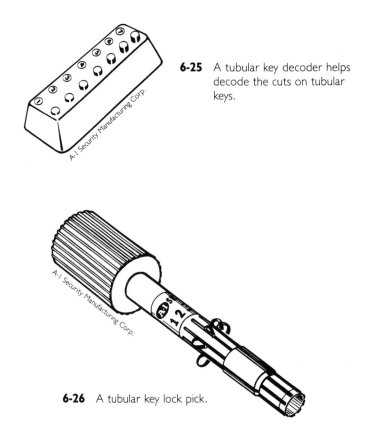

**6-25**   A tubular key decoder helps decode the cuts on tubular keys.

**6-26**   A tubular key lock pick.

**6-27**   A tubular key lock saw is used to drill the pins in a tubular lock.

A-1 Security Manufacturing Corp.

**6-28**   Wedges are used when opening locked vehicles.

A-1 Security Manufacturing Corp.

*Chapter* **7**

# Key-in-knob, deadbolt, and cylinder key mortise locks

$T$he information in this chapter is for products manufactured by Schlage Lock Company. But much of it also applies to products manufactured by other lock companies. The information will help you install and service most standard key-in-knob, deadbolt, and cylinder key mortise locks.

## HANDING OF DOORS

Door handing—placement on the right or left hand—is an important consideration for installing most locks. Some locks are nonhanded, and can be used with doors of any hand; others are for a specific hand. When a lock is installed on a door of the wrong hand, the lock will be upside down. This not only looks unprofessional, but could cause damage to the lock.

The hand of a door is based on the location and direction of swing of the door's hinges from the exterior side of the door. For example: If you're standing outside of a door that has hinges on the left side and that opens inward, you'd be looking at a left-hand door. Figure 7-1 shows the four types of door hands.

## KEY-IN-KNOB LOCK

A key-in-knob lock is one that is operated by inserting a key into its knob (FIG. 7-2). The knobs of key-in-knob locks come in a variety of designs. Figure 7-3 illustrates one of these.

**Single doors**          **Pairs of doors**

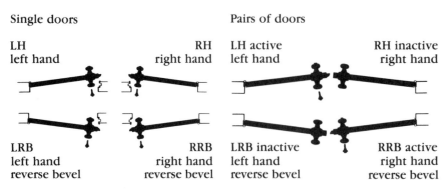

| LH | RH | LH active | RH inactive |
| left hand | right hand | left hand | right hand |

| LRB | RRB | LRB inactive | RRB active |
| left hand | right hand | left hand | right hand |
| reverse bevel | reverse bevel | reverse bevel | reverse bevel |

**7-1**   Consider door handing when installing locks. Schlage Lock Company

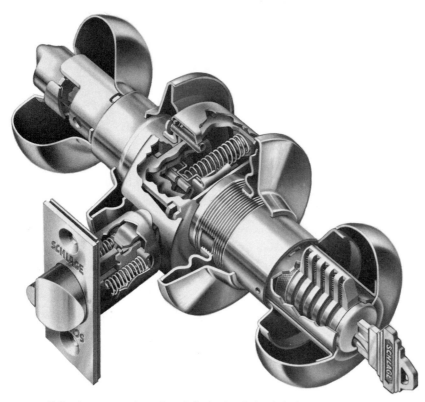

**7-2**   A cutaway view of an A Series key-in-knob lock. Schlage Lock Company

Schlage Lock Company invented this type of lock in 1925. The company's A Series key-in-knob locks are very popular among locksmiths. They are installed in the following way:

1. Use the template packaged with the lock to mark the door. Mark height line (center line of latchbolt) on edge of door. Suggested

Schlage Lock Company

**7-3** A popular knob design.

height from floor: 38 inches. Mark center point of door thickness. Position center line of template on height line. Hold in place and mark center point for $2^1/8$-inch hole (FIG. 7-4).

2. Bore a $2^1/8$-inch hole in the door. To avoid splintering the door, drill on both sides of the door to make the hole. Bore a $7/8$- or 1-inch hole, depending on latch housing diameter, straight into door edge to intersect with center of $2^1/8$-inch hole (FIG. 7-5). Then use latchfront faceplate to pattern for cutout. Front should fit flush with door surface. Install latch with screws.

For circular latch installation, drill a 1-inch-diameter latchbolt hole. Place a wooden block against bolt. Apply enough pressure to depress bolt. Tap block with hammer or mallet to drive latch into hole. Surface of latch faceplate should be flush with edge of door (FIG. 7-6).

3. Mark vertical line and height line on jamb exactly opposite center point of latch hole. Clean out hole and install strike.

For T strike, bore two $7/8$-inch holes, $^{11}/16$ inch deep in frame on vertical line $3/8$ inch above and below height line.

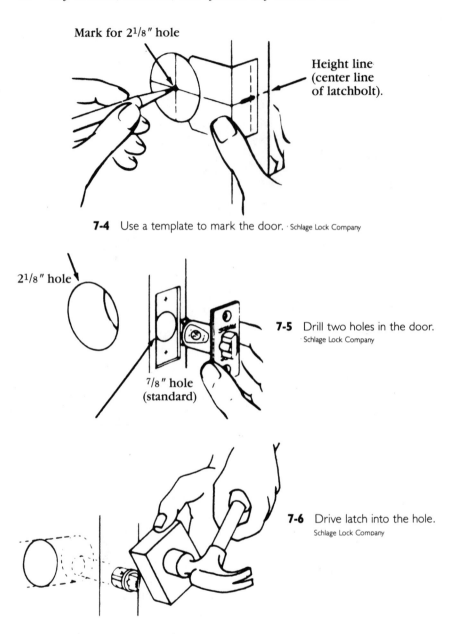

**7-4**   Use a template to mark the door.  · Schlage Lock Company

**7-5**   Drill two holes in the door.
  · Schlage Lock Company

**7-6**   Drive latch into the hole.
  Schlage Lock Company

For full lip strike, mark screw holes for strike so that screws lie on same vertical center line as latch screws. Cut out frame providing for clearance of latch bolt and strike tongue. Install strike (FIG. 7-7).

4. Remove inside trim by depressing knob catch, sliding knob off spindle, and removing appropriate rose design (FIG. 7-8).

5. Adjust rose by rotating it 1/16 inch short of housing for 13/8-inch-thick door. Rotate out to 3/16 inch for 17/8-inch door (FIG. 7-9).

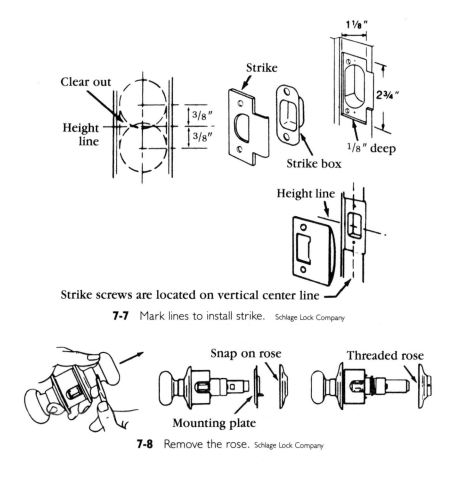

**7-7** Mark lines to install strike. Schlage Lock Company

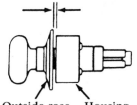

**7-8** Remove the rose. Schlage Lock Company

Rotate rose 1/16″ short of housing for 1 3/8″ thick door.

**7-9** Adjust the rose. Schlage Lock Company

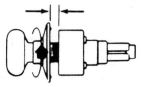

Outside rose    Housing

Rotate out to 3/16″ for 1 7/8″ door. This is the maximum adjustment.

6. Latch unit must be in place before installing lock. Be sure lock housing engages with latch prongs and retractor interlocks with latch bar (FIG. 7-10).

   For proper installation, deadlocking plunger on latch bolt must stop against strike, preventing forcing when door is closed. Do not attempt to mount lock unit with door closed.

7. Snap on rose. Slip mounting plate over spindle and fasten securely with two machine screws. Snap rose over spring clip on mounting plate (FIG. 7-11). Depress knob catch slide all the way onto spindle so catch engages into slot.

For a threaded rose, slip rose over spindle and screw onto threaded hub. Turn clockwise and tighten with spanner wrench (FIG. 7-12).

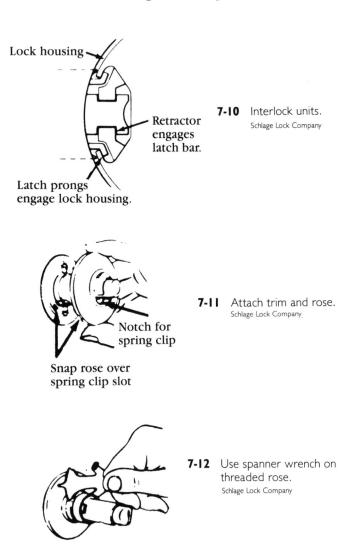

Lock housing

Retractor engages latch bar.

Latch prongs engage lock housing.

**7-10**  Interlock units.
Schlage Lock Company

Notch for spring clip

Snap rose over spring clip slot

**7-11**  Attach trim and rose.
Schlage Lock Company

**7-12**  Use spanner wrench on threaded rose.
Schlage Lock Company

## Servicing

When a key-in-knob lock isn't working properly, you might need to disassemble it to find out what the problem is. You might find a worn or broken spring. Or perhaps the lock needs to be cleaned and lubricated. The exploded lock views in FIGS. 7-13 through 7-16 shows the parts you'll find in typical key-in-knob locks. TABLE 7-1 is the list of part names corresponding to the numbered parts in FIGS. 7-13 through 7-16.

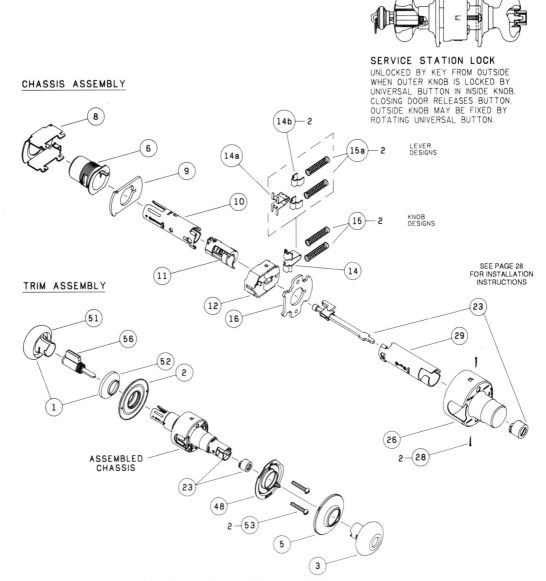

**SERVICE STATION LOCK**
UNLOCKED BY KEY FROM OUTSIDE
WHEN OUTER KNOB IS LOCKED BY
UNIVERSAL BUTTON IN INSIDE KNOB.
CLOSING DOOR RELEASES BUTTON.
OUTSIDE KNOB MAY BE FIXED BY
ROTATING UNIVERSAL BUTTON.

**7-13** Exploded view of Model A55PD lock. Schlage Lock Company

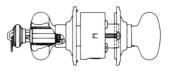

**CLASSROOM LOCK**
OUTSIDE KNOB LOCKED AND UNLOCKED BY
KEY. INSIDE KNOB ALWAYS UNLOCKED.

## CHASSIS ASSEMBLY

## TRIM ASSEMBLY

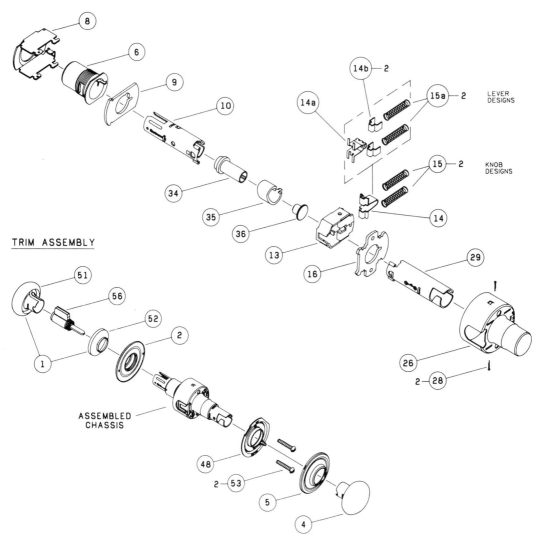

**7-14** Exploded view of Model A70PD lock. Schlage Lock Company

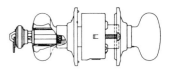

**STOREROOM LOCK**
OUTSIDE KNOB FIXED. ENTRANCE
BY KEY ONLY. INSIDE KNOB ALWAYS
UNLOCKED.

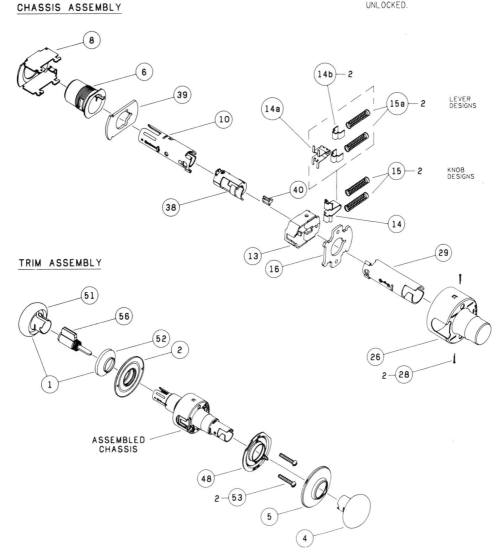

CHASSIS ASSEMBLY

TRIM ASSEMBLY

ASSEMBLED
CHASSIS

LEVER
DESIGNS

KNOB
DESIGNS

**7-15** Exploded view of Model A80PD lock. · Schlage Lock Company

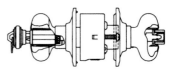

## HOTEL/MOTEL LOCK

OUTSIDE KNOB FIXED. ENTRANCE BY KEY ONLY. PUSH-BUTTON IN INSIDE KNOB ACTIVATES VISUAL OCCUPANCY INDICATOR, ALLOWING ONLY EMERGENCY MASTERKEY TO OPERATE. ROTATING INSIDE KNOB OR CLOSING DOOR RELEASES VISUAL OCCUPANCY INDICATOR. ROTATION OF INSIDE SPANNER-BUTTON PROVIDES LOCKOUT FEATURE BY KEEPING INDICATOR THROWN.

## CHASSIS ASSEMBLY

LEVER
DESIGNS

KNOB
DESIGNS

SEE PAGE 28
FOR INSTALLATION
INSTRUCTIONS

## TRIM ASSEMBLY

ASSEMBLED
CHASSIS

**7-16**  Exploded view of Model A85PD lock. Schlage Lock Company

**Table 7-1  Part one of Parts Index for Schlage A series locks.**

| SYM. | NO. | DESCRIPTION | A10S | A20S | A25D | A30D | A40S | A43D | A44S | A53PD | A55PD | A70PD | A73PD | A79PD | A80PD | A85PD |
|---|---|---|---|---|---|---|---|---|---|---|---|---|---|---|---|---|
| ① | 01-018** | KNOB & SLEEVE, CYLINDER | | | | | | | ● | ● | ● | ● | ● | ● | ● | ● |
| ② | 01-001** | ROSE, OUTSIDE | ● | ● | | ● | ● | ● | ● | ● | ● | ● | ● | ● | ● | ● |
| ③ | 01-009** | KNOB/LEVER, OPEN | | | | ● | ● | ● | ● | ● | | ● | | | | ● |
| ④ | 01-008** | KNOB/LEVER, CLOSED | ● | ● | ● | ● | | | | | ● | | | | ● | |
| ⑤ | 01-002** | ROSE, INSIDE | ● | | ● | ● | ● | | ● | ● | ● | ● | ● | | ● | ● |
| ⑥ | A201-399* | HUB & CAP, OUTSIDE | | | | ● | | | ● | ● | ● | ● | ● | ● | ● | ● |
| ⑦ | A201-406* | HUB & CAP, OUTSIDE | ● | ● | | | ● | ● | ● | | | | | | | |
| ⑧ | A508-598 | FRAME | ● | ● | ● | ● | ● | ● | ● | ● | ● | ● | ● | ● | ● | ● |
| ⑨ | A508-399 | PLATE, OUTSIDE | ● | ● | ● | ● | ● | ● | ● | ● | ● | ● | ● | | | |
| ⑩ | A301-387 | SPINDLE & CATCH, OUTSIDE | ● | ● | | ● | ● | | ● | ● | ● | ● | ● | ● | ● | ● |
| ⑪ | A301-402 | CAM, OUTSIDE | | | | ● | | | ● | ● | ● | | | | | ● |
| ⑫ | A590-158 | SLIDE | | | | ● | ● | | ● | | ● | | | | | ● |
| ⑬ | A590-159 | SLIDE | ● | ● | ● | | | | ● | | | ● | | ● | ● | |
| ⑭ | A508-597 | SEAT, SPRING (KNOB DESIGNS) | ● | ● | ● | ● | ● | | ● | ● | ● | ● | ● | ● | ● | ● |
| ⑭ⓐ | A501-525 | SEAT (2), SPRING (LEVER DESIGNS) | ● | ● | ● | ● | ● | ● | ● | ● | ● | ● | ● | | ● | ● |
| ⑭ⓑ | A501-645 | SEPARATOR, SPRING (LEVER DESIGNS) | ● | ● | ● | ● | ● | ● | ● | ● | ● | ● | ● | | ● | ● |
| ⑮ | A501-311 | SPRING (2), SLIDE (KNOB DESIGNS) | ● | ● | ● | ● | ● | ● | ● | ● | ● | ● | ● | ● | ● | ● |
| ⑮ⓐ | A508-605 | SPRING (2), SLIDE (LEVER DESIGNS) | ● | ● | ● | ● | ● | ● | ● | ● | ● | ● | ● | | ● | ● |
| ⑯ | A501-305 | PLATE, INSIDE | ● | ● | ● | ● | ● | ● | ● | ● | ● | ● | ● | | ● | ● |
| ⑰ | 01-055*** | PLUNGER & BUTTON, INSIDE | | | | ● | | | | | | | | | | |
| ⑱ | 01-056*** | PLUNGER & BUTTON, INSIDE | | | | | ● | | ● | | | | | | | |
| ⑲ | 01-057*** | PLUNGER & BUTTON, OUTSIDE | | | | ● | | | | | | | | | | |
| ⑳ | 01-058*** | PLUNGER & BUTTON, OUTSIDE | | | | | | ● | | | | | | | | |
| ㉑ | 01-059*** | PLUNGER & BUTTON, OUTSIDE | | | | | | | ● | | | | | | | |
| ㉒ | 01-060*** | PLUNGER & BUTTON, INSIDE | | | | | | | | ● | | | | | | |
| ㉓ | 01-061*** | PLUNGER & BUTTON, INSIDE | | | | | | | | | ● | | | | | |
| ㉔ | 01-062*** | PLUNGER & BUTTON, INSIDE | | | | | | | | | | | ● | | | |
| ㉕ | 01-063*** | PLUNGER & BUTTON, INSIDE | | | | | | | | | | | | | | ● |

 \*    SPECIFY FINISH.
 \*\*   SPECIFY DESIGN AND FINISH.
 \*\*\*  SPECIFY DESIGN, FINISH AND DOOR THICKNESS.

Schlage Lock Company

### Table 7-1   Part two of Parts Index for Schlage A series locks.

| SYM. | NO. | DESCRIPTION | A10S | A20S | A25D | A30D | A40S | A43D | A44S | A53PD | A55PD | A70PD | A73PD | A79PD | A80PD | A85PD |
|---|---|---|---|---|---|---|---|---|---|---|---|---|---|---|---|---|
| 26 | A301-403* | HOUSING & CAP. 1 3/8" - 1 7/8" DRS. | ● |  | ● | ● | ● |  | ● | ● | ● | ● | ● |  | ● | ● |
|  | A301-404* | HOUSING & CAP. 2" - 2 1/4" DRS. | ● |  | ● | ● | ● |  | ● | ● | ● | ● |  |  | ● | ● |
|  | A301-405* | HOUSING & CAP. 2 1/2" DRS. | ● |  | ● | ● | ● |  | ● | ● | ● | ● |  |  | ● | ● |
|  | A301-406* | HSG. THREADED 1 3/8" - 1 7/8" DRS. | ● |  | ● | ● | ● |  | ● | ● | ● | ● | ● |  | ● | ● |
| 27 | A508-600 | HOUSING |  | ● |  |  | ● |  |  |  |  |  | ● |  |  |  |
| 28 | C503-008 | COTTER PIN (2) | ● | ● | ● | ● | ● | ● | ● | ● | ● | ● | ● | ● | ● | ● |
| 29 | A301-386 | SPINDLE & CATCH, I/S, 1 3/8" - 1 7/8" DRS. | ● |  | ● | ● | ● |  | ● | ● | ● | ● | ● |  |  |  |
|  | A301-409 | SPINDLE & CATCH, I/S, 2" - 2 1/4" DRS. | ● |  | ● | ● | ● |  | ● | ● | ● |  |  |  | ● | ● |
|  | A301-410 | SPINDLE & CATCH, I/S, 2 1/2" DRS. | ● |  | ● | ● | ● |  | ● | ● | ● |  |  |  | ● | ● |
| 30 | A501-633 | SPINDLE, I/S, 1 3/8" - 1 7/8" DRS. |  | ● |  |  | ● |  |  |  |  |  |  |  |  |  |
|  | A500-001 | SPINDLE, I/S, 2" - 2 1/4" DRS. |  | ● |  |  | ● |  |  |  |  |  |  |  |  |  |
|  | A500-002 | SPINDLE, I/S, 2 1/2" DRS. |  | ● |  |  | ● |  |  |  |  |  |  |  |  |  |
| 31 | A501-498 | HUB, 1 3/8" - 1 1/2" DRS. |  |  | ● |  |  |  |  |  |  |  |  |  |  |  |
|  | A501-499 | HUB, 1 7/8" - 2 1/2" DRS. |  |  | ● |  |  |  |  |  |  |  |  |  |  |  |
| 32 | A301-391 | CAM |  |  |  | ● |  |  |  |  |  |  |  |  |  |  |
| 33 | A201-421 | SLIDE |  |  |  |  | ● |  |  |  |  |  |  |  |  |  |
| 34 | A501-721 | PLUG, CAM |  |  |  |  |  |  |  |  | ● | ● | ● |  |  |  |
| 35 | A501-776 | SPIRAL CAM |  |  |  |  |  |  |  |  | ● | ● | ● |  |  |  |
| 36 | A501-791 | SPACER, CAM |  |  |  |  |  |  |  |  | ● |  | ● |  |  |  |
| 37 | A501-768 | PLATE, INSIDE |  |  |  |  |  |  |  |  |  |  | ● |  |  |  |
| 38 | A201-370 | CAM |  |  |  |  |  |  |  |  |  |  |  | ● |  |  |
| 39 | A501-901 | PLATE, OUTSIDE |  |  |  |  |  |  |  |  |  |  |  |  | ● | ● |
| 40 | A501-615 | WEDGE |  |  |  |  |  |  |  |  |  |  |  |  | ● | ● |
| 41 | A201-782 | SLEEVE, CAM |  |  |  |  |  |  |  |  |  |  | ● |  |  |  |
| 42 | A201-558 | ROSE, OUTSIDE* |  | ● |  |  |  |  |  |  |  |  |  |  |  |  |
| 43 | A201-688 | TURN & PLATE, INSIDE* |  | ● |  |  | ● |  |  |  |  |  |  |  |  |  |
| 44 | A501-766 | ROSE, INSIDE** |  |  |  |  |  |  |  |  |  |  |  | ● |  |  |
| 45 | A501-767 | PLATE, MOUNTING |  |  |  |  |  |  |  |  |  |  |  | ● |  |  |

\*   SPECIFY FINISH.
\*\*  SPECIFY DESIGN AND FINISH.
\*\*\* SPECIFY DESIGN, FINISH AND DOOR THICKNESS.

Schlage Lock Company

**Table 7-1  Part three of Parts Index for Schlage A series locks.**

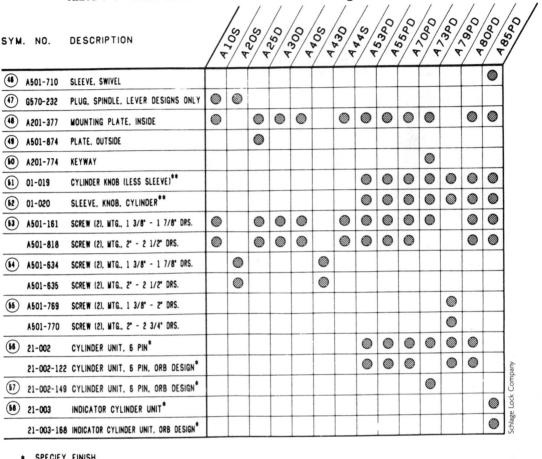

| SYM. | NO. | DESCRIPTION | A10S | A20S | A25D | A30D | A40S | A43D | A44S | A53PD | A55PD | A70PD | A73PD | A79PD | A80PD | A85PD |
|---|---|---|---|---|---|---|---|---|---|---|---|---|---|---|---|---|
| 46 | A501-710 | SLEEVE, SWIVEL | | | | | | | | | | | | | | ● |
| 47 | G570-232 | PLUG, SPINDLE, LEVER DESIGNS ONLY | ● | ● | | | | | | | | | | | | |
| 48 | A201-377 | MOUNTING PLATE, INSIDE | ● | | ● | ● | ● | | ● | ● | ● | ● | ● | | ● | ● |
| 49 | A501-874 | PLATE, OUTSIDE | | | ● | | | | | | | | | | | |
| 50 | A201-774 | KEYWAY | | | | | | | | | | | ● | | | |
| 51 | 01-019 | CYLINDER KNOB (LESS SLEEVE)** | | | | | | | | ● | ● | ● | ● | ● | ● | ● |
| 52 | 01-020 | SLEEVE, KNOB, CYLINDER** | | | | | | | | ● | ● | ● | ● | ● | ● | ● |
| 53 | A501-161 | SCREW (2), MTG., 1 3/8" - 1 7/8" DRS. | ● | | ● | ● | ● | | ● | ● | ● | ● | | | ● | ● |
|  | A501-818 | SCREW (2), MTG., 2" - 2 1/2" DRS. | ● | | ● | ● | ● | | ● | ● | ● | | | | ● | ● |
| 54 | A501-634 | SCREW (2), MTG., 1 3/8" - 1 7/8" DRS. | | ● | | | | ● | | | | | | | | |
|  | A501-635 | SCREW (2), MTG., 2" - 2 1/2" DRS. | | ● | | | | ● | | | | | | | | |
| 55 | A501-769 | SCREW (2), MTG., 1 3/8" - 2" DRS. | | | | | | | | | | | | ● | | |
|  | A501-770 | SCREW (2), MTG., 2" - 2 3/4" DRS. | | | | | | | | | | | | ● | | |
| 56 | 21-002 | CYLINDER UNIT, 6 PIN* | | | | | | | | ● | ● | ● | ● | ● | ● | |
|  | 21-002-122 | CYLINDER UNIT, 6 PIN, ORB DESIGN* | | | | | | | | | ● | ● | ● | | ● | ● |
| 57 | 21-002-149 | CYLINDER UNIT, 6 PIN, ORB DESIGN* | | | | | | | | | | | ● | | | |
| 58 | 21-003 | INDICATOR CYLINDER UNIT* | | | | | | | | | | | | | | ● |
|  | 21-003-168 | INDICATOR CYLINDER UNIT, ORB DESIGN* | | | | | | | | | | | | | | ● |

\* SPECIFY FINISH.
\*\* SPECIFY DESIGN AND FINISH.
\*\*\* SPECIFY DESIGN, FINISH AND DOOR THICKNESS.

## DEADBOLT LOCK

Figure 7-17 illustrates a typical deadbolt lock. A deadbolt is installed in much the same way as a key-in-knob lock. Figures 7-18 through 7-21 are exploded views of several types of deadbolt locks. TABLE 7-2 is the list of part names corresponding to the numbered parts in FIGS. 7-18 through 7-21.

**7-17**   Schlage deadbolt.

Schlage Lock Company

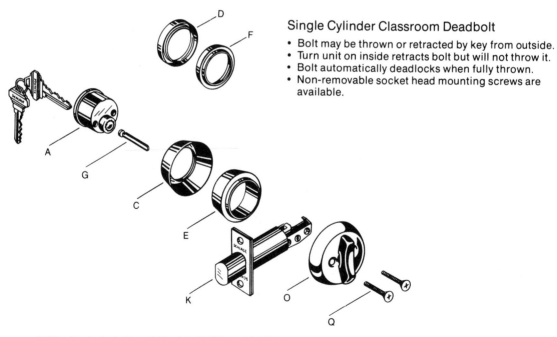

### Single Cylinder Classroom Deadbolt

- Bolt may be thrown or retracted by key from outside.
- Turn unit on inside retracts bolt but will not throw it.
- Bolt automatically deadlocks when fully thrown.
- Non-removable socket head mounting screws are available.

**7-18**   Exploded view of Models B263P and B463P locks. Schlage Lock Company

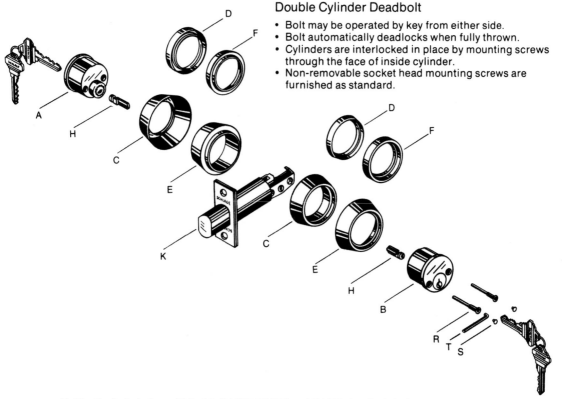

### Double Cylinder Deadbolt

- Bolt may be operated by key from either side.
- Bolt automatically deadlocks when fully thrown.
- Cylinders are interlocked in place by mounting screws through the face of inside cylinder.
- Non-removable socket head mounting screws are furnished as standard.

**7-19** Exploded view of Models B162P, B262P, and B462P deadbolt locks. Schlage Lock Company

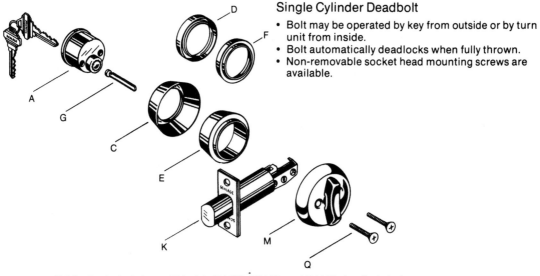

### Single Cylinder Deadbolt

- Bolt may be operated by key from outside or by turn unit from inside.
- Bolt automatically deadlocks when fully thrown.
- Non-removable socket head mounting screws are available.

**7-20** Exploded view of Models B160P, B260P, and B460P deadbolt locks. Schlage Lock Company

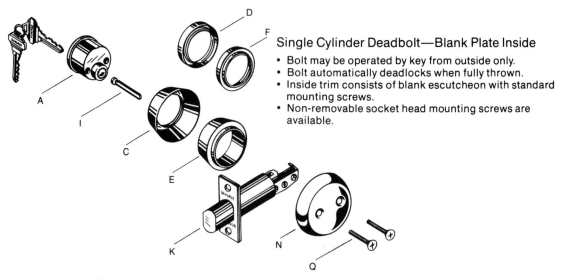

**Single Cylinder Deadbolt—Blank Plate Inside**

• Bolt may be operated by key from outside only.
• Bolt automatically deadlocks when fully thrown.
• Inside trim consists of blank escutcheon with standard mounting screws.
• Non-removable socket head mounting screws are available.

**7-21**   Exploded view of Models B261P and B461P deadbolt locks.   Schlage Lock Company

## CYLINDER KEY MORTISE LOCK

Commercial and industrial buildings are common places cylinder key mortise locks are used (FIG. 7-22). Those manufactured by Schlage Lock Company (L Series) are installed in the following way:

1. Measure desired height from finished floor, both sides and edge of door. Mark horizontal lines on door and door edge (FIG. 7-23).

2. Align template on edge of door with applicable horizontal at height line. Mark drill points (FIG. 7-24).

   Caution: The outside and inside of door might require different preparation. Read instructions thoroughly and use proper template for outside (exterior or corridor side) and inside of door.

3. Mortise door edge 1 inch. Drill $4^{1}/_{2}$ inches deep by $6^{3}/_{4}$ inches high (FIG. 7-25).

4. Drill proper holes for trim and lock function (FIG. 7-26).

   Note: For through holes, drill halfway from each side of the door to prevent splintering or otherwise damaging doors.

5. Recess to dimensions of lock face (FIG. 7-27).

6. Align strike template on jamb. Be sure to match center line on both strike and lock trim templates (FIG. 7-28). Bore 1-inch-diameter holes into door edge, $1^{1}/_{8}$ inch deep. Recess $5/_{32}$ inches for flush fit of strike and box. (Additional recess is required when using strike reinforcement.)

   Caution: Be sure auxiliary latch does not enter strike opening.

**Table 7-2  Part one of Parts Index for Schlage B series locks.**

| SYM. | NO. | DESCRIPTION | 160 260 460 | 261 461 | 162 262 462 | 263 463 | 264 464 | 180 280 480 | 560 | 562 | 250 | 251 | 252 | 270 |
|---|---|---|---|---|---|---|---|---|---|---|---|---|---|---|
| A | 22-017 | 5 Pin Cylinder Unit, Outside | • | • | • | • | | | | | • | • | • | |
| | 22-019 | 6 Pin Cylinder Unit, Outside | • | • | • | • | | | | | • | • | • | |
| B | 22-018 | 5 Pin Cylinder Unit, Inside | | | • | | • | | | | | | • | |
| | 22-020 | 6 Pin Cylinder Unit, Inside | | | • | | • | | | | | | • | |
| C | 36-067 | 7/16″ Trim Ring | • | • | • | • | • | | | | • | • | • | |
| D | 36-066 | 1/8″ Trim Ring | • | • | • | • | • | | | | • | • | • | |
| E | 36-069 | 7/16″ Security Insert, Std. B400 Series | • | • | • | • | • | | | | | | | |
| F | 36-068 | 1/8″ Security Insert, Std. B400 Series | • | • | • | • | • | | | | | | | |
| G | B202-323 | Cyl. Bar, 5 or 6 Pin, 1⅜″ or 1¾″ Dr.,Std. | • | | | • | | | | • | | • | | |
| H | B202-453 | Cyl. Bar, 5 Pin, 1⅜″ or 1¾″ Dr., (2 Req.), Std. | | | • | | | | | | | | • | |
| | B202-269 | Cyl. Bar, 6 Pin, 1⅜″ Dr. (2 Req.) | | | • | | | | | | | | | |
| | B202-369 | Cyl. Bar, 6 Pin, 1¾″ Dr. (2 Req.) | | | • | | | | | | | | • | |
| I | B202-267 | Cyl. Bar, 5 or 6 Pin, 1⅜″ or 1¾″ Dr.,Std. | | • | | | | | | | | | • | |
| J | E205-204 | Cyl. Bar, 5 Pin, 1⅜″ or 1¾″ Dr., Std. | | | | | • | | | | | | | |
| | B202-269 | Cyl. Bar, 6 Pin, 1⅜″ Dr. | | | | | • | | | | | | | |
| | E205-204 | Cyl. Bar, 6 Pin, 1¾″ Dr. | | | | | • | | | | | | | |
| K | 12-181 | Deadbolt, 5/8″ Throw, 2⅜″ BS., B100 Std. | • | | • | | | • | | | | | | |
| | 12-073 | Deadbolt, 1″ Throw, 2⅜″ BS., B400/B500 Std. | • | • | • | • | • | • | • | • | | | | |
| | 12-193 | Deadbolt, 5/8″ Throw, 2⅜″ BS., B200 Std. | • | • | • | • | • | • | | | | | | |
| L | 12-100 | Deadlatch, 9/16″ Throw, 2⅜″ BS., Std. | | | | | | | | | • | • | • | • |
| M | B202-321 | Rose & Turn, No Holdback | • | | | | | • | | | | | | |
| N | B502-815 | Blank Rose | | • | | | | | | | | • | | |
| O | B202-322 | Rose & Turn, One Way | | | | • | | | | | | | | |
| P | B202-320 | Rose & Turn, With Holdback | | | | | | | | | • | | | • |
| Q | B520-086 | Mtg. Screw, #10-32 x 2¼″, 1⅜″-1¾″ Dr., 5 or 6 Pin (2 Req.) | • | • | | • | | | | | • | • | | |
| R | B520-092 | Mtg. Screw, #10-32 x 2½″, 1⅜″ or 1¾″ Dr., 5 Pin (2 Req.) | | | • | | | | | | | | • | |
| | B520-094 | Mtg. Screw, #10-32 x 3″, 1⅜″ or 1¾″ Dr., 6 Pin (2 Req.) | | | • | | | | | | | | • | |

Schlage Lock Company

### Table 7-2   Part two of Parts Index for Schlage A series locks.

| SYM. | NO. | DESCRIPTION | 160 260 460 | 261 461 | 162 262 462 | 263 463 | 264 464 | 180 280 480 | 560 | 562 | 250 | 251 | 252 | 270 |
|------|-----|-------------|----|----|----|----|----|----|----|----|----|----|----|----|
| S | B502-894 | Drive Screw, (2 Req.) | | | ● | | ● | | | | | | ● | |
| T | B502-472 | Wrench | | | ● | | ● | | | | | | ● | |
| U | B520-112 | Bar Turn, 1⅜″ or 1¾″ Dr. | | | | | | ● | | | | | | ● |
| V | B502-409 | Backplate | | | | | | ● | | | | | | |
| X | F506-359 | Screw, Backplate, (2 Req.) | | | | | | ● | | | | | | |
| Y | B502-711 | Support | | | | | | | | | | | | ● |
| Z | B502-821 | Mtg. Screw, Wood Drs. | | | | | | ● | | | | | | ● |
| Z | B502-823 | Mtg. Screw, Metal Drs. | | | | | | ● | | | | | | ● |
| AA | B520-090 | Mtg. Screw, #10-32 x 1¾″, 5 & 6 Pin | | | | | ● | | | | | | | |
| AB | B502-497 | Anchor | | | | | ● | | | | | | | |
| BB | B202-317 | Cylinder Housing & Trim, Outside | | | | | | | ● | ● | | | | |
| BC | B520-098 | Mounting Plate | | | | | | | ● | ● | | | | |
| BD | B520-097 | Cylinder Housing, Inside | | | | | | | | ● | | | | |
| BE | B202-319 | Rose & Turn | | | | | | | ● | | | | | |
| BF | B520-101 | Trim, Housing, Inside | | | | | | | | ● | | | | |
| BG | B520-103 | Cylinder Guard | | | | | | | ● | ● | | | | |
| BH | B520-104 | Cyl. Guard Retainer Clip | | | | | | | ● | ● | | | | |
| BI | 22-002 | Cylinder, 6 Pin, (No Cylinder Bar) | | | | | | | ● | | | | | |
| BK | B202-269 | Cyl. Bar, 1⅝″ to 2⅛″ Dr., Std. (2 Req.) | | | | | | | | ● | | | | |
| BL | B520-102 | Cyl. Retainer, Outside | | | | | | | ● | ● | | | | |
| BM | B520-108 | Screw, Cylinder Retainer, (2 Req.) | | | | | | | ● | ● | | | | |
| BN | B520-107 | Screw, Mtg. Plate (4 Req.) | | | | | | | ● | ● | | | | |
| BO | B520-110 | Screw, Rose & Turn (2 Req.) | | | | | | | ● | | | | | |
| BP | A501-634 | Screw, Cyl. Hsg., Inside (2 Req.) | | | | | | | | ● | | | | |
| BQ | B520-105 | Plug, Cyl. Hsg. Screw (2 Req.) | | | | | | | | ● | | | | |
| BR | B520-111 | Screw, Hsg. Trim, Inside (1 Req.) | | | | | | | | ● | | | | |
| BS | 22-043 | Cylinder, 6 pin (No Cylinder Bar) | | | | | | | | ● | | | | |

FUNCTIONS

Schlage Lock Company

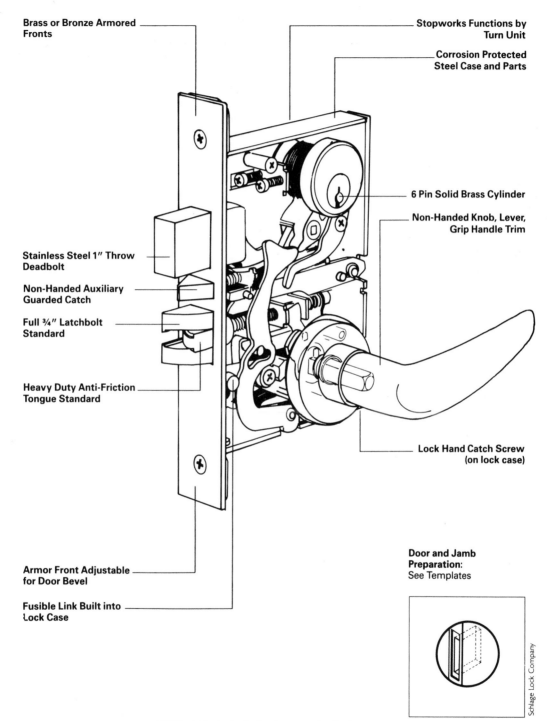

**Brass or Bronze Armored Fronts**

**Stopworks Functions by Turn Unit**

**Corrosion Protected Steel Case and Parts**

**6 Pin Solid Brass Cylinder**

**Non-Handed Knob, Lever, Grip Handle Trim**

**Stainless Steel 1″ Throw Deadbolt**

**Non-Handed Auxiliary Guarded Catch**

**Full ¾″ Latchbolt Standard**

**Heavy Duty Anti-Friction Tongue Standard**

**Lock Hand Catch Screw (on lock case)**

**Armor Front Adjustable for Door Bevel**

**Fusible Link Built into Lock Case**

**Door and Jamb Preparation:** See Templates

Schlage Lock Company

**7-22**   L Series cylinder key mortise lock.

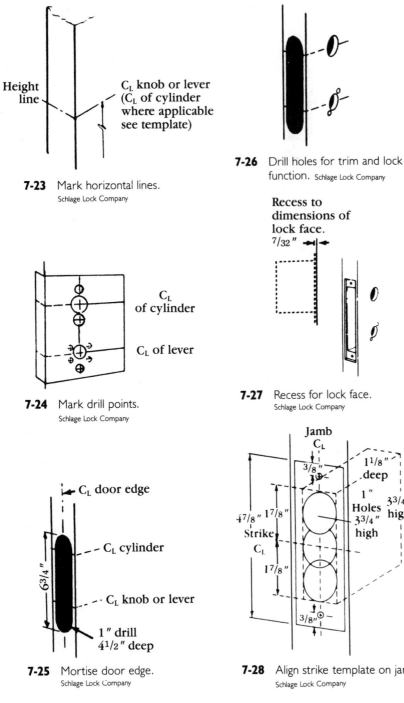

Height line

$C_L$ knob or lever ($C_L$ of cylinder where applicable see template)

**7-23** Mark horizontal lines.
Schlage Lock Company

$C_L$ of cylinder

$C_L$ of lever

**7-24** Mark drill points.
Schlage Lock Company

$C_L$ door edge

$C_L$ cylinder

$C_L$ knob or lever

6 3/4 "

1 " drill
4 1/2 " deep

**7-25** Mortise door edge.
Schlage Lock Company

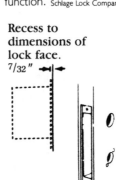

**7-26** Drill holes for trim and lock function. Schlage Lock Company

**Recess to dimensions of lock face.**
7/32 "

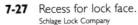

**7-27** Recess for lock face.
Schlage Lock Company

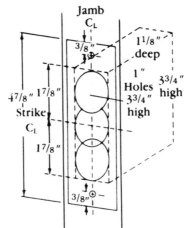

Jamb
$C_L$

3/8 "

1 1/8 " deep

1 " Holes 3 3/4 " high

4 7/8 " 1 7/8 "

Strike
$C_L$

3 3/4 " high

3 3/4 " high

1 7/8 "

3/8 "

**7-28** Align strike template on jamb.
Schlage Lock Company

7. Adjust for door bevel by loosening two screws on top and bottom of lock case. Position faceplate tabs at correct angle and tighten screws (FIG. 7-29).

   Insert lock case into mortise cutout (armor front removed) and fasten to door (FIG. 7-30). Catch screw must always be on inside of lock chassis.

8. Insert and install inside and outside spindle and springs so that pin stop is in contact with the hub of lock case (FIG. 7-31).

9. Trim installation (rose trim) levers. Thread screw posts onto outside rose. Place spring cage onto screw posts with arrows pointing in direction of lever rotation. Install outside trim onto spindle with screw posts through holes in lock case (FIG. 7-32).

10. Install inside lever trim. Join spring cage and mounting plate. Slide mounting plate over screw posts. Insert and tighten mounting screws. Place inside rose over mounting plate. Slide lever into position and tighten bushing with spanner wrench (FIG. 7-33).

11. Thread posts onto outside rose mounting plate (FIG. 7-34).

12. Install inside knob trim. Slide mounting plate onto spindle. Insert and tighten mounting screws. Place inside rose over mounting plate. Slide knob into position and tighten bushing with spanner wrench (FIG. 7-35).

13. Install cylinders (exposed type). Exposed cylinders are installed after escutcheon trim (FIG. 7-36). Tighten set screws against cylinders. Install armored front.

14. Install cylinders (concealed type). Cylinder plug must project approximately $7/16$ inch from face of door. Tighten set screws against cylinders. Install armored front.

15. Install escutcheon trim (lever trim only). Position spring cages on spindles with arrows pointing in direction of lever rotation. Slide outside trim unit onto spindle and mounting posts through door. Hold unit in place.

    Place inside trim unit onto spindle. Insert and tighten mounting screws (inside of door only). Turn piece must be vertical when bolt is retracted (FIG. 7-37).

16. Install turn piece and project deadbolt. Align screw holes on vertical and fasten with screws (FIG. 7-38).

17. Install emergency button by snapping into place (refer to template).

18. Install "Do Not Disturb" indicator plate (hotel function) and cylinder collar. Fasten with provided screw. Refer to template ($9/16$-inch-diameter hole).

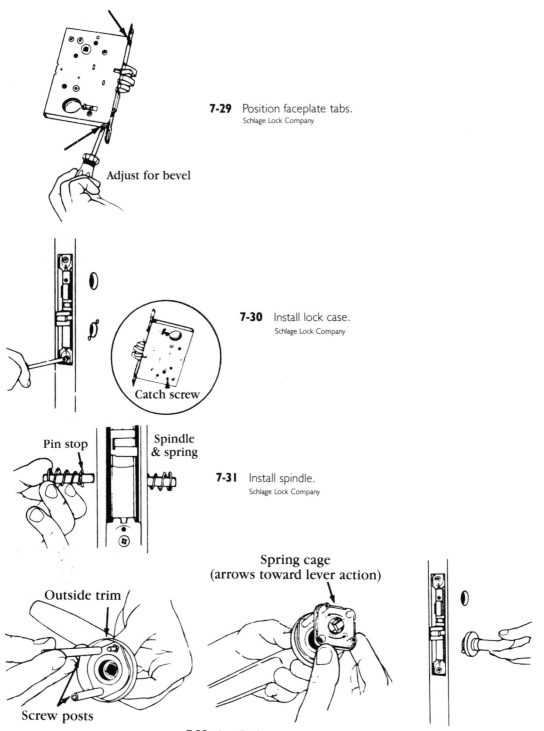

**7-29**  Position faceplate tabs.
Schlage Lock Company

Adjust for bevel

**7-30**  Install lock case.
Schlage Lock Company

Catch screw

Pin stop

Spindle & spring

**7-31**  Install spindle.
Schlage Lock Company

Spring cage
(arrows toward lever action)

Outside trim

Screw posts

**7-32**  Install trim. Schlage Lock Company

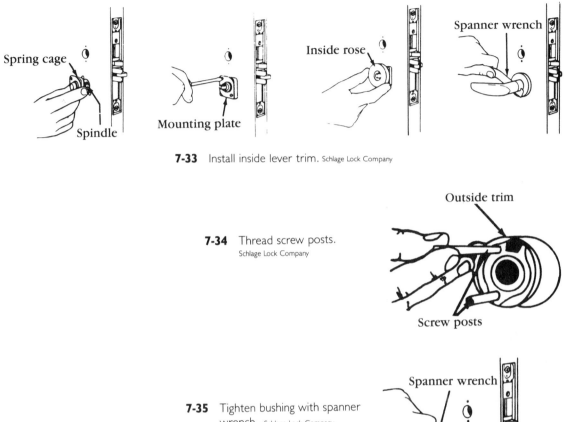

**7-33** Install inside lever trim. Schlage Lock Company

**7-34** Thread screw posts.
Schlage Lock Company

**7-35** Tighten bushing with spanner wrench. Schlage Lock Company

**7-36** Install cylinder.
Schlage Lock Company

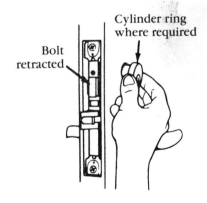

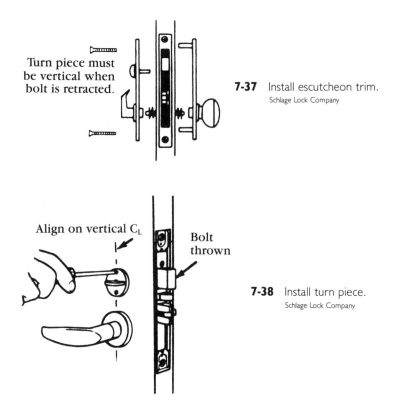

Turn piece must
be vertical when
bolt is retracted.

**7-37**   Install escutcheon trim.
Schlage Lock Company

Align on vertical C_L

Bolt
thrown

**7-38**   Install turn piece.
Schlage Lock Company

## Changing lock hand

1. Change latch handing: Remove armor front. Pull anti-friction tongue and latch bolt away from chassis and rotate complete unit, adjusting for door handing. Reinstall armor front (FIG. 7-39).

2. Change chassis handing: Remove catch screw from one side of chassis and install on opposite side. Catch screw must always be on inside of latch chassis for proper functioning of lock (FIG. 7-40).

## Grip handle installation

1. Insert outside spindle and inside spindle and spring. Inside spindle pin stop should be in contact with hub of lock case (FIG. 7-41).

2. Thread mounting posts and screws onto inside of grip handle. Lift thumbpiece while grasping handle and place on outside door (FIG. 7-42). Install wood screw at bottom of grip handle after completing assembly of inside trim. For metal door tap $^{7}/_{64}$-inch-diameter pilot hole.

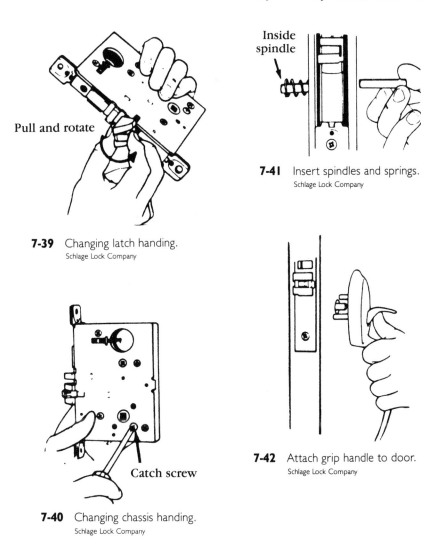

**Pull and rotate**

**7-39** Changing latch handing.
Schlage Lock Company

**Catch screw**

**7-40** Changing chassis handing.
Schlage Lock Company

**Inside spindle**

**7-41** Insert spindles and springs.
Schlage Lock Company

**7-42** Attach grip handle to door.
Schlage Lock Company

# Chapter 8

# High-security cylinders

$A$ lock cylinder is a complete operating unit that usually consists of a plug, shell, tumblers, springs, a plug, a cam/tailpiece or other actuating device, and other necessary operating parts. It contains a keyway, and is installed into locks to allow them to be operated with keys. Most high-security cylinders are designed to be used in locks made by many different manufacturers.

A high-security cylinder is one that has special features to thwart attempts to operate it without a proper key. Most high-security cylinders have built-in safeguards against picking, impressioning, and drilling. Such cylinders are usually operated by special keys, and provide various levels of key control.

The cylinders covered here reveal the wide range of differences among high-security cylinders. This chapter also discusses how the cylinders are constructed, rekeyed, serviced, and installed. Much of the information applies to other high-security cylinders.

## THE CORKEY SYSTEM

CorKey Control Systems, Inc. has patented kits that are used as replacement doorknobs and cylinders on most major brands of deadbolt, key-in-knob, and rim locks. The kits turn standard mechanical locks into card operated access control systems, usually without modifying the locks.

The *card operated replacement* kits (or Cor-Kits, for short) have been used at the U.S. Mint building in San Francisco, at the Honolulu International Airport, and at numerous hotels and universities throughout the world. Figure 8-1 shows a Cor-Kit used with a deadbolt.

Corkey Control Systems, Inc.

**8-1**   A Cor-Kit can be used with a deadbolt lock.

Corkey Control Systems, Inc.

**8-2**   A Cor-Kit, with key, installed on a key-in-knob lock.

With a Cor-Kit, locks can be opened by sliding a key into a slot at the top of the knob or cylinder. This slot then becomes the lock's new keyway. Figure 8-2 shows a Cor-Kit replacement knob and key. The key, called a CorKey, can be coded to operate one or several locks within a system.

The CorKey is flat and looks like a military dog tag. It is a magnetically coded card usually encased in a steel housing. Plastic housings are available for light-duty applications.

The keys and locks within a CorKey system can be decoded and recoded as often as the user chooses to do so. Coding of the locks and keys is controlled by the user of a Cor-Kit system. The manufacturer provides encoding equipment with the original setup.

## Installing Cor-Kits on deadbolt locks

The Cor-Kit 400 Series models are used to operate deadbolt locks and panic locks. The following instructions are for installing model 400R on rim cylinder lock hardware such as deadbolts and panic locks:

If you are installing a new deadbolt lock, follow the lock manufacturer's instructions to locate and drill the hole through the door for the cylinder. Do not mount the lock or strike. Go on to Step 1 below.

If you are replacing the cylinder on a lock remove the lock from the door. Then follow these simple instructions to mount the Cor-Kit on the outside of the door:

1. Place the two mounting screws through the two outside holes in the round steel plate that has three holes in it (FIG. 8-3).
2. Feed the screws through the hole in the door from the inside.
3. Hold the Cor-Kit on the outside of the hole, feeding the tailpiece through the center hole of the round steel plate.
4. Position Cor-Kit in hole and press to locate point to drill a 1/4-inch hole for pin, which secures the lock on the door against wrenching. Drill hole for pin 1/2-inch into door.
5. Screw the two screws into the threaded holes of the Cor-Kit until it is snug on the door—but not too tight.
6. Place hardware in position on the inside of the door, engaging Cor-Kit tailpiece. You will need the steel plate or washer that originally secured the cylinder.
7. If hardware was previously installed, make sure you can replace it in the same position as before—otherwise, it might be necessary to turn Cor-Kit housing or to file inside of hole in door to accommodate the two new screws.
8. Tighten Cor-Kit to the door.
9. Secure so there is no binding of Cor-Kit tailpiece.
10. If you are mounting the deadbolt lock for the first time, secure it to inside of door, keeping alignment with tailpiece of Cor-Kit so it will not bind. Then mount strike to match bolt location as the lock manufacturer directs.

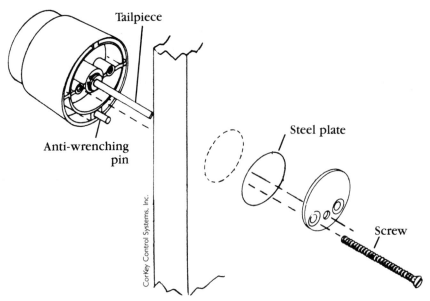

**8-3** An exploded view of a Cor-Kit for a rim lock.

## Operating the Cor-Kit

Follow these steps to operate a Cor-Kit:

1. Insert CorKey, arrow side out into slot.
2. When it meets resistance, start to turn complete front cylinder while applying added pressure on CorKey.
3. CorKey will go further into slot engaging tailpiece mechanism. Continue to turn to actuate bolt.
4. CorKey may be inserted or removed from any point around housing. Cor-Kit freespins whenever CorKey is released or removed.

## Installing Cor-Kits on Kwikset tubular deadbolts

The Cor-Kit Model 400K is used to operate Kwikset 800 Series tubular deadbolts; Model 485K is used for the Kwikset 885 Series. If you are installing a new deadbolt, follow the manufacturer's instructions to locate and drill the holes through the door for the cylinder, bolt, and strike. The following installation instructions apply to both series:

1. Remove inside thumbturn by removing two screws in rose, or if 885 Series (double cylinder unit), remove inside cylinder.
2. Back out two mounting screws until front cylinder and ring can be removed.
3. File or drill notch in bottom of hole in outside of door to accommodate anti-wrench projection on 400K housing for single cylinder, and on both outside and inside for double cylinder.

If you are putting Cor-Kit on outside of lock with thumb-turn on inside, continue to Step 4. If you have Model 485, skip to Step 8. (Note: For the double cylinder you must use the new type bolt shown in FIG. 8-4.)

4. Determine if your Kwikset is a new or old type by inspecting tail-piece hole in the bolt. If you have new type, insert two spacer bars to prevent tailpiece from turning in large hole (FIG. 8-5).

5. Install Cor-Kit in outside hole in door, feeding tailpiece through hole in bolt and into thumbturn hole. If too long, cut off. Remove thumbturn.

6. Secure Cor-Kit with mounting screws and plate, just as original cylinder.

7. Replace thumbturn and rose on inside, secure with two screws.

If you are installing Kwikset deadbolt follow manufacturer's instruction up to "Install Exterior Mechanism." Also do the following:

8. Identify inside and outside Cor-Kits. Outside unit has snap ring on center coupling, inside unit has hole in housing bottom for set screw (FIG. 8-6).

9. To disassemble for coding or installation, remove snap ring on outside unit. Pull out Cor-Kit cylinder and push out cylinder from inside unit. Be sure key works Cor-Kits.

Old type =          New type =

**8-4** Two types of Kwikset bolts. CorKey Control Systems, Inc.

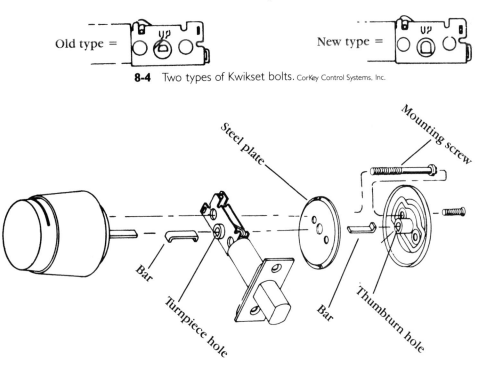

**8-5** A Model 400K Cor-Kit. CorKey Control Systems, Inc.

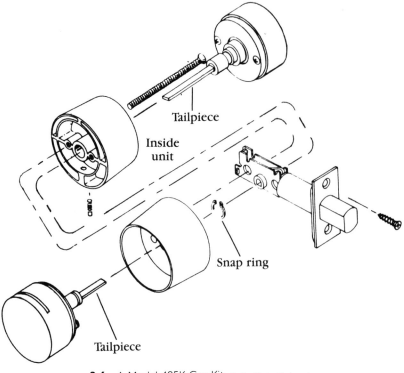

Tailpiece

Inside unit

Snap ring

Tailpiece

**8-6** A Model 485K Cor-Kit. CorKey Control Systems, Inc.

10. Reassemble outside unit, install snap ring.

11. Use Kwikset's instructions to install bolt.

12. Place Cor-Kit in outside hole so tailpiece goes through bolt drive hole. Mark tailpiece 1/4 inch from bolt body, cut off excess.

13. Temporarily assemble inside Cor-Kit. Hold in inside door hole with tailpiece through bolt drive hole. Mark it 1/4 inch from bolt body. Cut off excess.

14. Replace outside Cor-Kit in hole with tailpiece in bolt.

15. Remove inside cylinder from its housing.

16. Place eight 32-x-5/16-inch Nylok Allen screws on 5/16-inch Allen wrench. Feed carefully into hole in bottom of housing, into second hole in bottom of housing, and into second hole in hub. This starts Allen screws.

17. Place housing in inside hole and secure it to outside housing, through bolt with two screws provided.

18. Insert inside Cor-Kit, feeding tailpiece through bolt hole. Push Cor-Kit firmly all the way into housing.

19. Use Allen wrench and tighten screw.

20. Use Kwikset's instructions to install door strike.

## DOM IX KG SYSTEM CYLINDER

The cylinders in the DOM IX KG system (FIG. 8-7) have a horizontal keyway and two rows of teardrop-shaped and mushroom pin tumblers that are virtually impossible to pick. The tumblers are arranged in an offset position to one another. The heads of these pins are in the form of cutaway, bevelled half-discs, which are stabilized in their lateral position by the elliptically shaped cross-sectional design of the pin tumblers.

Special dimple keys are used to operate the cylinders (FIG. 8-8). The keys are designed to make unauthorized duplication very difficult. Key control is through DOM's registration certificate program; duplicate keys can be obtained from the factory with proper identification.

The keys include a patented floating ball that is integrated with the key. No locking is possible without that ball. Only after a deflection pin inside the keyway has been overridden by the floating ball is it possible for the ball to actuate the tenth blocking pin. The ball's mobility in its ball cage is just enough to enable it to override the deflection pin.

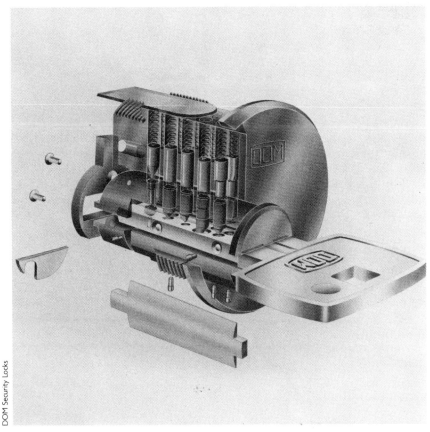

DOM Security Locks

**8-7** A DOM IX KG cylinder is extremely burglar resistant.

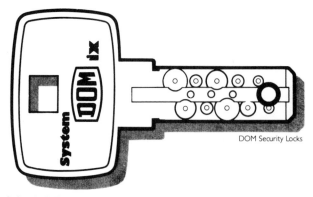

DOM Security Locks

**8-8**   A DOM IX KG key can be duplicated only by the factory.

When the key is completely inserted, the floating ball is pressed down by the lifting pin and operates the blocking pin. Only when all pin tumblers and profile control pins are in the right position and when the blocking pin has been operated correctly is it possible to operate the cylinder.

The profile pins, which are not controlled by springs and which operate laterally and vertically, are guided into position. When the key is inserted the pins each fall into their own respective borings, either from above or from the side. When all pin tumblers are in position and when all profile control pins are resting correctly in their key borings, the cylinder can be operated.

A rigid instrument such as a lock pick cannot operate the blocking mechanism of this lock, because the deflection pin protruding into the keyway prevents such an instrument from being inserted. In addition to this safety measure, each row of tumblers is equipped with two tapered core pins, which jam whenever an opening attempt is made with a picking instrument, thus rendering it virtually impossible to turn the core of the cylinder.

## Construction key

The construction key for the DOM IX KG cylinder (FIG. 8-9) allows temporary operation during the construction period. These cylinders are fitted with an insert that prevents the use of the permanent keys during the construction period. Only the construction key can be used during that phase. When the building is finished and handed over to the owner, the temporary inserts are removed so that only the original keys operate the cylinder. All construction keys become inoperative at this time.

## Split key

Any DOM IX KG cylinder can be equipped with a split key (FIG. 8-10). The split key is in two separate parts, which requires two persons to operate

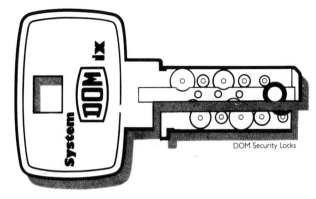

DOM Security Locks

**8-9** Construction keys for DOM IX KG cylinders allow temporary operation of cylinders.

the cylinder together. The "two-party compulsory locking" guarantees that the cylinder cannot be opened or closed by one person alone. This is especially useful to protect places like drug cabinets, banks, computer rooms, and evidence rooms.

## KABA GEMINI

Lori Corporation's Kaba Gemini cylinder is operated by dimple keys that have cuts drilled at precise angles. Figure 8-11 shows several types of such cylinders. They are UL listed and extremely difficult to pick or impression.

The cylinder can be used in deadbolt locks, key-in-knob locks, and padlocks. It fits the DIN standard profile cylinders (Hahn or Euro types), as well as American key-in-knob cylinder configurations. The American style cylinder is manufactured at the Lori plant in Connecticut.

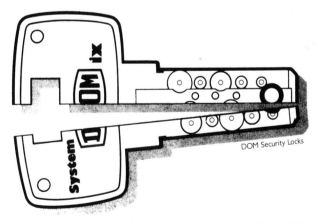

DOM Security Locks

**8-10** Any DOM IX KG cylinder can be equipped with a split key.

Lori Corp.

**8-11**    Kaba Gemini cylinders are available for a variety of lock types.

### Excerpt: Kaba High-Security Manual

*Note: Lori Corporation and Anthony "A. J." Hoffman, have given permission for the section of the Kaba High Security Manual (Copyright 1987) that pertains to Kaba Gemini cylinders to be reprinted here. A. J. Hoffman, CML, is the author of the manual. Slight modifications have been made to allow the information to fit the format of this book. The following information is, in substance, a reprint of section CST-2 of the manual.*

The Kaba Gemini Cylinder has two rows of side pins (FIG. 8-12). Keys for the cylinder have cuts made at 15-degree angles. Most dimple key machines are not capable of cutting Kaba Gemini keys. (Lori Corporation offers a machine designed to cut Kaba keys.) Many key machines that have 15-degree vises can't maintain the accuracy in spacing and depth required for consistent results on Gemini.

The edge has three active cut depths plus a #4 cut depth used in master keying (FIG. 8-13). Four depths are used on the sides. Remember, #1 depth is the deepest and #4 the shallowest. The increment is .35 millimeter (.0138 inch) for side and edge cuts.

There are ten edge positions available, but because of their close spacing, no two can be drilled adjacent to one another in the cylinder. The #1 position is not used at this time.

Stock cylinders (FIG. 8-14) are all drilled at right-hand or odd positions (3, 5, 7, 9) on the edge. This means that any stock Gemini cylinder may be rekeyed to any stock Gemini key. In order for the factory to guarantee that

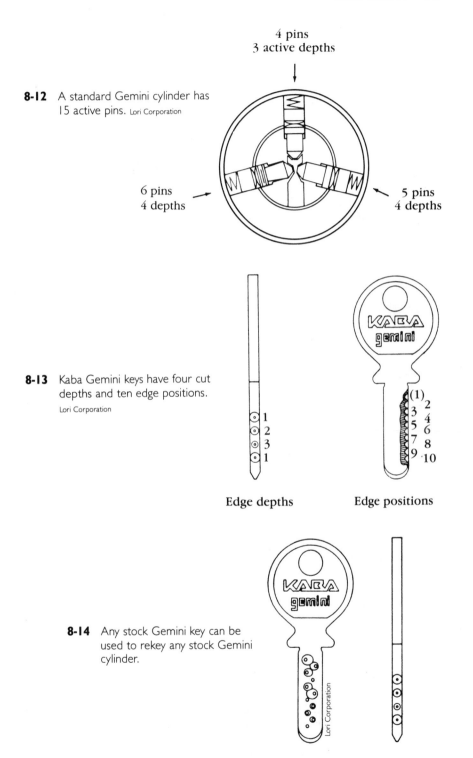

**4 pins**
**3 active depths**

**8-12** A standard Gemini cylinder has 15 active pins. Lori Corporation

**6 pins**
**4 depths**

**5 pins**
**4 depths**

**8-13** Kaba Gemini keys have four cut depths and ten edge positions.
Lori Corporation

1
2
3
1

(1)
2
3
4
5
6
7
8
9 ·10

**Edge depths**

**Edge positions**

**8-14** Any stock Gemini key can be used to rekey any stock Gemini cylinder.

Lori Corporation

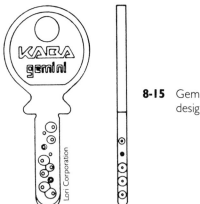

**8-15**   Gemini key for factory-designed keying system.

no stock Gemini key will ever operate a cylinder in a factory-designed keying system, cylinders furnished for keying systems (FIG. 8-15) are normally drilled at left-hand or even positions (2, 4, 6, 8, 10) on the edge. This means that persons servicing factory keying systems as well as stock cylinders will need both hands of the cylinders.

Gemini architectural hardware cylinders are UL listed. They are furnished with hardened steel pins as standard. This provides a high degree of drill resistance to the cylinder. The drill resistance runs the entire length of a Kaba Gemini cylinder, rather than only on the front portion. A minimum of two and a maximum of four mushroom bottom pins are used in each Gemini cylinder to provide resistance to both picking and impressioning. All pins have a hard electroless nickel plating to prevent corrosion.

Because the cylinder diameters are small, the drivers are compensated. The longer the bottom, the shorter the top pin (FIG. 8-16). For this reason, it isn't practical to service the cylinder with a plug follower.

When combining a cylinder, remember that the edge bottom pins are different from the side bottom pins. The same top pins, however, are used for both sides and the edges.

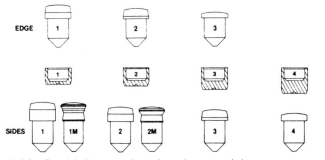

**8-16**   Gemini pins come in various shapes and sizes. Lori Corporation

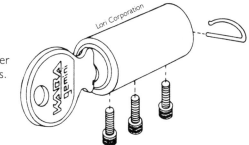

**8-17** A Gemini key-in-knob cylinder body uses three bible screws.

Because it isn't practical to use a plug follower, a special holding fixture is used for combinating and sleeving these cylinders. With practice it should take you about 5 minutes to combinate a Kaba Gemini cylinder.

The replacement cylinder product line for American architectural hardware is built around two basic units: the cylinder body and the core. The cylinder body (FIG. 8-17) is used for all key-in-knob cylinders. It is recognizable by the three tapped holes in the cylinder shell which accept "bible" screws, and by the horizontal hole in the rear of the plug. The core (FIG. 8-18) is used for all rim and mortise cylinders and some tubular deadlock cylinders. It has a hole in the shell to receive the core retaining screw.

## Kaba Gemini key-in-knob cylinders

All Kaba Gemini key-in-knob cylinders are built around one cylinder body. Bibles, adapters, and tailpieces of various shapes and sizes are attached to the cylinder body (FIG. 8-19), enabling it to assume many forms. This chameleon effect greatly reduces the investment required to be able to retrofit a wide variety of locksets with high-security cylinders.

Here's how this "build-a-cylinder" concept works: If most of your retrofit high-security sales are for "brand X" key-in-knob locks, you can stock only Kaba cylinders designed for that brand. Then if a customer needs a high-security cylinder for another brand, you can take the necessary parts from the build-a-cylinder kit and attach them to one of your cylinder's body. There's no need to wait for distributor or factory shipments or to stock dozens of different types of cylinders.

**8-18** A Gemini core.

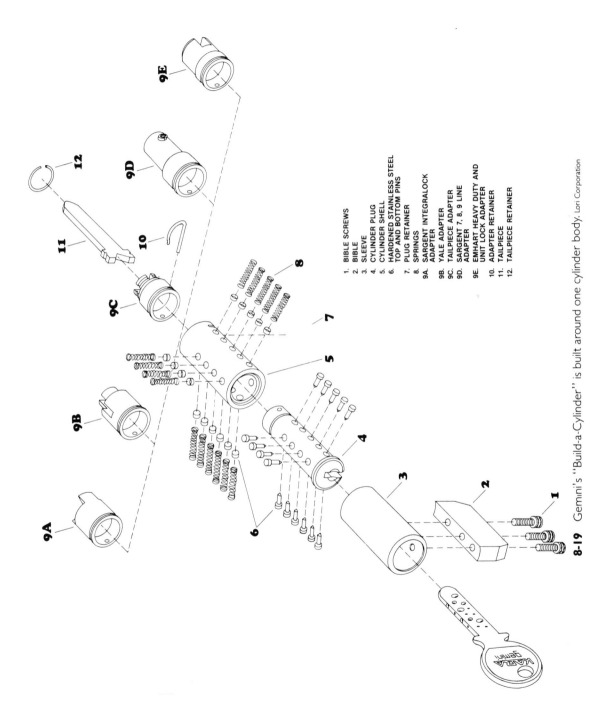

1. BIBLE SCREWS
2. BIBLE
3. SLEEVE
4. CYLINDER PLUG
5. CYLINDER SHELL
6. HARDENED STAINLESS STEEL
   TOP AND BOTTOM PINS
7. PLUG RETAINER
8. SPRINGS
9A. SARGENT INTEGRALOCK
   ADAPTER
9B. YALE ADAPTER
9C. TAILPIECE ADAPTER
9D. SARGENT 7, 8, 9 LINE
   ADAPTER
9E. EMHART HEAVY DUTY AND
   UNIT LOCK ADAPTER
10. ADAPTER RETAINER
11. TAILPIECE
12. TAILPIECE RETAINER

**8-19** Gemini's "Build-a-Cylinder" is built around one cylinder body. Lori Corporation

### Using Kaba Gemini cylinders in padlocks

When padlocks are needed with Gemini cylinders, Master Lock Company's 29 Series may be used. If you use the company's original cylinder retaining plate on the bottom of the padlock, the plate must be altered slightly.

Because the bible attaches to the Gemini cylinder at the bottom of the keyway rather than at the top, the plate must be installed 180 degrees from the position in which it was designed to be installed. There is a chamfer only halfway around the inside edge of the plate. The nonchamfered portion keeps the plate from being installed in the opposite position. You simply need to finish the chamfer, using a grinder or belt sander, then install the plate in a rotated position for the Kaba Gemini cylinder.

### Kaba Gemini core cylinders

Although technically not an "interchangeable core," this core is identical for all mortise, rim, and tubular deadlock cylinders (FIG. 8-20). In order to use the same core in various types and lengths of cylinders, a cam driver is used. This piece makes up for differences in the types of cams and tailpieces among cylinders.

The core is held into the housing by several screws, including a hardened retaining screw. This makes the core resistant to slide hammer attacks.

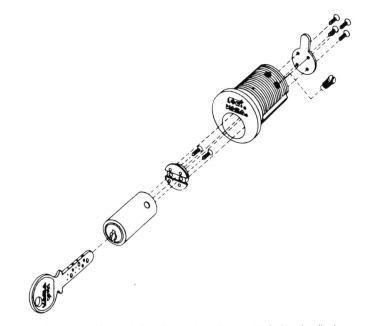

**8-20** Gemini's core is identical for all mortise, rim, and tubular deadlocks. Lori Corporation

## MEDECO CYLINDERS

Since the 1960s, Medeco Security Locks, Inc. has been a leading manufacturer of high-security locks and cylinders. The company's cylinders are UL listed and designed to resist picking, impressioning, hammering, wrenching, and prying. They are available for a wide assortment of lock types, including mortise, rim, deadbolt, and key-in-knob (FIG. 8-21).

### Principles of operation

Figure 8-22 is an exploded view of a typical Medeco cylinder, showing the important elements of Medeco's dual locking system. The first important element is a set of pin tumblers that must be elevated by the cuts of the key. The second element is the sidebar within the cylinder that requires the pin tumblers to be rotated to specific positions by the key. This rotation aligns a slot in each pin tumbler. When the pin tumblers have been properly elevated and rotated, the "fingers" or projections on the sidebar can enter the pin tumbler slots, which frees the plug to rotate to the locked or unlocked position (FIG. 8-23).

For additional security, Medeco cylinders are equipped with hardened steel inserts that protect strategic areas of the cylinder from drilling and surreptitious entry. Anti-drill rods surround the keyway and hardened inserts protect the shear line and sidebar areas of the cylinder.

A Medeco cylinder requires a key that must be made on a special key machine. This limits the availability of duplicate keys.

### Medeco Biaxial cylinders

The Biaxial cylinder is an improved version of Medeco's original cylinder. The two cylinders look similar and operate on the same principle. The primary differences between them are the angles of the bottom pins and the corresponding angles of the keys used to operate the cylinders.

**8-21**  Medeco cylinders are available for a variety of lock types.

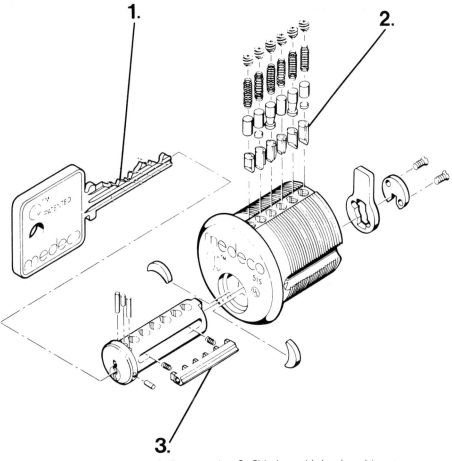

**8-22** 1. Medeco key. 2. Bottom pins. 3. Side bar with hardened insert.
Medeco Security Locks

Like the original Medeco cylinder, the Biaxial is UL listed, extremely burglar-resistant, and is available for a wide variety of locks. The Biaxial offers over 1 billion possible key changes—more than 50 times the changes the original cylinder design offered.

## SCHLAGE PRIMUS CYLINDERS

Manufactured by Schlage Lock Company, the Schlage Primus Security Cylinder offers a new dimension in security cylinders. It is operated by a unique "patent protected" key.

These cylinders are used for highly sensitive government installations, public and private institutions, as well as for commercial and residential applications. They are available for Schlage's A, B, C/D, E, H, and L Series locks, and allow for simple, cost-effective retrofitting.

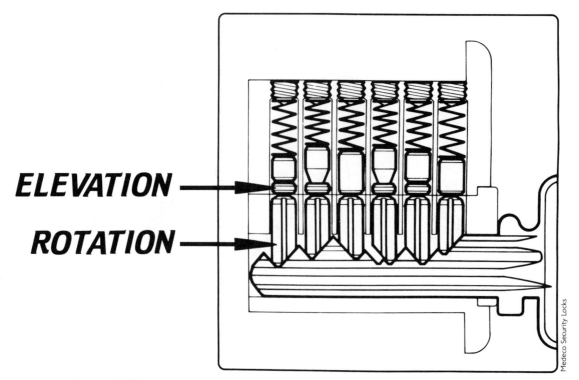

**ELEVATION**

**ROTATION**

**8-23**   Pins in a Medeco cylinder must be elevated and rotated by a key.

## Construction

Figures 8-24 and 8-25 show exploded views of Primus cylinders. The Primus is a six-pin cylinder, precision built to extremely close tolerances. It is machined to accept a side bar and a set of five finger pins.

The side bar and finger pins, in conjunction with Schlage's conventional six-pin keying system, provide two independent locking mechanisms that are operated simultaneously by a specially designed Primus key. Hardened steel pins are incorporated in the cylinder plug and housing to protect it from drilling and other forceful attacks.

## Primus key control

To meet a wide variety of security needs, Schlage has created four levels of control for Primus keys. Each level requires special registration and an identification card for duplication of the Primus key. Figure 8-26 shows the differences among the levels.

## Assembling Primus cylinders

To assemble a Schlage Primus cylinder, do the following:

1. Insert two sidebar springs into cylinder plug. Insert cylinder plug upside down into a plug holding fixture, with sidebar engaging slot in holding fixture (FIG. 8-27).

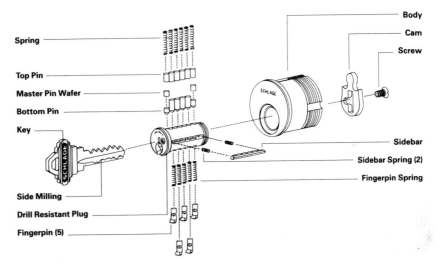

**8-24**   Exploded view of a Schlage Primus mortise cylinder. Schlage Lock Company

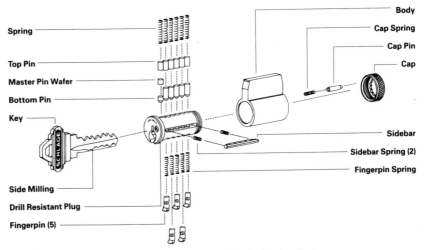

**8-25**   Exploded view of a Schlage Primus key-in-knob cylinder. Schlage Lock Company

2. Insert five finger pin springs into holes at the bottom of the plug (FIG. 8-28).

3. Determine the correct finger pins for each position as required by side milling of keys (FIG. 8-29).

4. Insert finger pins while advancing key upside down into cylinder plug in order to capture pins in position (FIG. 8-30). Maintain light pressure on finger pins while inserting key.

5. With key fully inserted, rotate plug slightly to ensure correct selection of finger pins (FIG. 8-31). Plug will not rotate if wrong finger pin(s) have been installed.

| SECURITY LEVEL ONE | ■ Keys Stocked Locally<br>■ Keying Locally Controlled<br>■ Standard Side Bit Milling<br>■ Positive I.D. Required<br>■ Serviced by Primus I Centers |
| --- | --- |
| SECURITY LEVEL TWO | ■ Keys Locally Stocked<br>■ Factory Masterkeying Available<br>■ Factory Side Bit Milling<br>■ Positive I.D. Required<br>■ Serviced by Schlage Primus II Centers |
| SECURITY LEVEL THREE | ■ Keys Factory Controlled<br>■ Random Selection of Side Bit Milling<br>■ Positive I.D. Required |
| SECURITY LEVEL FOUR | ■ Keys Factory Controlled<br>■ Restricted Side Bit Milling<br>■ Positive I.D. Required |

**8-26** Four levels of key control are available for Schlage cylinders. Schlage Lock Company

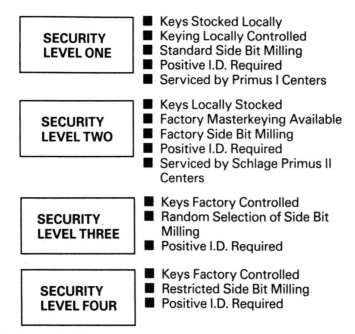

Slot    Sidebar    Springs

Cylinder plug
upside down

**8-27** Insert plug upside down.
Schlage Lock Company

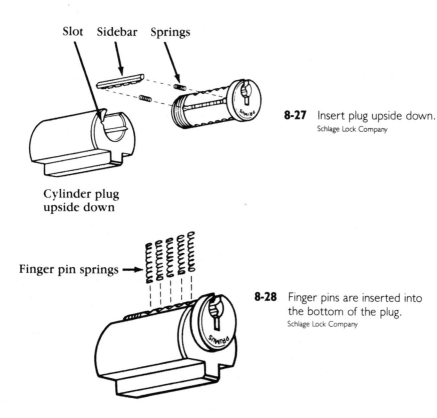

Finger pin springs ➔

**8-28** Finger pins are inserted into the bottom of the plug.
Schlage Lock Company

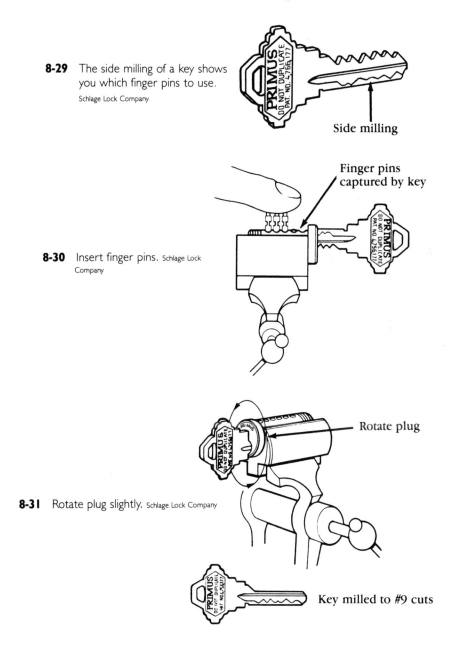

**8-29**  The side milling of a key shows you which finger pins to use.
Schlage Lock Company

Side milling

Finger pins captured by key

**8-30**  Insert finger pins. Schlage Lock Company

Rotate plug

**8-31**  Rotate plug slightly. Schlage Lock Company

Key milled to #9 cuts

Note: When master keying, use a key with the deepest cuts to complete plug and body assembly. If no one single key allows all bottom and master pins to be at or below the shear line, cut a key to all #9 cuts (deepest depth) for final assembly.

6. Select bottom and master pins from key bitting list and load cylinder plug. Add a small pinch of graphite into plug holes (FIG. 8-32).

7. Use chart to determine stack height of combined bottom and master pins and select correct top pins (FIG. 8-33, top). Insert top pins and springs into appropriate holes in loading rod (bottom).

8. Slide sleeve into loading rod. Depress springs with knife tip and advance sleeve along rod to guide groove (FIG. 8-34).

9. Slide cylinder body onto loading rod. Keep cylinder body slightly rotated (about 15 degrees) and push sleeve off until body is in line with groove in loading rod. Remove sleeve (FIG. 8-35).

10. Rotate cylinder body so that holes in loading rod align with holes in cylinder body. Use knife to transfer pins and springs into cylinder body. Rotate cylinder body to capture pins and springs (FIG. 8-36).

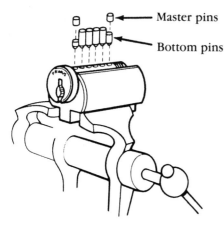

Master pins

Bottom pins

**8-32**  Select bottom and master pins. Schlage Lock Company

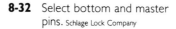

| Top Pin | 1 | 2 | 3 |
|---|---|---|---|
| Stack Height | 0-3 | 4-6 | 7-9 |

**8-33**  Use chart to determine stack height (top) and insert top pins (bottom). Schlage Lock Company

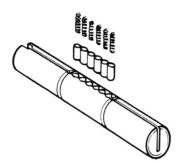

11. Remove cylinder plug from holder, with assembly key fully inserted, by maintaining pressure on side bar (FIG. 8-37). Keep cylinder body slightly rotated and slide body onto cylinder plug. When assembling mortise cylinders use opposite end of loading rod.

12. Rotate key and plug to align bottom and top pins. Maintain finger pressure on cylinder plug face while removing key (FIG. 8-38).

13. Complete cylinder assembly with cap, pin spring, cap pin, driver bar, and cap or mortise cylinder cam (FIG. 8-39).

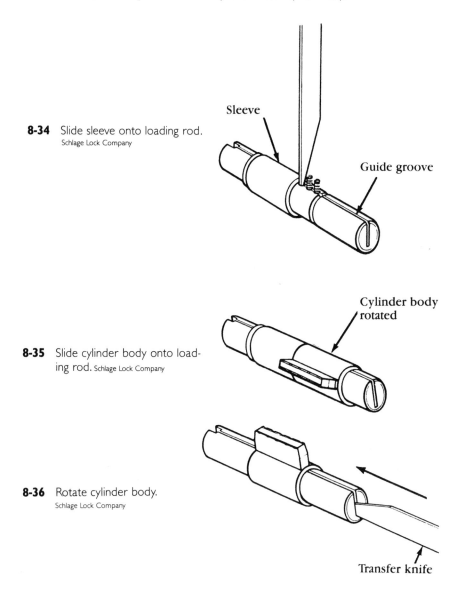

Sleeve

Guide groove

**8-34** Slide sleeve onto loading rod.
Schlage Lock Company

Cylinder body rotated

**8-35** Slide cylinder body onto loading rod. Schlage Lock Company

**8-36** Rotate cylinder body.
Schlage Lock Company

Transfer knife

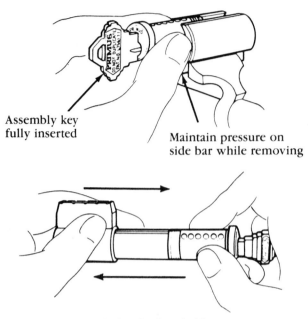

**Assembly key
fully inserted**

**Maintain pressure on
side bar while removing**

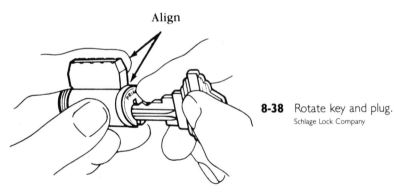

**8-37**  Remove cylinder plug from holder. Schlage Lock Company

**Align**

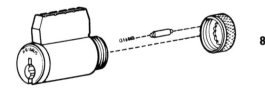

**8-38**  Rotate key and plug.
Schlage Lock Company

**Carefully
remove key**

**8-39**  Complete cylinder assembly.
Schlage Lock Company

*Chapter* **9**

# Pushbutton combination locks

*A* pushbutton combination lock is operated by pushing a series of buttons (usually labeled with letters or numbers) in unison or in sequence. It might or might not also be operated by a key. If a key is used, it is usually an override feature for emergencies only.

These locks are most often designed for commercial applications such as employee entrances, dormitories, and apartment lobbies, but some are available for residential applications. Modern pushbutton combination locks are ideal for a high-traffic business, because they offer thousands (and in some cases, millions) of possible combinations. It is virtually impossible for someone to randomly guess the right combination to operate one of the locks.

The combinations on pushbutton locks can usually be changed quickly and easily by authorized personnel. These locks can greatly reduce or even eliminate costs of issuing, collecting, and reissuing keys.

Simplex Access Controls Corporation is a major manufacturer of pushbutton combination locks. This chapter reviews how to install and service some of the company's most popular models.

## 1000 SERIES LOCKS

Simplex's Unican 1000 Series locks (FIG. 9-1) are completely mechanical; they don't require batteries or electrical wiring. The locks offer thousands of possible combinations, and any combination can be changed in seconds by any authorized person.

Simplex Access Controls Corp.

**9-1**   Simplex 1000 Series pushbutton combination locks are used in high-security areas.

To install the Models 1000-1 and 1000-2 (those without passage set) on 1⁵/₈- to 1⁷/₈-inch doors, proceed as follows:

1. Because every installation is unique, first carefully check windows, frame, door, etc. to make sure these procedures will not cause damage.

2. Carefully fold the paper template (FIG. 9-2) as indicated. Tape the template securely to the door, so that all indicated folds are properly aligned with the edge of the door. Using an awl (or similar marking tool), make the marks for drilling the four holes at precisely the points indicated on the template (FIG. 9-3). Remove the template.

    Caution: When a metal frame features an existing strike, be sure to locate the strike template, so that the latch hole center is directly aligned with the center of the strike cutout.

3. A hole saw with center guide drill bit should be used for drilling the 2¹/₈-inch (54mm) hole and the two 1-inch (25mm) holes. Use standard drilling bits for the two ¹/₄-inch (6mm) holes (FIG. 9-4).

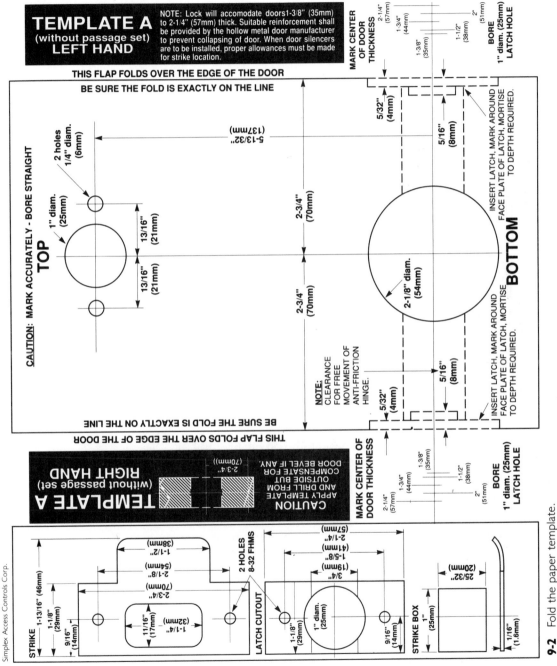

## TEMPLATE A
(without passage set)
### LEFT HAND

**NOTE:** Lock will accomodate doors1-3/8" (35mm) to 2-1/4" (57mm) thick. Suitable reinforcement shall be provided by the hollow metal door manufacturer to prevent collapsing of door. When door silencers are to be installed, proper allowances must be made for strike location.

THIS FLAP FOLDS OVER THE EDGE OF THE DOOR

BE SURE THE FOLD IS EXACTLY ON THE LINE

CAUTION: MARK ACCURATELY - BORE STRAIGHT

**TOP**

2 holes 1/4" diam. (6mm)

1" diam. (25mm)

13/16" (21mm)

13/16" (21mm)

2-3/4" (70mm)

5-13/32" (137mm)

2-3/4" (70mm)

**NOTE:** CLEARANCE FOR FREE MOVEMENT OF ANTI-FRICTION HINGE.

2-1/8" diam. (54mm)

**BOTTOM**

MARK CENTER OF DOOR THICKNESS

2-1/4" (57mm)
1-3/4" (44mm)
1-3/8" (35mm)
2" (51mm)
1-1/2" (38mm)

**BORE** 1" diam. (25mm) **LATCH HOLE**

5/32" (4mm)
5/16" (8mm)

INSERT LATCH, MARK AROUND FACE PLATE OF LATCH, MORTISE TO DEPTH REQUIRED.

5/32" (4mm)
5/16" (8mm)

INSERT LATCH, MARK AROUND FACE PLATE OF LATCH, MORTISE TO DEPTH REQUIRED.

BE SURE THE FOLD IS EXACTLY ON THE LINE

THIS FLAP FOLDS OVER THE EDGE OF THE DOOR

## TEMPLATE A
(without passage set)
### RIGHT HAND

2-3/4" (70mm)

**CAUTION** APPLY TEMPLATE AND DRILL FROM OUTSIDE BUT COMPENSATE FOR DOOR BEVEL IF ANY.

MARK CENTER OF DOOR THICKNESS

2-1/4" (57mm)
1-3/4" (44mm)
1-3/8" (35mm)
2" (51mm)
1-1/2" (38mm)

**BORE** 1" diam. (25mm) **LATCH HOLE**

**STRIKE**

1-13/16" (46mm)
1-1/8" (29mm)
9/16" (14mm)

1-1/2" (38mm)
2-1/8" (54mm)
2-3/4" (70mm)

11/16" (17mm)
1-1/4" (32mm)

2 HOLES 8-32 FHMS

**LATCH CUTOUT**

1-1/8" (29mm)
9/16" (14mm)

1" diam. (25mm)

2-1/4" (57mm)
1-5/8" (41mm)
3/4" (19mm)

**STRIKE BOX**

25/32" (20mm)
1" (25mm)

1/16" (1.6mm)

**9-2** Fold the paper template.

Simplex Access Controls Corp.

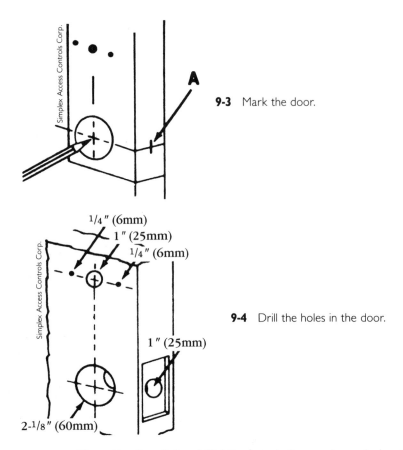

**9-3**  Mark the door.

**9-4**  Drill the holes in the door.

Place the tip of the drill bit against the mark made by the awl. Apply pressure evenly until the circular blade cuts the first side of the door and the tip of the pilot bit emerges through the other side, then stop. Drill through the other side until the 2$^1$/$_8$-inch circular hole is completed. Then use the same procedure to drill the 1-inch hole in door face.

Using a $^1$/$_4$-inch bit, drill the two holes. The final 1-inch hole is drilled through the edge of the door. Carefully drill until the hole saw is visible through the 2$^1$/$_8$-inch hole, then stop.

Caution: Make a mortise cutout for anti-friction hinge of latch.

**For Metal Doors:** Cut opening according to latch face plate. Mount top and bottom brackets to accept two latch mounting screws.

**For Wood Doors:** Insert latch into the 1-inch hole until the face plate abuts the door edge. Draw a line around the face plate. Then remove the latch from the door. Using a sharp 1-inch wood chisel, remove approximately $^1$/$_8$ inch (3mm) of material, or enough for the face plate to be perfectly flush with the edge of the door (FIG. 9-5).

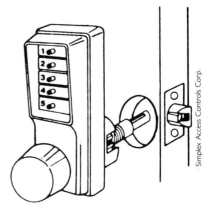

**9-5** Make latch cutout on wood doors.

4. Insert the latch into the 1-inch door edge hole until the latch face plate is flush with the edge of the door. Secure the latch face plate with two Phillips head screws, which are provided (FIG. 9-6).

5. Remove the inside knob by depressing the spring-loaded retaining pin that is visible through the poke hole. Pull the handle as the pin is depressed. Release the retaining pin. Continue to pull the handle. The handle has now been separated from the lock body.

6. Properly align the latch unit and the cylindrical drive unit by depressing the latch bolt slightly. Referring to FIG. 9-7: Make sure that the latch case clips (A) engage the front opening of cylinder drive unit cover (B). Then engage the end of the tailpiece (C) with the shoe retracting hood hooks (D).

7. Test the operation of the lock by depressing the factory-set combination: Press buttons 2 and 4 simultaneously, then press button 3 (FIG. 9-8).

Turn the outside knob clockwise to stop position. As the knob is turned, the latch is retracted until it is flush with the latch face plate. Release the outside knob. The latch will return

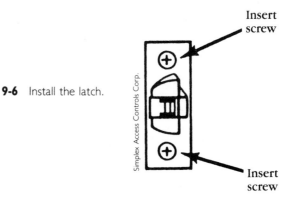

**9-6** Install the latch.

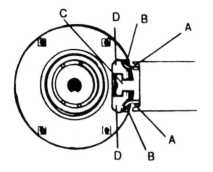

**9-7**  Install the lock. Simplex Access Controls Corp.

to the fully extended position. Turn the inside sleeve clockwise or counterclockwise to stop position. The latch will retract until it is flush with the latch face plate. After the inside sleeve is released, the latch bolt will return to the fully extended position.

8. Position the reinforcing plate onto the sleeve as illustrated in FIG. 9-9. Insert the two mounting screws through the reinforcing plate, and into the two 1/4-inch holes. Tighten the screws to secure the front lock housing to the door.

9. Position the trim plate over the reinforcing plate. Gently screw the threaded ring onto the cylindrical drive unit (FIG. 9-10). Tighten the threaded ring using the spanner wrench provided.

   Note: If you have difficulty screwing the ring onto the drive unit, remove the cover and readjust the reinforcing plate: Do not force the ring.

10. Locate the retaining pin through the poke hole in the collar of the knob (FIG. 9-11). If the threaded ring in the trim plate covers this poke hole, then loosen the ring and align the hole in the ring with the hole in the collar. Depress the knob retaining pin using the pointed tip of the spanner wrench. Slide the inside knob until the retaining pin snaps back into place.

11. Retighten the threaded ring. Turn the knob clockwise to stop position. The latch should retract smoothly until it is flush with

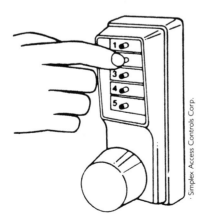

**9-8**  Test operation of the lock.

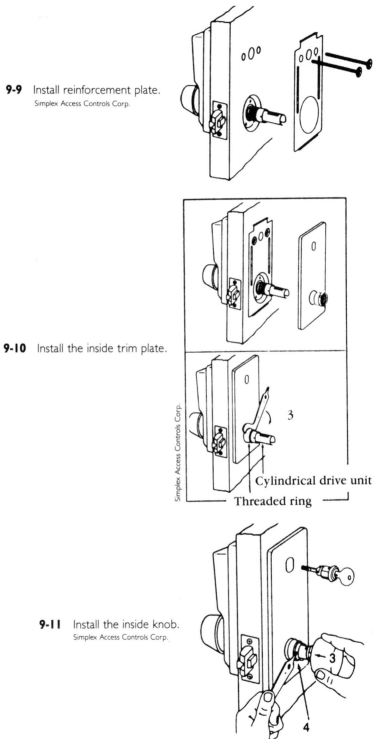

**9-9**   Install reinforcement plate.
Simplex Access Controls Corp.

**9-10**   Install the inside trim plate.

Cylindrical drive unit

Threaded ring

Simplex Access Controls Corp.

**9-11**   Install the inside knob.
Simplex Access Controls Corp.

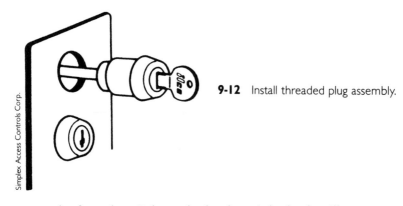

Simplex Access Controls Corp.

**9-12**  Install threaded plug assembly.

the face plate. Release the knob and the latch will return to the fully extended position.

12. Insert the correct key into the cylinder of the threaded plug assembly (FIG. 9-12). Engage the lock screw through the hole in the trim cover. The mounting stud is located in the 1-inch diameter cutout. If necessary, shorten the lock screw.

    Turn the key clockwise until the threaded plug assembly abuts the trim plate. The key can only be removed in either the vertical or horizontal position.

13. Mark location of the strike on the door frame. Make sure the line through the screw holes of the strike is well aligned with the line through the screw holes on the face of the latch when the door is closed (FIG. 9-13).

14. Mortise frame for strike box, a minimum depth of ³/₄ inches (19mm). This will guarantee that the latch can be fully extended

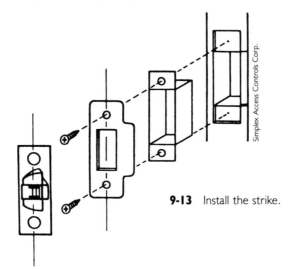

Simplex Access Controls Corp.

**9-13**  Install the strike.

into the door jamb. Place the strike box in the mortised cutout. Secure the strike plate with the screws provided. If necessary, draw a line around the strike. Use this line as a guide to cut out a minimum of $1/16$ inch (2mm) of material, or enough to make the strike plate flush with the door jamb.

Caution: Check the operation of the latch by making sure that the latch deadlocking plunger stops against the strike plate, and does not slip into the strike opening when the door is closed. If this situation occurs, then a total lockout can result. This will cancel the warranty of the complete lock mechanism.

If there is a gap between the edge of door and frame (or in the case of double doors, the edge of door and the edge of door) of more than $1/4$ inch, the latch will fail to engage the strike jamb. If necessary, adjust the gap using the rubber bumpers included in the lock box.

### Installation on $1^3/8$- to $1^1/2$-inch doors

The lock has been preset at the factory to accommodate $1^5/8$- to $1^7/8$-inch doors. For $1^3/8$- to $1^1/2$-inch doors, adjust the lock as follows (FIG. 9-14):

1. Remove the back plate and the cylindrical drive unit from the back plate.
2. Remove and discard the spacer between the back plate and the cylindrical drive unit. Remount cylindrical drive unit onto back plate using the shorter screws provided.
3. Remove cross pin from position B (refer to FIG. 9-14). Place new cross pin in position C. Reinstall back plate onto the front of lock case.

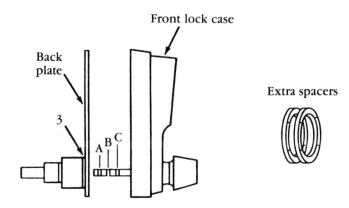

**9-14** Adjust the lock for varying door thickness. Simplex Access Controls Corp.

### Installation on 2- to 2¹/₄-inch doors

To adjust the lock to accommodate doors 2 to 2¹/₄ inches, proceed as follows:

1. Remove the back plate and the cylindrical drive unit from the back.

2. Insert the extra spacer provided in the accessory pack between the cylindrical drive unit and the back plate. Be sure that the sharp edges are adjoining.

3. Remount the cylindrical drive unit onto the back plate using longer screws provided. Add a cross pin in position A (refer to FIG. 9-14).

4. Remount back plate onto the front lock case.

   Warning: Damage may result if the knob hits against the wall or wall stop. If such is the case, all warranties are null and void.

### Reversing lock location

Unless otherwise stated in the manufacturer's literature included with a lock, all locks are factory-assembled for left-hand operation. To reverse the hand of a lock, proceed as follows (FIG. 9-15):

1. Remove the back plate, which contains the cylindrical drive unit, from the lock case. Unscrew the cylindrical drive unit from the back plate.

2. Turn the cylindrical drive unit so that the cutout for the latch faces in the opposite direction.

3. Reattach cylindrical drive unit to the back plate. Remount back plate onto the lock case.

4. Tighten all screws securely.

5. Test the lock to make sure it is still working properly.

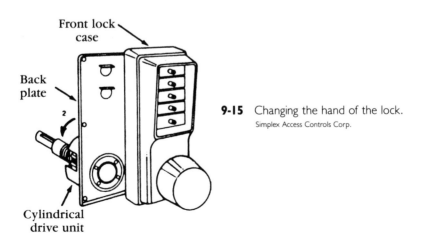

**9-15**  Changing the hand of the lock.
Simplex Access Controls Corp.

## SERIES 3000

The Simplex Series 3000 pushbutton combination locks (FIG. 9-16) are made for narrow stile doors. The locks are fully mechanical, and can be recombined in less than a minute by authorized personnel.

A key may be used to deactivate the pushbutton combination to prevent unauthorized reentry after hours, and to reactivate the combination for the next day's shift. Management's key may also activate the latch hold-back feature, which removes all security from the door.

For a complete installation, you'll need both a lock housing and a drive assembly; they are sold separately. Those narrow stile doors already equipped with Adams Rite hardware (latches, key cylinders, egress devices) need only the Simplex Series 3000 lock for a complete installation.

### Lock housing assembly

You'll need a large Phillips head screwdriver and a small and medium flat blade screwdriver to install a 3000 Series lock. The installation procedure is as follows:

1. Remove mounting plate assembly (FIG. 9-17) by removing one round head screw and one flat head screw.
2. Install mortise cylinder by turning cylinder until cam contacts lock in plate. Turn cylinder counterclockwise less than one turn until the key is positioned in pull position, also aligning to cylinder set screw grooves.

**9-16** Simplex 3000 Series locks are fully mechanical.

3. Assemble two cylinder positioning screws. Make sure the heads of these screws are below the underside of the cylinder cam. Reassemble mounting plate assembly.

4. Interposer arm should be positioned central to cylinder cam. If it is not, adjust cylinder depth. Assemble cam stop screw.

    (Note: Interposer stud to be centered as shown in FIG. 9-17 before mounting combination lock housing.)

## Mounting lock to stile

To mount a Series 3000 lock to the stile, proceed as follows:

1. Place combination lock over lock-in studs and slide downward until a stop position is attained—approximately ⅛ inch of movement (FIG. 9-18). Top of lock housing and top of drive assembly should be flush when properly positioned.

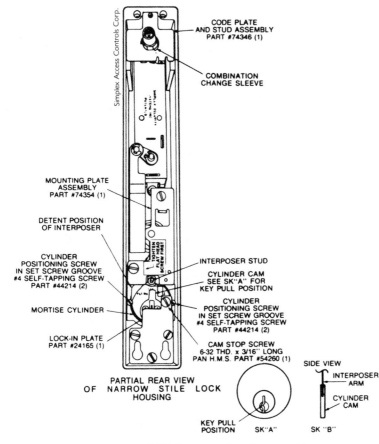

**9-17**　Simplex 3000 Series lock housing assembly.

2. Fix two each long round-head machine screws as shown in FIG. 9-18.

3. Using key, thread control lock assembly into combination change sleeve (FIG. 9-19) until trim plate is snug against stile. Key can only be removed in vertical or horizontal position. Before closing door, refer to operating instructions to check the operation of all lock functions.

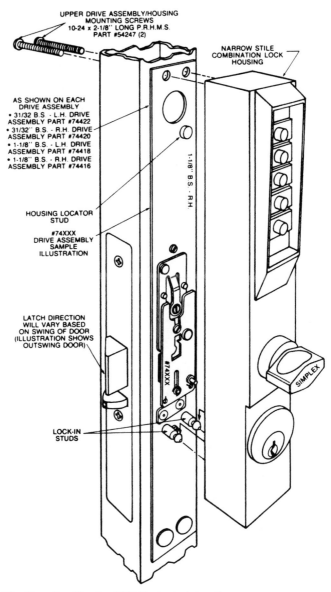

UPPER DRIVE ASSEMBLY/HOUSING MOUNTING SCREWS 10-24 x 2-1/8" LONG P.R.H.M.S. PART #54247 (2)

NARROW STILE COMBINATION LOCK HOUSING

AS SHOWN ON EACH DRIVE ASSEMBLY
• 31/32 B.S. - L.H. DRIVE ASSEMBLY PART #74422
• 31/32" B.S. - R.H. DRIVE ASSEMBLY PART #74420
• 1-1/8" B.S. - L.H. DRIVE ASSEMBLY PART #74418
• 1-1/8" B.S. - R.H. DRIVE ASSEMBLY PART #74416

1-1/8" B.S. - R.H.

HOUSING LOCATOR STUD

#74XXX DRIVE ASSEMBLY SAMPLE ILLUSTRATION

LATCH DIRECTION WILL VARY BASED ON SWING OF DOOR (ILLUSTRATION SHOWS OUTSWING DOOR)

#74XXX

SIMPLEX

LOCK-IN STUDS

**9-18** Mounting Simplex 3000 Series lock to stile. Simplex Access Controls Corp.

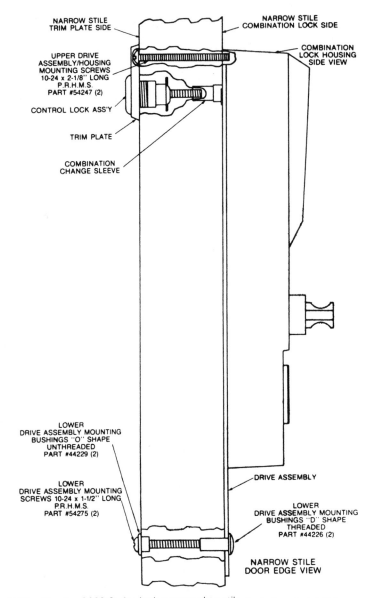

NARROW STILE
TRIM PLATE SIDE

NARROW STILE
COMBINATION LOCK SIDE

COMBINATION
LOCK HOUSING
SIDE VIEW

UPPER DRIVE
ASSEMBLY/HOUSING
MOUNTING SCREWS
10-24 x 2-1/8" LONG
P.R.H.M.S.
PART #54247 (2)

CONTROL LOCK ASS'Y

TRIM PLATE

COMBINATION
CHANGE SLEEVE

LOWER
DRIVE ASSEMBLY MOUNTING
BUSHINGS "O" SHAPE
UNTHREADED
PART #44229 (2)

LOWER
DRIVE ASSEMBLY MOUNTING
SCREWS 10-24 x 1-1/2" LONG
P.R.H.M.S.
PART #54275 (2)

DRIVE ASSEMBLY

LOWER
DRIVE ASSEMBLY MOUNTING
BUSHINGS "D" SHAPE
THREADED
PART #44226 (2)

NARROW STILE
DOOR EDGE VIEW

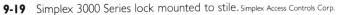

**9-19**   Simplex 3000 Series lock mounted to stile. Simplex Access Controls Corp.

*Chapter* **10**

# Lock picking, impressioning, and forced entry

*L*ocksmiths must be able to enter locked areas quickly while causing the least possible amount of damage. Few people would appreciate a locksmith who gained entry into a locked home by kicking the door in or breaking a window.

Lock picking, key impressioning, and forced entry techniques are commonly used by locksmiths to enter locked buildings and automobiles. This chapters covers all those methods in detail.

## LOCK PICKING

You might have seen television shows in which people pick locks in less time than it takes to blink your eyes. Television detectives simply insert a tool into a keyway and the lock opens. The picking process they use, however, is greatly simplified. It usually takes from 3 to 5 minutes for an experienced locksmith to pick a standard lock. Sometimes it takes much longer.

In theory, any mechanical lock that is operated by inserting a key into a keyway can be picked. A tool can be made to simulate the mechanical action of any key. But some locks are specially designed to thwart picking attempts so picking them usually requires special tools. Before you can learn to professionally pick locks, you need to be familiar with the internal constructions of the basic lock types. That information was given in Chapters 2, 5, 7, and 8.

## Tools

The only tools needed for picking most types of locks are picks and torque (or turning) wrenches. The pick is used to align the tumblers to a position that will allow the plug to be turned and the torque wrench to rotate the plug. In some cases, a lock pick can also act as a torque wrench.

There is a wide variety of types, styles, and sizes of picks and torque wrenches (FIGS. 10-1 and 10-2). Which are ''best'' is largely a matter of personal preference. Sometimes the size of a lock's keyway dictates the kind of pick that can be used. It's a good idea to have several types and sizes of lock picks and torque wrenches.

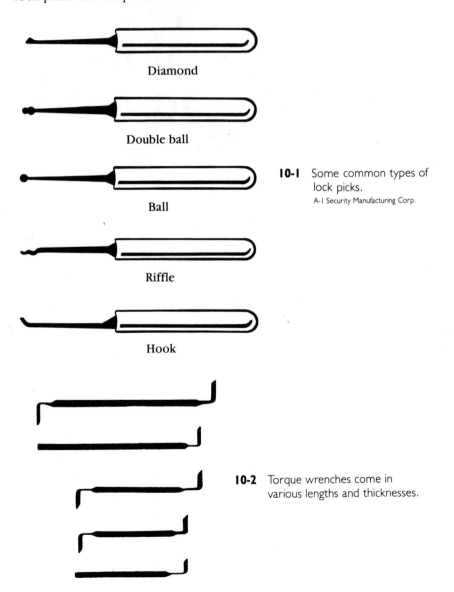

Diamond

Double ball

Ball

**10-1**  Some common types of lock picks.
A-I Security Manufacturing Corp.

Riffle

Hook

**10-2**  Torque wrenches come in various lengths and thicknesses.

### Picking pin tumbler locks

The ability to pick pin tumbler locks quickly is one of the most important skills a locksmith needs to have. Because the pin tumbler lock is the most commonly used lock, it's the one that locksmiths most often have to pick.

There are several methods for picking pin tumbler locks. The "feel" method is the most professional; it's also the most difficult to learn. You should become proficient at using the feel method, before learning other methods. Otherwise, you might become dependent on the less precise methods.

**The feel method**   Many locksmiths use a diamond or hook pick (see FIG. 10-1) when using the feel method to pick a pin tumbler lock. Hold the pick as you would a pencil, and insert it all the way into the keyway with the pick's point or curve pointing upward toward the pins. The pick will then be under all the bottom pins.

Using your other hand, insert the small bent portion of the torque wrench into the keyway from the top or bottom of the keyway, whichever position is easier. Be sure the torque wrench isn't so far into the keyway that it's touching the pins. Using the hand that's holding the torque wrench, place your index finger on the end of the torque wrench farthest away from the plug. Use only that index finger to steady the torque wrench and to apply pressure.

Slowly lift the last bottom pin (the one farthest from the face of the cylinder) with your pick, while applying light pressure to the torque wrench with your index finger. When the top of that pin reaches the plug's shear line, your index finger will sense a little movement of the plug. Remember how much give you felt at that time. Release the pressure on the torque wrench, but don't allow it to fall out of the plug.

Move the pick toward the face of the cylinder and stop beneath the next bottom pin. Slowly lift that pin while applying slight pressure to the torque wrench. When the top of the pin reaches the shear line, you will feel the plug turn slightly; remember how much give you feel. Moving toward the face of the cylinder, repeat that process with each bottom pin. At this point you should know which pin allows the plug to rotate the most, and which the least.

Place the tip of the pick directly beneath the pin that allowed the plug to turn the least. Slowly lift that pin to the shear line again while applying slight pressure to the torque wrench. When the plug turns a little, release some (but not all) of the pressure you're applying to the torque wrench. Move the pick beneath the pin that now allows the plug to turn the least. Slowly lift that pin to the shear line while applying a little more pressure to the torque wrench. The plug should turn a little more. Release some of the pressure to the torque wrench. Move the tip of the pick directly beneath the pin that now allows the plug to turn the least. Repeat the procedure until the lock is picked.

**Principles of the feel method**   Understanding the principle behind the feel method is important for learning to perform it. The method works because of two things: First, the lower pin chambers of a pin tumbler cylinder are never perfectly aligned with their corresponding lower pin chambers. Second, the sets of upper and lower pin chambers are never perfectly aligned with one another; instead of forming a straight line, the sets of pin chambers form a slight zigzag pattern across the plug and cylinder case (FIG. 10-3).

Sometimes the variation from one set of pin chambers to the next set is as little as .0002 inch. There's always some variation, because mass production techniques can't create perfect locks. The higher the quality of the lock, the less variation the lock will have among its sets of pin chambers (and between its upper and lower pin chambers).

The slight misalignment within a cylinder allows the cylinder to be picked. When a bottom pin is lifted to the shear line while pressure is being applied to the torque wrench, the top pin that's resting on that bottom pin is also being lifted. As the top pin is being lifted, it leans against a wall within its upper pin chamber. When that top pin reaches the shear line (which occurs the same time the bottom pin reaches the shear line), the plug is able to turn slightly. When it turns, a small ledge is created on the plug. This ledge prevents the top pin from falling back into the lower pin chamber. Even when the pick is moved away from the corresponding bottom pin, and the bottom pin falls back into the plug, that top pin will stay on that ledge at the shear line as long as adequate pressure is being applied to the plug.

As each top pin is set on the ledge, the ledge gets bigger because the plug is able to turn more. When all the top pins are sitting on the ledge, all the bottom pins will be at or below the shear line. The plug is then free to be rotated into position to operate the lock.

How can you tell when a bottom pin has reached the shear line? When a ledge is formed by the plug to prevent the top pin from dropping into the lower pin chamber, a slight ceiling is formed below the upper pin chamber that makes it a little more difficult for the bottom pin to go into the upper pin chamber. If you're lifting the bottom pin slowly enough, you will feel when it bumps against that ceiling; that's when you stop lifting the pin.

But lifting each pin exactly to the shear line is only part of the task of picking a pin tumbler cylinder. You also have to apply just the right

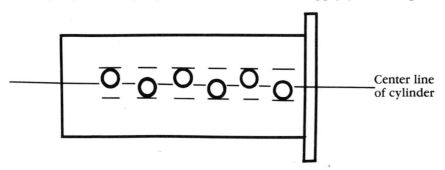

Center line of cylinder

**10-3**   Pin chambers within a cylinder are not perfectly aligned with one another.

amount of turning pressure to the plug. The proper amount of pressure varies among the sets of pin chambers within a cylinder. While lifting one bottom pin you might need to use an amount of pressure that differs from the amount you used with another bottom pin.

If too much turning pressure is applied to the plug, the pins will be bound between the upper and lower pin chambers, and you won't be able to lift the pins to the shear line. If too little pressure is applied, the ledge on the plug will be too small to prevent the top pins from dropping into the lower pin chambers.

Applying too little turning pressure to the plug is seldom the problem. In most cases, people apply far too much pressure. A very small ledge is needed to hold the top pins in place.

**Practicing the feel method**   Even though you know the principle behind the feel method and the steps required to perform it, you will need to practice a great deal before you can use the method proficiently. It will probably seem very difficult at first—perhaps even impossible! But with enough proper practice, you will become an expert at picking locks.

You'll learn this method faster if you first practice picking a lock that has only one set of pin tumblers (one bottom pin and one corresponding top pin and spring). After you feel comfortable picking the lock, add another set of pin tumblers to it. Continue adding sets of pins until you can pick the lock with five sets of pin tumblers in it.

When practicing, don't try to rush the learning process. It takes awhile to develop the feel you need to pick pin tumbler locks. Just visualize what's happening in the lock as you're picking it, and take your time. Just as you developed the feel to ride a bike, you can develop the feel to pick locks.

**The rake method**   The most common method used to pick locks is the rake method (also called raking). The rake method is based primarily on luck. People who only use the rake method to pick locks have never taken the time to learn the feel method.

Raking is done by inserting a pick (usually a half diamond or a riffle) into a keyway past the last set of pin tumblers, and then quickly moving the pick in and out of the keyway while varying tension of the torque wrench. The scrubbing action of the pick causes all the pins to jump up to (or above) the shear line, and the varying pressure of the torque wrench helps catch the top pins in place. Raking sometimes works very quickly.

Many times locksmiths use both the rake method and the feel method to pick a lock. You can rake a few times first to bind some top pins into position, and then switch to the feel method to finish the job. With experience, you'll know which method is best to use in a given situation.

### Picking lever tumbler and disc tumbler locks

Lever tumbler and disc tumbler locks are picked in much the same way as pin tumbler locks. The same types of picks are used for all three lock types. If you're able to pick pin tumbler locks, you'll have no trouble picking locks of the other two types.

When picking a lever tumbler lock, you'll be lifting lever tumblers instead of pins. When picking disc tumbler locks, you'll need to pull the discs tumblers downward, instead of lifting them up.

When picking double-bitted disc tumbler locks, you might need to use a two-sided pick. If so, use the rake method.

The side bar wafer lock is a type of disc tumbler lock that is difficult to pick using common picking tools. Bringing the tumblers to the shear line isn't hard, but until all of them are aligned at once, the side bar will prevent the lock from being operated.

To pick a side bar wafer lock, apply pressure to the side bar to force it against the tumblers so that each tumbler you put in place with your pick will stay in place until all the tumblers have been properly positioned. One way to do that is to drill a small hole in the face of the cylinder beside the side bar, and insert a thin wire into the hole to press the side bar against the tumblers.

## Using a lock picking gun

A lock picking gun (FIG. 10-4) is a tool that is shaped like a gun. It can't automatically open locks but it can sometimes make the job of picking easier.

When using a lock picking gun, you'll also need to use a torque wrench. Just insert the pick end of the gun into the keyway beneath the last bottom pin. Insert the torque wrench. Squeeze the trigger a few times while varying the pressure of the torque wrench. The gun pick end of the gun will slap the bottom pins, causing them to jump up. The torque wrench catches the pins in place—much like the rake method of lock picking.

## Lock picking tips

Locks with worn tumblers are usually easier to pick than new locks. When picking a lock on an automobile, work on the one that is least often used (such as the one on the passenger door).

Thinner picks are usually easier to work with than are thicker picks.

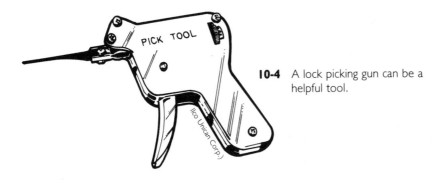

**10-4** A lock picking gun can be a helpful tool.

## KEY IMPRESSIONING

The ability to impression keys can make many locksmithing tasks easier. However, a great deal of practice is required to develop the skill.

Impressioning is done by inserting a properly prepared blank into a keyway and moving the blank so the tumblers scratch the bitting surface of the blank. When tumblers scratch the blank, they leave small marks. Those marks indicate where and how deep to cut the blank. In most cases, the process of preparing, inserting, and cutting the blank must be attempted several times before the blank can be made into a key that will operate the lock.

Although the principle behind key impressioning is the same for all basic types of locks, the specific steps differ depending on the types of locks.

### Impressioning keys for warded locks

To impression a key for a warded padlock you'll need a pair of vise grips, the correct key blank, a lighted candle, and a small warding (square cut) file. Proceed as follows:

1. Clamp the vise grips to the bow of the proper blank. Using the vise grips as a handle, hold the blank over the candle flame until the blade is covered with soot on both sides.

2. Insert the blank into the lock and gently twist the blank left to right a few times. Remove from the lock, being careful not to rub the soot off the blade. You will see several marks on the blade; they show where the cuts need to be made. The marks should be on both sides and on both cutting edges of the blade.

3. Use a corner of the file to scratch the center of each mark. You need the scratches to show where the cuts need to be made, because much of the soot will come off the blade while you're handling the blank.

4. Use your warding file to make a 90-degree (squared) cut at each scratch you made—on both sides of the blank. Make all the cuts about 1/4 inch deep.

5. You should now have a key that will operate the lock. If it doesn't, you might need to widen the cuts or make them a little deeper. Use the vise grips and candle to put more soot on the key, and then twist it in the lock again. The new marks will tell you where the key needs to be cut again.

### Impressioning keys for pin tumbler locks

The pin tumbler lock is the most difficult type of lock to impression a key for. The tumbler marks left on a key blank are usually very small, and can sometimes be misleading. It takes a lot of practice to learn to notice the marks and to understand what they indicate.

The procedure for impressioning a key for a pin tumbler lock is as follows:

1.  Find out which cut depths and spaces are used for keys that operate the lock you're impressioning. That information can be obtained from depth and space charts.

2.  Select the proper brass blank and use a key coding machine (or a caliper and a #4 swiss file) to place the shallowest cuts possible (usually #0 or #1 cuts) at each proper space along the length of the blade. This will show you how to properly space subsequent cuts you'll make while impressioning.

3.  File the length of the bitting edge of the blade to a knife point. Do not decrease the width of the blade, only the thickness of the blade.

4.  Clamp vise grips to the bow of the blank, and using the vise grips as a handle, insert the blank into the lock's keyway. Gently twist the plank from left to right a few times, and gently rock it up and down a few times.

5.  Remove the blank from the lock and notice the small marks on or near the shallow cuts on your blank. Find the cut that has the heaviest tumbler mark near it—the pins don't always leave marks in the center of proper spaces. Use your file and caliper to deepen that cut to the next depth—usually #1 or a #2 on the depth chart—for the lock you're working on. Don't deepen any other cuts.

6.  Use your file to replace the knife edge on the blank. Then insert it back into the lock and gently twist and rock the blank again. Remove the blank and notice the tumbler marks left on the cutting edge of the blade. If the heaviest mark is on the cut you just deepened, deepen the cut to the next depth (based on the chart). If the heaviest mark is near another cut, deepen that cut to its next depth.

7.  Replace the knife edge on the blank and reinsert it into the lock. Look at the marks and find the heaviest one. If the heaviest one is near the last cut you deepened, then deepen that cut to its next depth. But if the heaviest mark is near any other cut you've deepened, don't deepen that cut. Once you've stopped deepening a cut and started deepening other ones, never deepen that cut again.

8.  Continue replacing the knife edge on the blank, reinserting it into the lock, deepening the heaviest mark, etc.—until you have a working key. Remember to never file a cut deeper once it stops getting the heaviest tumbler mark, even if it later gets the heaviest mark.

When filing cuts, you need to make them the same shape and angle of the original key that was made for the lock you're working on. Figure 10-5 shows the three types of cuts most commonly found on keys. If the

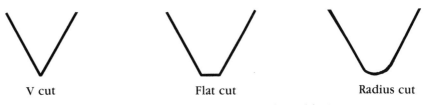

| V cut | Flat cut | Radius cut |

**10-5** V, flat, and radius grooves are commonly used for key cuts.

cut shapes and angles are made improperly, the key you've impressioned will not allow the pin tumblers to seat properly in the cuts; the pins will either seat too high or too low. Many depth and space charts (see Appendix C) have drawings that show the shapes and angles key cuts should be.

## FORCED ENTRY TECHNIQUES

Most locksmiths consider it unethical to use forcible entry techniques, except in emergencies when professional techniques won't work, or when professional techniques would not be cost effective. It usually isn't sensible to waste time trying to pick a lock open, for example, if someone inside a building needs immediate medical attention. Nor is it in a customer's best interest for a locksmith to impression a key to open an inexpensive lock.

### Loiding

Locks with spring-loaded latch bolts, such as most key-in-knob locks and some rim locks, can be opened by using a thin piece of plastic or metal (a credit card or a knife) to press out of a strike and back into the lock. This process is called *loiding*. The card or knife is inserted between the door and jamb at the beveled edge of the latch bolt, and is pushed against the latch bolt.

Loiding is more difficult to do with a spring-loaded latch bolt that has a deadlatching bar attached to it. When the lock is properly installed, the bar is depressed against the strike plate and prevents the bolt from being depressed. You have to free the deadlatching bar before loiding will work on the lock.

### Jimmying

Many deadbolt locks and most spring-loaded locks (including those with deadlatching bars) can be opened by jimmying. To jimmy a lock simply insert a prying bar between the door and jamb near the lock's bolt. Pry the door far enough away from the jamb to allow the bolt to come out of the strike. The more loosely a door is fitted and the smaller the lock's bolt, the easier it is to jimmy a lock.

### Drilling pin tumbler locks

Most pin tumbler locks can be easily drilled open by drilling through the top pins. Position a $1/8$-inch or $3/32$-inch drill bit on the face of the cylinder about $1/8$ inch above the shear line, directly in line with the top pins. As you drill through the cylinder, you will feel the drill jerk forward each time you go through a pin. After you've drilled through the last pin, use a key blank or a screwdriver to rotate the plug.

Some pin tumbler cylinders have hardened pins or plates protecting the pins. They too can be drilled through, but drilling them takes a little longer. A tungsten carbide drill bit can make it easier to drill through hardened pins and plates.

Locksmithing supply houses sell jigs for properly aligning a drill bit to a cylinder. The jigs can be useful, but aren't necessary for drilling through a cylinder.

## Removing mortise cylinders

A mortise cylinder can be forcibly removed by first prying off the cylinder collar, then using a wrench to twist the cylinder out. When the cylinder is wrenched out, it will need to be replaced because its threads will be stripped. Many locksmith supply houses sell mortise cylinder removal tools; these work in much the same way as wrenches.

## Opening padlocks

The most common way of opening an inexpensive padlock is to cut the shackle with bolt cutters. The process is quick, and allows a locksmith to earn extra money selling new padlocks. When cutting a shackle, make sure people aren't standing too close. Sometimes the lock will fly off its hasp. To open a high-quality padlock, it might be in the customer's best interest for you to drill out the cylinder plug or pick the lock.

Another way to enter a padlocked area is to use a bolt cutter to cut the hasp loop (also called a staple) that the padlock's shackle is in. Sometimes you can simply unscrew the hasp and remove both it and the padlock intact.

# Chapter 11

# Master keying

*C*ontrary to a popular myth, there is no master key that can open every lock (or even most locks). The closest thing to such a mythological key is a skeleton key. As explained in Chapter 4, a skeleton key can be used to operate a large number of warded locks. A master key is simply a key that has been cut to operate two or more locks that have different key combinations. It can only operate locks that have been modified to allow it to operate them.

To *master key* means to modify a group of locks in such a way that each can be operated by a uniquely bitted key that can't operate the others—and all can be operated by a master key. A key that operates only one lock—or two or more locks keyed alike—within a master system is called a *change key*.

Some people confuse master keying with locks that are keyed alike. The latter all have the same keying combination and are operated by the same key. In a house, for example, the front and back door might be keyed alike so that a single key can be used to operate both locks. Two master keyed locks, however, are combinated differently from each other; neither's change key can operate the other lock.

Master keying is frequently done on locks in hotels and apartment complexes. In an apartment complex, for example, a master key system can allow each tenant to have a change key to operate only the lock for his or her apartment, and the apartment manager can have a master key that operates all the locks in the complex.

Some master key systems consist of more than one level. In a multi-level system, different locks are modified to be operated by different master keys, and all the locks within the system can be operated by the top-level master key. For example, a hotel might have a system that allows each guest to have a change key, each maid to have a master key to open one group of rooms, and the manager to have a top-level master key that can open all the doors in the hotel. Figure 11-1 shows how a multi-level master key system works.

Often some locks within a master keyed system are *maison keyed*. That is, they are modified to be operated by two or more change keys. The front door of an apartment building, for example, may be maison keyed so that each tenant's change key can unlock that door.

Both maison keying and master keying systems are used primarily for convenience. Without them, an apartment or hotel would need to have and control the use of many more keys. However, both types of keying systems reduce security. Whenever a lock is modified to be operated by more than one key, the lock becomes less secure. Likewise, a lock that can be operated by two levels of master keys is less secure than one that can be operated by only one master key.

How a group of locks are master keyed depends on the type and model of the locks. Some general master keying principles apply to each type of lock, but certain models with each type require special treatment. This is especially true for many high-security pin tumbler locks that have unique features.

## WARDED LOCKS

Whenever a group of warded locks needs to be master keyed, it is done by the manufacturer. Master keying is done by placing wards of various sizes and shapes within each lock. A change key is cut in such a way that it will bypass the wards of one of the locks, so it can operate that lock but none of the others. A master key is cut to bypass the wards in all the locks, so it can operate all of them.

As explained in Chapter 4, warded locks offer little security. They are seldom used in situations that require master keying, and therefore are seldom master keyed.

## LEVER TUMBLER LOCKS

Lever tumbler locks are also usually master keyed by the manufacturer. Master keying such locks is time-consuming, and seldom practical for a locksmith to do.

A lever tumbler lock can be master keyed by using double-gated levers. One set of gates is operated by the change key; the other set by the master key.

Another way a lever tumbler lock can be master keyed is by using a control lever tumbler that has a pin running through it, connecting the tumbler to all the other tumblers in the lock. The master key is cut to operate the lock by lifting the control tumbler.

## DISC TUMBLER LOCKS

A disc tumbler lock that isn't part of a master keyed system usually has a rectangular cutout in each tumbler. The lock is master keyed by using

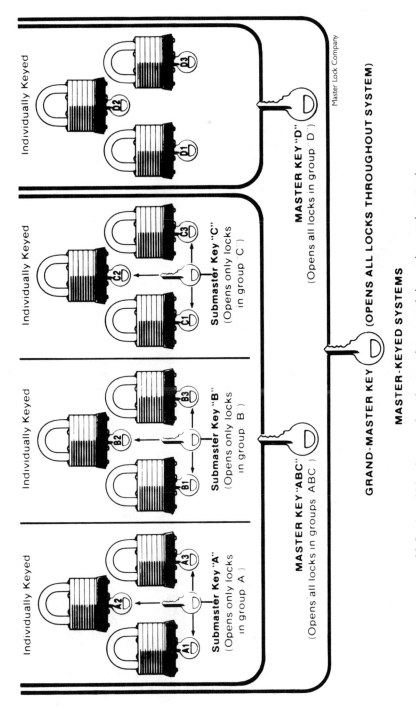

**MASTER-KEYED SYSTEMS**

11-1    A multi-level master keyed system has several change keys and master keys.

tumblers with cutouts shaped to allow more than one key combination to operate it. One side of the tumbler is operated by the change key, the other side by the master key. In addition to having different cuts, the keys have different (reversed) keyway grooves. The keyway of a master keyed disc tumbler lock is designed to accept two different keyway groove patterns.

## PIN TUMBLER LOCKS

Although most other types of locks are usually master keyed by lock manufacturers, locksmiths are frequently called upon to master key pin tumbler locks. Such locks offer many more keying possibilities than do other types, and can be master keyed in several different ways.

Some pin tumbler locks have keyways that are specially designed for master keying. Such keyways are called *sectionals*. They allow certain keys within a master keyed system to fit into them, and prevent certain keys within the system to fit into them. With sectional keyways (FIG. 11-2) a person can combinate two or more locks the same and issue a distinct change key for each lock. Even though the change keys have the same spacing depths and cuts, they cannot enter the other's lock—but another key that is specially designed can enter both locks. Sectional keyways are especially useful for a large master key system because they can provide many master key levels.

Another way pin tumbler locks are master keyed is by creating more than one shear line within each cylinder. This is done by placing master pins between the top and bottom pins. A master keyed pin tumbler lock

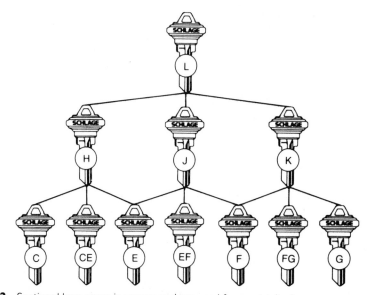

**11-2**  Sectional keys come in groups and are used for master keying.  Schlage Lock Company)

has one shear line for a change key, and others for master keys (FIG. 11-3). In a well-designed system, none of the change keys will bring the pins of any lock to the shear line used by a master key; likewise, none of the master keys will bring the pins of any lock to the shear line used by other master keys.

Key interchange is when a key in a master keyed system causing the pins of a lock to reach a shear line that the key wasn't meant to use. That condition is undesirable, and is easy to avoid by properly planning the master keying system.

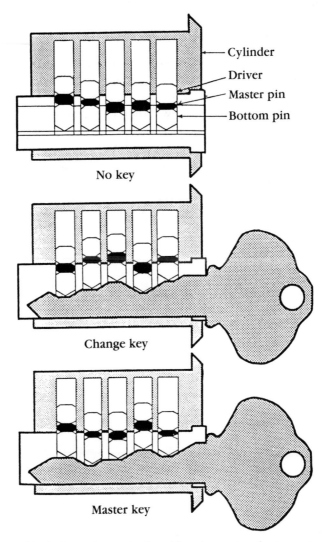

**11-3** Pin tumbler locks can be master keyed by using master pins to create more than one shear line in each lock.

You can easily master key two pin tumbler locks without preplanning a system. Assuming each lock is already operated by a distinctly bitted key of the same length and keyway grooves, proceed as follows:

1.  Select a cylinder key of the same length and keyway grooves of the other two (the key can be taken from another lock). Make sure the key bitting differs from the bitting of the other two keys.

2.  Using a plug follower to keep the top pins in place, remove the plug from the cylinder.

3.  Using various size bottom and master pins (the master pins should set on top of the bottom pins), rekey the lock so that the original key causes the pins to create one shear line, and the master key causes them to form a second shear line. Reassemble the cylinder and make sure both keys operate it. Set the original key for that lock aside.

4.  Repeat Steps 2 and 3 with the other lock, using that lock's original key and the master key. You will then have a simple master keyed system.

Locksmith supply houses sell charts for designing more complicated master key systems. Some of the charts are available on computer discs. To use them you need to know how many pins are in the locks you're using, how many change keys you want, how many levels of master keys you want, and how many master keys you want for each level.

*Chapter* **12**

# Special door hardware

*L*ocks represent a small part of the hardware commonly used on or with doors. Other hardware includes hinges, strikes, and door reinforcers. A locksmith who is familiar with the wide range of door hardware can profit greatly by selling and installing the products.

## BUTTS AND HINGES

A door that binds or sags can be an eyesore and can hinder the performance of locks and door closers. Often these problems can be resolved by simply replacing the door's butts or hinges.

Butts and hinges come in a variety of shapes, styles, and types to be used for many different functions and applications. A butt or hinge usually consists of three basic parts: two metal leaves or plates, each with knuckles on one edge; and one pin that fits through the knuckles of both leaves. When the knuckles of both leaves are joined, they form the barrel of the butt or hinge.

The pin may be removable or nonremovable. A removable pin can be pulled out of the barrel. A nonremovable pin is retained by a small retaining pin or a set screw. Ordinarily, the retaining pin or set screw is concealed when the door is closed.

Nonremovable pins are especially useful for exterior doors where the barrel of the butt or hinge is exposed to the outside. Without such pins, a person could enter a door simply by pulling the pins out of the door's hinges.

Fast or rivet pins have both ends machined on and are factory sealed. This type of nonremovable pins is often used in prisons and psychiatric hospitals.

Many people use the terms "butt" and "hinge" synonymously. However, there is a distinction between the two. All butts are hinges, but all hinges are not butts. Hinges designed for applications in which their leaves normally abut each other—such as on the edge of a door and door jamb—are called butt hinges or butts (FIG. 12-1).

Usually one or both leaves of a butt are swagged, or bent slightly at the knuckles. That brings the leaves into closer contour with the barrel and allows for a tighter fit.

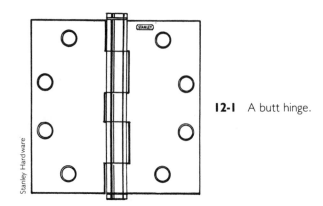

**12-1**   A butt hinge.

## Classification

Butts and hinges are classified in three ways: by screw hole pattern, by type of installation, and by function. Those that have standard screw hole sizes and patterns are called template hinges. They're used mostly on metal doors and pressed metal frames. Non-template hinges are those with staggered screw hole patterns; such hinges are used on wooden doors and wood frames. Blank face or plain hinges have no predrilled screw holes. These are used when it's necessary to field drill screw holes or weld hinges into place.

In addition to screw hole pattern, butts are also classified by type of installation. Full-mortise butts are installed by mortising both leaves (FIG. 12-2). Full-surface butts are installed by surface mounting both leaves. Butts that have one leaf mounted to the door frame and the other mor-

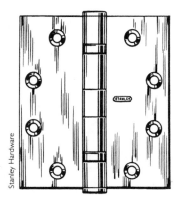

**12-2**   A full-mortise template hinge.

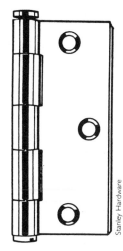

**12-3** A half-surface hinge is installed by mortising the jamb leaf.

Stanley Hardware

tised into the edge of the door are half-mortise butts. Half-surface butts are installed in the reverse manner (FIG. 12-3).

Sometimes butts and hinges are identified by how they function. For example, a clear swing hinge is one that allows a door to swing clear of the passageway (usually a 180-degree swing), permitting full use of the door opening.

Regardless of the door type, weight, or size, a butt can usually be used on it. Butts come in four standard sizes, based on barrel length: 3-inch, 3½-inch, 4-inch, and 5-inch. These are made in a variety of finishes to match other door hardware.

- Simple butts are non-handed; they can be used on both left-hand and right-hand doors. These buts have nonremovable pins.

- Loose pin butts also have nonremovable pins. Their pins are fixed into one leaf. Loose pin butts are designed to allow the door to be removed without disturbing the pin or unscrewing the leaves. The door is removed by lifting it up so the leaves attached to the pins will clear the fixed pins. Loose pin butts are handed.

- Rising butts are also handed. They have knurled knuckles and are designed to lift a door up as it swings open. They can be used to allow a door to clear a heavy carpet.

- Ball bearing butts turn on two or more lubricated ball bearings instead of on pins. The lubricant and ball bearings are housed in "ball bearing raceways" that look like small knuckles resting between the knuckles of the barrel. Ball bearing butts are used on heavy doors.

- Concealed hinges are used on folding doors, and look very different from most other types of hinges. They're installed by drilling two proper size parallel holes, one in each meeting edge of the door. One cylindrical end of the concealed hinge is inserted into one hole, and the other end into the other hole. Then both ends are screwed into place, resulting in a completely concealed hinge.

- Gravity pivot hinges allow a door to swing either way. Some have a hold-open feature. Double-acting hinges, as the name implies, permit folding doors to swing either way (FIG. 12-4).

- Spring-loaded hinges can be used as door closers on fire doors and large screen doors. Some models have adjustable spring tension (FIG. 12-5).

- Pivot reinforced hinges don't require a door frame. Such hinges are used for recessed, flush, or overlay doors.

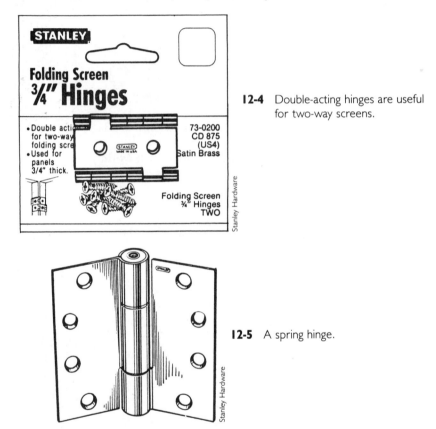

**12-4**  Double-acting hinges are useful for two-way screens.

**12-5**  A spring hinge.

### J-U-5 replacement hinges

Manufactured by Brookfield Industries, Inc., J-U-5 hinges are designed for replacement and new installation on store fronts, public buildings, apartment buildings, and other places where the doors are frequently used.

J-U-5 hinges come as part of a kit. The kit includes the following: 2 hinges, 12 aluminum expanders, 12 steel fasteners, 4 security strips, 1 expanding tool, and 1 instruction sheet. The kit makes installation of J-U-5 hinges easy. To install a J-U-5 hinge, proceed as follows:

1. Strip the door and jamb of old worn hinges, broken pivots, or slip-ins. Chuck the door up, leaving equal clearance on the top

**12-6** Use the hinge as a template.

Brookfield Industries, Inc.

and bottom and side to side. To allow for vertical door adjustment, set hinge barrel gap between $1/8$ and $3/16$ inch.

2. Using the hinge as a template, set the barrel of the hinge in the center line of the clearance between the door and jamb approximately 6 inches from both top and bottom (FIG. 12-6). On doors where butts or slippings were removed, cover the old cutout with the new hinge.

3. Scribe the center hole of the leaf with a pencil, and drill a $1/4$-inch hole (FIG. 12-7).

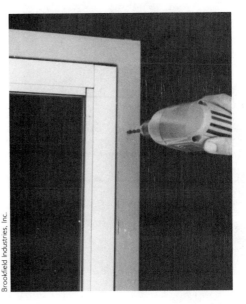

**12-7** Drill a $1/4$-inch hole.

Brookfield Industries, Inc.

4. Insert a blind threaded fastener through both hinge and frame, expanding the fastener with the tool provided (FIG. 12-8). Make sure the hinge is straight.

5. Drill the center hole and insert the fastener on the opposite leaf. Then drill the remaining four holes and expand the fasteners (FIG. 12-9). Repeat this procedure on the lower hinge.

6. Install 12 steel screws (supplied with the kit) and tighten (FIG. 12-10).

7. Clean the hinges. Apply self-adhering cover strips, remove chucks, and test the door for a free swing. Balance up and down with the vertical adjustments screw, if necessary. The installation is now complete (FIG. 12-11).

## HIGH-SECURITY STRIKE PLATES

When customers go to a locksmithing shop to buy a good lock they are often shocked to hear the price of a lock that's highly resistant to picking, impressioning, and drilling. A less expensive lock combined with a high-security strike plate can be sold to such customers as an alternative. Although such plates are relatively inexpensive, they provide resistance to the most common form of forcible entry: kick-ins (FIG. 12-12).

**12-8** Insert a blind threaded fastener.

Brookfield Industries, Inc.

Burglary statistics show that picking, drilling, and impressioning are rarely used for gaining entry into residences. Most standard strike plates are held to a frame by two small wood screws, and can easily be popped out of the frame by a well-placed kick.

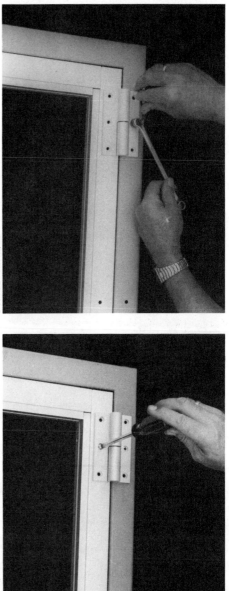

**12-9**   Expand the fasteners.

Brookfield Industries, Inc.

**12-10**   Install steel screws.

Brookfield Industries, Inc.

Brookfield Industries, Inc.

**12-11** A completed J-U-5 hinge kit installation.

M.A.G. Eng. & Mfg., Inc.

**12-12** Kicking in a door is the leading method of forced entry.

### Strike 3

Manufactured by M.A.G. Eng. & Mfg., Inc., the Strike 3 is a popular high-security strike. It's made of heavy-gauge steel (one-piece construction), and fits all standard door frames. Four hardened screws anchor it to the stud. When properly installed, the Strike 3 makes the frame as strong as the deadbolt (FIG. 12-13). To install a Strike 3, proceed as follows:

1. Mark center of bolt on frame.
2. Locate center mark on template with center mark on frame (FIG. 12-14). Tape template in place.

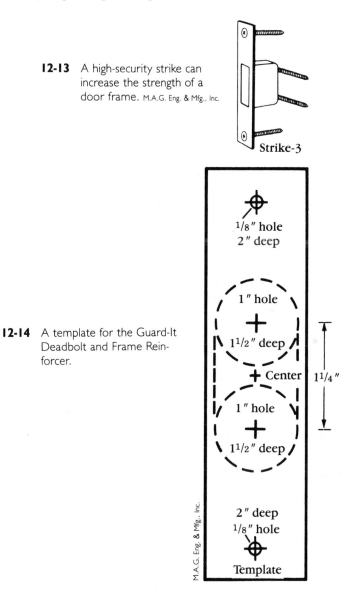

**12-13** A high-security strike can increase the strength of a door frame. M.A.G. Eng. & Mfg., Inc.

Strike-3

**12-14** A template for the Guard-It Deadbolt and Frame Reinforcer.

1/8″ hole
2″ deep

1″ hole
1 1/2″ deep

+ Center

1 1/4″

1″ hole
1 1/2″ deep

2″ deep
1/8″ hole

Template

M.A.G. Eng. & Mfg., Inc.

3. Draw around outside of template. Using an awl, mark four holes per template. Drill the holes.

4. Chisel out between 1-inch holes and fit strike box.

5. Cut ⅛ inch deep for strike (FIG. 12-15).

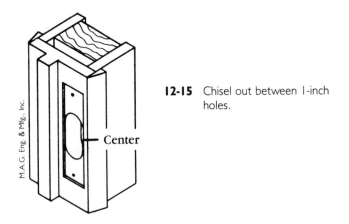

**12-15** Chisel out between 1-inch holes.

6. Install strike and faceplate with top and bottom screws. Be sure two holes in strike box are toward center of frame (FIG. 12-16).

7. Drill two ⅛-inch pilot holes at bottom of strike box toward center of frame and install screws. Angle screws as shown in FIG. 12-16.

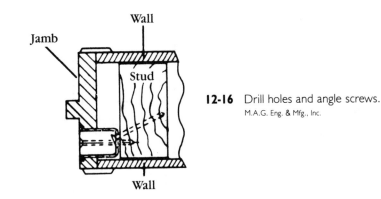

**12-16** Drill holes and angle screws.
M.A.G. Eng. & Mfg., Inc.

## INSTALL-A-LOCK DOOR REINFORCERS

Manufactured by M.A.G. Eng. & Mfg., Inc., the Install-A-Lock door reinforcer makes a door as strong as the door's lock. In addition to guarding against kick-in attacks, the door reinforcer is excellent for lock conversions and misdrilled holes (FIG. 12-17).

### Original series

The original standard series Install-A-Lock is used with recessed latches. The 2000 series is used with both recessed and drive-in latches. Both

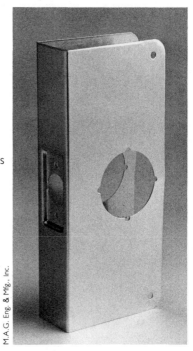

**12-17** An Install-A-Lock reinforces the strength of a door.

M.A.G. Eng. & Mfg., Inc.

types are available in several popular finishes, and in sizes to fit all doors. To install an indented unit Install-A-Lock, proceed as follows:

1. First make sure the unit you're using matches the backset, door thickness, and hole size for your installation. Drill holes in door using instructions furnished with lockset purchased, or remove old lockset from door.

2. Cut out door edge (FIG. 12-18). For double units the "A" dimensions or centers between lock holes (FIG. 12-19) is 3⅝ or 4 inches.

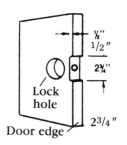

**12-18** Cut out door edge.
M.A.G. Eng. & Mfg., Inc.

3. Slide unit over door.

4. Install latch with ⅜-inch machine screws provided. Then install lockset. Tighten lockset to hold Install-A-Lock in position. Push unit flush against edge of door, then tighten lockset. Install four mounting screws.

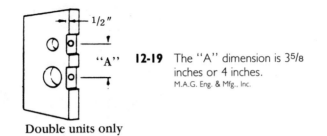

**12-19**  The "A" dimension is 3$^5$/$_8$
inches or 4 inches.
M.A.G. Eng. & Mfg., Inc.

Double units only

Note for Tight Fitting Doors: If Install-A-Lock hits jamb when
closing, remove and cut out $^1$/$_{32}$ inch on door edge so unit fits
flush with door edge.

## 2000 Series

To install a 2000 Series (flat edge unit) Install-A-Lock, proceed as follows:

1. Drill holes in door using instructions furnished with lockset purchased, or remove old lockset from door. Remove latch screws,
   but don't remove the latch.
       If door is damaged, install clips on latch as shown in FIG.
   12-20. Place latch in Install-A-Lock, and insert 1$^1$/$_2$-inch screws.
2. Slide unit over door (FIG. 12-21).

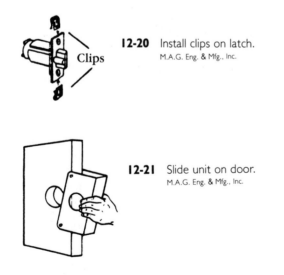

**12-20**  Install clips on latch.
M.A.G. Eng. & Mfg., Inc.

**12-21**  Slide unit on door.
M.A.G. Eng. & Mfg., Inc.

3. Replace latch screws with 1$^1$/$_2$-inch furnished screws (FIG. 12-22).
4. Install four mounting screws, and the installation is complete (FIG.
   12-23).

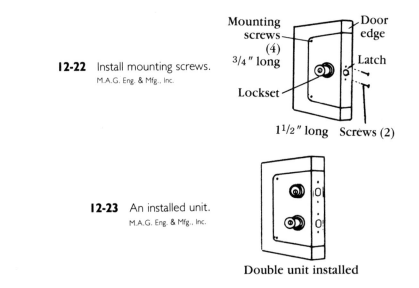

**12-22** Install mounting screws.
M.A.G. Eng. & Mfg., Inc.

**12-23** An installed unit.
M.A.G. Eng. & Mfg., Inc.

Double unit installed

### Uni-Force door reinforcer

The Uni-Force is another type of door reinforcer manufactured by M.A.G. Eng. & Mfg. Like the company's Install-A-Lock, the Uni-Force greatly increases a door's strength. The Uni-Force is for doors with deadbolt and key-in-knob locks. It is installed on the edge of a door, but not underneath the door's lock (FIG. 12-24). To install the Uni-Force, proceed as follows:

1. Remove two latch screws, but not the latch (FIG. 12-25).

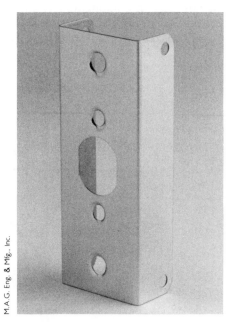

**12-24** The Uni-Force is a door reinforcer for deadbolt and key-in-knob locks.

M.A.G. Eng. & Mfg., Inc.

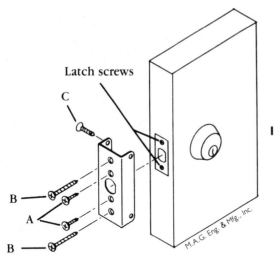

Latch screws

C

B

A

B

M.A.G. Eng. & Mfg., Inc.

**12-25**  Uni-Force installation.

2.  Place unit over door.

3.  Replace latch screws with two 1½-inch-long screws provided.

4.  Drill ⅛-inch pilot holes 2 inches deep for screws. Install 2-inch screws.

5.  Install four ¾-inch-long screws.

## MISCELLANEOUS PRODUCTS

In addition to the products previously mentioned in this chapter, lock-smiths can profit from selling all types of door hardware. Hasps, for example, are used in conjunction with padlocks by many people (FIG. 12-26). Because hasps are very easy to install, you probably won't make much money installing them (FIG. 12-27).

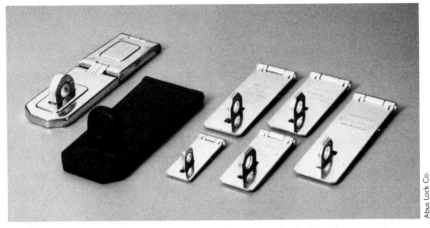

Abus Lock Co.

**12-26**  An assortment of hasps are good products for locksmiths to stock.

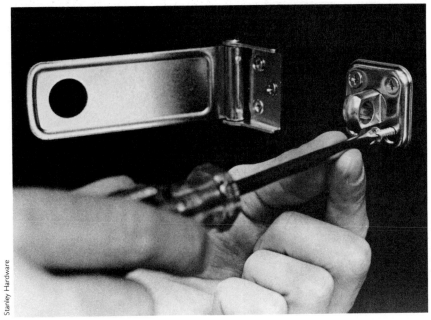

Stanley Hardware

**12-27** Hasps are easy to install.

Likewise, door stops (FIG. 12-28), door holders (FIG. 12-29), and door viewers (FIG. 12-30) are also easy to install. Smart locksmiths will stock all these items in their store front shops, because people often go to a locksmith shop when they need such products.

**12-28** Doorstops help prevent the wall from being damaged by door knobs.

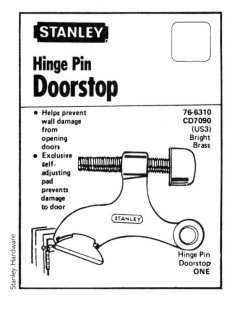

Stanley Hardware

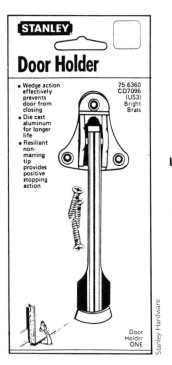

**12-29** A door holder is used to prop a door open.

**12-30** A door viewer lets you see who's knocking at a door.

*Chapter* **13**

# Emergency exit door devices

*F*ire and building codes often require institutions and commercial establishments to have doors that can be quickly and easily opened at any time from the inside. But the businesses need to be able to prevent those doors from being used for unauthorized entry.

Most institutions and commercial establishments use emergency exit door devices as a cost-effective way to handle both matters (FIG. 13-1).

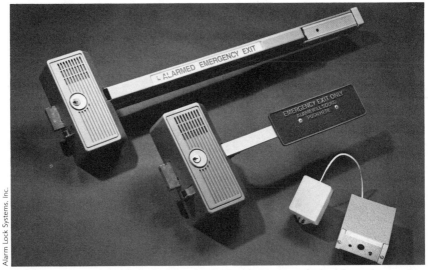

Alarm Lock Systems, Inc.

**13-1** Emergency exit door devices are used to keep doors locked from the outside while allowing easy egress from the inside.

Such devices are easy to install and offer excellent money-making opportunities for locksmiths.

This chapter explains how to install, operate, and service some of the models manufactured by Alarm Lock Systems, Inc. Much of the information also applies to most other popular models.

In normal use of a properly installed emergency exit device, when the door is closed, a key is used to project a lock bolt to secure the door. Depressing the device's clapper plate or pushing a bar located on the inner side of the door releases the deadbolt and deadlatch to open the door.

Some models provide outside key and pull access when an outside cylinder and door pull is installed. In these models, entry remains unrestricted from both sides of the door until the deadbolt is relocked by key from inside or outside the door.

Many emergency exit door devices feature an alarm that will sound when a door is opened without a key. The better alarms are dual piezo (double sound).

Other useful features to consider on an emergency exit door device include: For which hand it is installed (nonhanded models are the most versatile); the length of the deadbolt (a 1-inch throw is the minimum desirable length); and special security features (such as a hardened insert in the deadbolt).

## PILFERGARD MODEL PG-10

One of the most popular emergency exit door devices is the Pilfergard PG-10 (FIG. 13-2). It has a dual piezo alarm, can be armed and disarmed from inside or outside a door, and is easy to install on single or double doors.

**13-2** The Pilfergard PG-10 is a popular emergency exit door device among locksmiths.

Alarm Lock Systems, Inc.

The device is surface mounted—approximately 4 to 6 feet from the floor—on the interior of the door, with a magnetic actuator on the frame (or vice versa). It is armed or disarmed with any standard mortise cylinder (which is not supplied). Opening the door, removal of the cover, or any attempt to defeat the device with a second magnet, when armed, activates the alarm. To install the PG-10 Pilferguard, proceed as follows:

1. Remove cover by depressing test button and lifting cover out of slot.
2. Mark and drill holes per template directions and drill sizes (5 for alarm unit, 2 for magnetic actuator).
3. For outside cylinder installation, follow Steps A through E. Otherwise go on to Step 4.
   a. Drill a 1¼-inch hole as shown on template.
   b. Install a rim type cylinder through the door and allow flat tailpiece to extend 1 inch inside door.
   c. Position cylinder so that keyway is vertical (horizontal if PG-10 is installed horizontally).
   d. Hold PG-10 in position over mounting holes and note that outside cylinder tailpiece is centered in clearance hole in base of PG-10 (rotate cylinder 180 degrees if not).
   e. Tighten outside cylinder mounting screws.
4. Install PG-10 and magnetic actuator with seven screws.
5. Install threaded 1¼-inch mortise cylinder in PG-10 cover using hardware supplied (FIG. 13-3). Keyway must be horizontal so that tailpiece extends toward center of unit when key is turned.

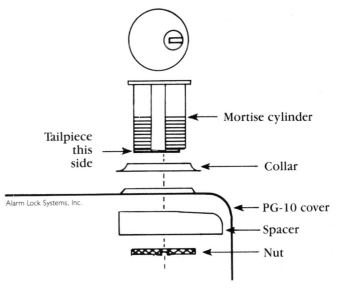

**13-3** Install mortise cylinder.

6. Move slide switch to off (FIG. 13-4). Connect battery. Hook cover on end slot and secure with two cover screws. Note: One of these screws acts as tamper alarm trigger, so be sure screws are fully seated.

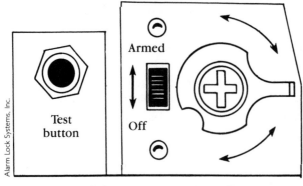

**13-4**  Move slide switch to off.

Caution: When installing the PG-10 on a steel frame, it might be necessary to install a non-magnetic shim between the magnetic actuator and the frame. This prevents the steel frame from absorbing the magnet's magnetic field, which could cause a constant alarm condition or occasional false alarms. The shim should be $1/2 \times 2^{1}/2 \times 1/8$ inch thick, and may be constructed from plastic, bakelite, or aluminum.

## Checking out and operating

You can check out and operate the PG-10 by doing the following:

1. With slide switch in off position, depress test button. Horns should sound.
2. To test using magnetic actuator:
   - Close door.
   - Arm PG-10 by turning key clockwise 170 degrees.
   - Open door, alarm should sound.
   - Close door, alarm should remain sounding.
   - Silence alarm by turning key counterclockwise until it stops.
3. Close door and rearm PG-10 by turning key clockwise until it stops.
4. Periodic test: Unit should be tested weekly using test button to insure battery is working. Test button only operates when PG-10 is turned off.

## PILFERGARD MODEL PG-20

The Pilfergard PG-20 (FIG. 13-5) is a sleek modern version of the PG-10. The PG-20 is designed to fit all doors, including narrow stile doors. It has a flashing LED on its alarm, and is as easy to install as the PG-10. To install the PG-20, proceed as follows:

1. Remove cover from mounting plate. Install mortise cylinder (which is not supplied), keeping key slot pointing down in the 6 o'clock position. Screw lock ring on cylinder.

**13-5**  The Model PG-20 emergency exit door device can be used on narrow stile doors.

Alarm Lock Systems, Inc.

2. Select proper template for specific type of door.

3. Mark and drill $9/64$-inch-diameter holes per template directions on the door and jamb. This will require four holes for mounting plate and two holes for magnetic actuator. Note: Certain narrow stile doors require only two holes for mounting plate.

4. For outside key control only: Drill $1^1/4$-inch-diameter hole as shown on template. Install a rim type cylinder (not supplied) through the door and allow the flat tailpiece to extend $5/16$ inch beyond the door. Position cylinder, keeping key slot pointing down, in the 6 o'clock position. Tighten cylinder screws.

5. Knock out the necessary holes from the mounting plate and install on the door with the No. 8 sheet metal screws (supplied). Make sure that rim cylinder tailpiece fits in cross slot of ferrule if outside cylinder is used.

6. Install the magnetic actuator on the door jamb with the two No. 8 sheet metal screws supplied.

   Note: It is sometimes necessary on steel frames to install a non-magnetic shim between the magnetic actuator and the frame. This is done to prevent the steel frame from absorbing the magnet's magnetic field, which could cause a constant or occasional false alarm condition. The shim should be $1/2$-x-$2^1/2$-x-$1/8$-inch-thick non-magnetic material such as plastic, bakelite, or rubber.

7. Make sure battery is connected.

8. Cut white jumper only on the side of unit that magnet will not be installed. Unit will not function otherwise.

9. Select options desired as follows:

   a. Cut yellow jumper for 15-second entry delay. (Alarm will sound 15 seconds after any entry through door if unit is in armed condition.) To avoid the alarm condition upon entry, reset the unit within 15 seconds. This feature is used for authorized entrance.

   b. Cut red jumper for 15-second exit delay. (Unit will be activated after 15 seconds each time unit is turned on with key.) This feature allows authorized non-alarmed exit.

   c. Place shunt jumper plug in position C or 2. If jumper is in position C, alarm will sound continuously until battery is discharged. If jumper is in position 2, alarm will be silenced after 2 minutes and unit will rearm. However, the LED will continue to flash, indicating an alarm has occurred.

10. Install cover on the mounting plate, making sure that the slide switch on PC board fits into the cam hole. Secure the cover with the four screws supplied.

## Testing the PG-20

To field test the PG-20, proceed as follows:

1. With the door closed, turn the key counterclockwise until it stops (less than one-quarter turn). Unit is now armed instantaneously or delayed.

2. Push test button and alarm should sound, verifying that the unit is armed.

3. Open the door, then close it. Alarm should sound instantaneously or delayed.

4. Note pulsating sounder and flashing LED, indicating an alarm.

5. To reset the unit, turn key clockwise until it stops (less than one-quarter turn). Unit is now disarmed. This can be done at any time.

## EXITGARD MODELS 35 AND 70

To install Exitgard Models 35 or 70 panic locks, proceed as follows:

1. With door closed, tape template to inside face of door with center line approximately 38 inches above the floor. Mark all hole locations with a punch or awl. Note that the keeper is surface installed—on jamb for single doors, on other leaf of pairs of doors.

2. If outside cylinder is being used, mark cylinder hole center. If alarm lock pull #707 is furnished, mark the four holes on the template for the pull, in addition to the cylinder hole.

3. Drill holes as indicated for lock and keeper. Drill holes for outside pull, if used.

4. Loosen hex head bolt holding cross bar to lock and pull bar off clapper arm. Remove lock cover screws. Depress latch bolt. Lift up and remove lock cover.

5. (Disregard this step if cylinder or thumb turn are already installed.) Remove screws holding cylinder housing (FIG. 13-6) to bolt cover. Install rim cylinder with keyway horizontal facing the front of the lock—opposite the slot in the rear of the cylinder support. Reinstall support while guiding the tailpiece into the slot cam.

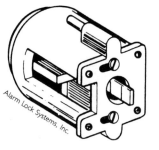

**13-6**   Remove screws holding cylinder.

6. Use key to try proper operation. Key should withdraw in either the fully locked or fully unlocked position of the deadbolt. If not, the cylinder and cam are mistimed. The cylinder housing must be removed, the cam turned a quarter turn to the right, and the cylinder housing reinstalled.

7. Attach the lock to the door. For single doors, remove the keeper cover and roller, install keeper, replace roller and cover. Double door keeper #732 is installed on the surface of the inactive leaf, as furnished.

8. Remove hinge pivot cover. Reinstall push bar section on the lock clapper as far forward as it will go. Tighten hex head bolt under lock clapper. Attach hinge side pivot assembly, using a level or tape to assure cross bar is level. If cross bar is too long, loosen hex head bolt on underside of clapper and remove pivot assembly. Cut bar to proper length, deburr edges, and reinstall pivot assembly on the cross bar. Only after pivot base has been installed should dogging screw (FIG. 13-7) be loosened and the pivot block removed and discarded.

9. Dog screw should face floor. If not, remove and reinstall from below. Replace pivot cover, with hole for dog screw facing the floor.

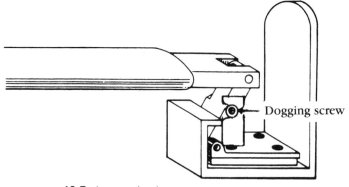

**I3-7**  Loosen dogging screw. Alarm Lock Systems, Inc.

10. Test lock operation by projecting deadbolt by key into keeper. Depress cross bar fully to retract deadbolt, then release latchbolt and open door.

11. On single doors, close the door and adjust the keeper so the door is tightly latched. After final adjustments, install holding screw in keeper to maintain position permanently. On pairs of doors, adjust plastic slide on keeper so door is tightly latched.

12. (If installing Model 35, disregard this step.) Connect power plug. Repeat Step 10. Horns should sound until deadbolt is projected by key. Attach self-adhesive sign to bar and door.

13. Replace lock cover.

## ALARM LOCK MODELS 250, 250L, 260, AND 260L

Models 250, 250L, 260, and 260L by Alarm Lock Systems, Inc. are used to provide maximum security on emergency exit doors. Each of the models feature sleek architectural design and finishes, dual piezo sounder, low battery alert, simple modular construction, selectable 2-minute alarm or constant alarm, and hardened insert in the deadbolt. These models are installed in the following way:

1. With the door closed, select the proper template. Tape it to the inside face of the door with the center line about 38 inches above the floor, according to template directions.

2. Mark and drill the holes.

3. Remove the lock cover and four screws holding the cylinder housing to the bolt cover. Install rim cylinder (CER) with the keyway horizontal, facing the front of the lock in the nine o'clock position. Cut the cylinder tailpiece $3/8$ inch beyond the base of cylinder. Reinstall the cylinder housing, guiding the tailpiece into the crosshole of the cam with the four screws.

4. Use the key to test for proper operation of deadbolt. The key should be able to be withdrawn from the lock in either the fully locked or fully unlocked position of the deadbolt. If not, the cylinder and the cam are misaligned and the cylinder housing must be removed. Turn the cam one-quarter turn to the right, and reinstall the cylinder housing.

   Note: The deadbolt can be projected into the keeper by turning the key counterclockwise. Likewise, it can be withdrawn from the keeper by turning the key clockwise one full turn.

5. For outside cylinder only (CER-OKC): Install the rim cylinder with the keyway horizontal facing the front of the door in the three o'clock position. Use the screws supplied. Cut the tailpiece $3/8$ inch beyond the inside face of the door.

6. For outside cylinder only: Guide the tailpiece of the outside cylinder into the crosshole of the cam.

7. Install the lock to the door with the four screws supplied.

8. For single doors only: Remove the keeper cover, roller, and pin. Install the keeper base on the door with two screws supplied. Reinstall the pin, roller, and cover with two screws.

9. For double doors only: Install the rub plate for a $1^3/4$-inch-wide door, from inside the door. Install the keeper with two screws. Do not tighten the screws fully because the keeper will require adjustment.

10. For single doors only: Close the door, project the deadbolt and adjust the keeper so that the door latches tightly. Retract the deadbolt, hold the keeper, release the latch, and open the door. Open the keeper cover and tighten the screws. Drill a .157-inch-diameter hole, as shown on the template, for the holding screw. Fasten the keeper with a screw. Reinstall pin, roller, and cover with two screws.

11. For double doors only: Close the door, project the bolt, and adjust the plastic slide on the keeper so the door is tightly latched. Tighten the screws.

## ELECTRONIC EXIT LOCK MODEL 265

Alarm Lock Systems, Inc. Model 265 emergency exit door device has the following features: a non-handed unit; a deadbolt with hardened steel insert that can be operated with an outside key; 15-second delay before door can be opened after clapper arm plate has been pushed; a lock that only requires 5 to 10 pounds of force to operate; a dual piezo horn; and a disarming beep when bolt is retracted with key (FIG. 13-8). To install the model 265's lock and keeper, proceed as follows:

1. With the door closed, select the proper template. Tape it to the inside face of the door with the center line approximately 38 inches above the floor, according to template directions.

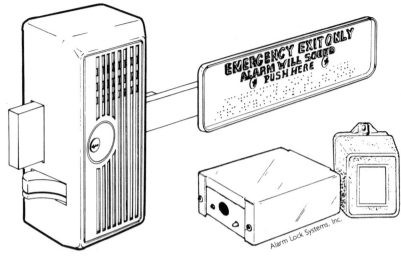

**13-8**   An electronic exit lock Model 265.

2. Mark and drill the following holes (see template for details):
   a. For single and double doors mark six .157-inch-diameter holes—four for the lock mounting plate and two for the keeper.
   b. Mark a ¹/₄-inch-diameter hole for the rub plate on double doors that are 1³/₄ inch thick.
   c. If outside cylinder (CER-OKC) is used, mark the center of the 1¹/₄-inch-diameter hole.

3. If mounting the lock on a hollow metal door and wires are run through the door, align the lock with the holes drilled in Step 2A above. Mark and drill a ³/₈-inch hole in the door to align with the hole in the baseplate near the terminal strip (FIG. 13-9).

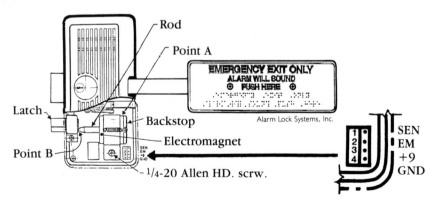

**13-9**   Connect one end of the 4-conductor cable to the control box.

4. Remove the lock cover and four screws holding the cylinder housing to the bolt cover. Install rim cylinder (CER) with the keyway horizontal, facing the front of the lock in the nine o'clock position. Cut the cylinder tailpiece $3/8$ inch beyond the base of the cylinder. Reinstall the cylinder housing, guiding the tailpiece into the crosshole of the cam with four screws.

5. Use the key to test for proper operation of the deadbolt. The key should be able to be withdrawn from the lock in either the fully locked or fully unlocked position of the deadbolt. If not, the cylinder and the cam are misaligned and the cylinder housing must be removed. Turn the cam one-quarter turn to the right, and reinstall the cylinder housing. (Note: The deadbolt can be projected into the keeper by turning the key counterclockwise one full turn.)

6. For outside cylinder only (CER-OKC): Install the rim cylinder with the keyway horizontal facing the front of the door in the three o'clock position. Use the screws supplied. Cut the tailpiece $3/8$ inch beyond the inside face of the door.

7. For outside cylinder only: Guide the tailpiece of the outside cylinder into the crosshole of the outside cam.

8. Install the lock to the door with four No. 10 screws that are supplied.

9. For single doors only: Remove the keeper cover, roller, and pin. Install the keeper base on the door with two screws that are supplied. Reinstall the pin, roller, and cover with two screws. Do not tighten the screws fully, because the keeper will require adjustment.

10. For single doors only: Close the door, project the deadbolt, and adjust the keeper so that the door is tightly latched. Retract the deadbolt, hold the keeper, release the latch, and open the door. Open the keeper cover and tighten the screws. Drill a .157-inch-diameter hole, as shown on the template, for the holding screw, and fasten the keeper with a No. 10 screw. Reinstall the pin, roller, and cover with two screws.

11. For double doors only: Close the door, project the bolt, and adjust the plastic slide on the 732 keeper so the door is tightly latched. Tighten the screws.

12. A fine adjustment in the latch and electromagnet mechanism might be necessary. With the door pulled fully closed, check to see that the backstop is in complete contact with the electromagnet. There will be a small gap, approximately $1/32$ inch or less, between the rod and latch. If not, loosen the Allen head screw and slide the electromagnet to the right or left until adjusted. Retighten the Allen head screw.

## Installing the control box

The control box for Model 265 is installed as follows:

1. Remove the control box cover.

2. Select a location for the control box on the hinge side of the door and mount it to the wall, using the three No. 10 × $3/4$-inch self-tapping screws.

## Wiring

Model 265 is wired in the following way:

1. A four-conductor No. 22-AWG cable is needed to connect the control box to the lock. There are two ways of bringing electric current from the hinge side of the door frame to the door: Use the Armored Door Loop Model 271 and Disconnect one of the 271 end boxes. Insert the loose end of the armored cable into the $1/2$-inch hole on the control box and secure it with the retaining clip. Or, use a continuous conductor hinge with flying leads.

2. Connect one end of the four-conductor cable to the control box terminal strip P2, and the other end of the cable to the lock terminal strip P3. The terminal strips are marked as follows:

   1) SEN        −sence
   2) EM         −electromagnet
   3) +9VDC
   4)             −ground

Do not cross wires.

3. Connect one end of an approved twin lead cable to the terminal strip at P1-1 and P1-2 (see FIG. 13-9), and connect the other end of the cable to the transformer provided. Do not plug in the transformer at this time.

4. Connect one end of another approved twin lead cable to the terminal strip at P1-3 and P1-4 (see FIG. 13-9). Connect the other end to the normally closed alarm relay contacts of one of the following: An approved supervised automatic fire detection system or an approved supervised automatic sprinkler system.

5. Install the control box cover.

## Operating the 265

Before installing the lock cover, do the following:

1. For continuous alarm, leave the black jumper plug on the terminal strip installed. For 2-minute alarm shutdown, remove the black jumper plug from the terminal strip.

2. Connect the 9-volt-battery to its connector. A short beep will sound, ensuring that the 265 is powered and ready.

3. Plug the transformer into a continuous 115 ac volt source. Note the red pilot lamp on the control box is lit.

4. Open the door and install the lock cover with the four screws that are supplied.

## Testing the unit

To test the 265, do the following:

1. Lock and unlock the door using the cylinder key. Note that a small beep sounds when the bolt is retracted, indicating a disarmed condition. The door can be opened by pushing the clapper plate, after a 15-second wait, without causing an alarm.

2. Lock the door again and push the clapper plate. Immediately the alarm will pulse loudly and the following will occur: The 265 latch will impede the opening of the door for approximately 15 seconds, then remain unimpeded until it is manually reset with the key. If continuous alarm was selected above, the piezo sounder will remain on for 2 minutes then reset.

3. In the event of a power failure, the impeding latch will be disabled, but the 9-volt battery will provide standby power for the alarm circuit. This can be tested by disconnecting the transformer from the 115-volt source and pushing the clapper plate. The door should now open immediately and the alarm should sound.

4. In the event of a fire panel alarm, the impeding latch will also be disabled. This can be tested by disconnecting the wire going to the control box at P1-3 and pushing the clapper plate. The door should open immediately and the alarm should sound.

The unit's battery can be tested by pushing and holding the test button on the control box and pushing the clapper plate to alarm the unit. If the piezo sounder is weak or doesn't operate at all, replace the 9-volt battery.

## ALARM LOCK MODELS 700, 700L, 710, AND 710L

Standard features for each emergency exit door device in Alarm Lock Systems, Inc. 700 and 710 Series include: a non-handed unit; a deadbolt with hardened steel insert that can be operated by outside key and can be used for single or double doors; a deadlatch for easy access from inside without alarm; loud dual piezo horn; a selectable continuous or 2-minute alarm shutdown; a disarming beep when bolt is retracted with key; a low-battery beep when the battery needs to be replaced; and a retriggerable alarm after 2-minute shutdown (this for Model 710 only). Figure 13-10 shows a Model 700.

The 700 and 710 series models are installed as follows:

1. With the door closed, select the proper template. Tape it to the inside face of the door, with the center line approximately 38 inches above the floor, according to template directions.

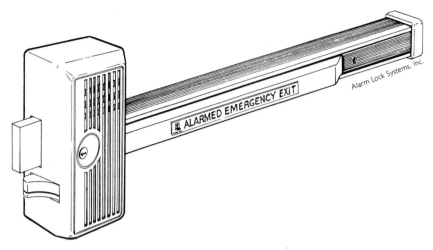

**13-10**   A panic exit alarm Model 700.

2. Mark and drill the following holes (see template for details):

   a. For single and double doors mark six .157-inch-diameter holes—four for the lock mounting plate and two for the keeper.

   b. Mark a $1/4$-inch diameter hole for the rub plate on double doors $1^3/4$ inch thick.

   c. If outside cylinder (CER-OKC) is used, mark the center of the $1^1/4$-inch-diameter holes.

   d. If outside pull Model 707 is used, mark the center of the four $1/4$-inch-diameter holes.

   e. If mounting the lock on a hollow metal door and wires are to be run through the door, drill hole X (see template). Note: If outside pull is used, drill four $1/4$-inch-diameter holes through the door from the inside. Then drill $3/4$-inch-diameter holes $1^1/4$ inch deep from the outside of the door.

3. Remove the lock cover and four screws holding the cylinder housing to the bolt cover. Install rim cylinder (CER) with the keyway horizontal, facing the front of the lock in the nine o'clock position. Cut the cylinder tailpiece $3/8$ inch beyond the base of the cylinder. Reinstall the cylinder housing, guiding the tailpiece into the crosshole of the cam with four screws.

4. Use the key to test for proper operation of the deadbolt. The key should be able to be withdrawn from the lock in either the fully locked or fully unlocked position of the deadbolt. If not, the cylinder and the cam are misaligned and the cylinder housing must be removed. Give the cam one-quarter turn to the right, and reinstall the cylinder housing. Note: The deadbolt can be projected into the keeper by turning the key counterclockwise and can be withdrawn from the keeper by turning the key clockwise one full turn.

5. For outside cylinder only (CER-OKC): Install the rim cylinder with the keyway horizontal facing the front of the door in the three o'clock position, using the screws supplied. Cut the tailpiece 3/8 inch beyond the inside face of the door.

6. For outside cylinder only: Guide the tailpiece of the outside cylinder into the crosshole of the outside cam.

7. Install the lock loosely to the door with four No. 10 screws as supplied. Do not tighten them at this time.

8. Insert the bar and channel assembly under the channel retainer bracket, which is mounted to the lock baseplate. Hold the bar and channel assembly horizontally against the door using a level. Slide the end cap bracket into the end of the channel and, using the bracket as a template, mark and drill the two .157-inch-diameter mounting holes on the door. If the channel is too long, cut the channel and channel insert to the proper length and deburr edges.

9. Attach the pushbar to the lock at the clapper arm hinge bracket, using the 1/2-inch screw and No. 10 internal tooth lockwasher provided.

10. Mount the end cap bracket to the door with two No. 10 screws provided, and tighten the lock securely to the door.

11. Attach the end cap to the end cap bracket using the 1/2-inch oval head screw provided.

12. For single doors only: Remove the keeper cover, roller, and pin. Install the keeper base on the door with two screws supplied. Reinstall the pin, roller, and cover with two screws.

    For double doors only: Install the rub plate for a 13/4-inch-wide door, from inside the door. Also install the 732 keeper with two No. 10 screws supplied. Do not tighten the screws fully because the keeper will require adjustment.

13. For single doors only: Close the door, project the deadbolt, and adjust the keeper so that the door is tightly latched. Retract the deadbolt, hold the keeper, release the latch, and open the door. Open the keeper cover and tighten the screws. Drill a .157-inch-diameter hole, as shown on template, for holding screw. Fasten the keeper with a No. 10 screw. Reinstall pin, roller, and cover with two screws.

    For double doors only: Close the door, project the bolt, and adjust the plastic slide on the 732 keeper so the door is tightly latched and tighten the screws.

## Operating 700 and 710 series models

Before installing the lock cover, do the following:

1. For continuous alarm, leave the blank jumper plug on the terminal strip, as installed. For 2-minute auto-alarm shutdown, remove the black jumper plug from the terminal strip.

2. Connect the battery connector to the 9-volt battery, observing the proper polarity. A short beep will sound, ensuring that the lock is powered and ready.

3. Install the lock cover with four screws supplied.

4. Close door.

## Testing the 700 and 710 series models

To test a 700 or 710 Series model, do the following:

1. Lock and unlock the door using the cylinder key. Note that a small beep sounds when the deadbolt is retracted, indicating a disarmed condition. The door can now be opened without an alarm, by pushing the push bar.

2. Lock the door again and push the push bar to open it. Immediately the alarm will pulse loudly and one of the following will occur: If continuous alarm has previously been selected, the alarm will sound until the lock is manually reset by locking the deadbolt with the key. If auto-alarm shutdown has been selected, the alarm will sound for 2 minutes then reset.

   When the battery becomes weak, the sounder will emit a short beep approximately once a minute, indicating the battery needs replacing.

## Special operations

If the unit has a retriggerable alarm (Model 260), after the initial 2-minute alarm and auto shutdown, the alarm will retrigger if the door is opened again. This function will remain retriggerable until the door is relocked with the key. Whenever an alarm is caused by opening the door, and the door is left open, the 2-minute alarm shutdown will be inhibited.

To use the unit's dogging operation, insert the 3/16-inch Allen wrench into the dogging latch through the hole in the channel insert. Turn the dogging latch counterclockwise a half turn, push in the pushbar, and turn the dogging latch clockwise a quarter turn until it stops. Release the pushbar and notice that it stays depressed and the door is unlatched.

## ALARM LOCK MODEL 715

Standard features of the Model 715 include: the unit is non-handed; a deadbolt with hardened steel insert that can be operated by outside key; a 15-second delay before door can be opened after pushbar has been pushed; a lock that only requires 5 to 10 pounds of force to operate and can be used for single or double doors; loud dual piezo horn; selectable, continuous, or 2-minute alarm shutdown; and a disarming beep when bolt is retracted with key.

The Model 715 electronic exit lock is installed in the following way:

1. With the door closed, select the proper template and tape it to the inside face of the door, with the center line approximately 38 inches above the floor.

2. Mark and drill the following holes (see template for details):

a. For single and double doors: Mark six .157-inch-diameter holes—four for the lock mounting plate and two for the keeper.

b. Mark a 1/4-inch-diameter hole for the rub plate on double doors 13/4 inch thick.

c. If outside cylinder (CER-OKC) is used, mark the center of the 11/4-inch-diameter hole.

d. If mounting the lock on a hollow metal door and wires are run through the door, align the lock with the holes drilled in step 2A. Mark and drill a 3/8-inch hole in the door to align with the hole in the baseplate near the terminal strip P3 (FIG. 13-11).

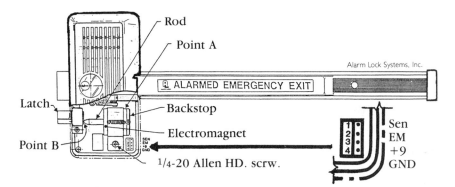

**13-11**   Connect one end of the 4-conductor cable to the control box.

3. Remove the lock cover and four screws holding the cylinder housing to the bolt cover. Install rim cylinder (CER) with the keyway horizontal, facing the front of the lock in a nine o'clock position. Cut the cylinder tailpiece 3/8 inch beyond the base of the cylinder. Reinstall the cylinder housing, guiding the tailpiece into the crosshole of the cam with four screws.

4. Use the key to test for proper operation of the deadbolt. The key should be able to be withdrawn from the lock in either the fully locked or fully unlocked position of the deadbolt. If not, the cylinder and the cam are misaligned and the cylinder housing must be removed. Turn the cam a quarter turn to the right, and reinstall the cylinder housing. Note: The deadbolt can be projected into the keeper by turning the key counterclockwise and can be withdrawn from the keeper by turning the key clockwise one full turn.

5. For outside cylinder only (CER-OKC): Install the rim cylinder with the keyway horizontal facing the front of the door in the three o'clock position. Use the screws supplied. Cut the tailpiece 3/8 inch beyond the inside face of the door.

6. For outside cylinder only: Guide the tailpiece of the outside cylinder into the crosshole of the outside cam.

7. Install the lock loosely to the door with four No. 10 screws, as supplied. Do not tighten them at this time.

8. Insert the bar and channel assembly under the channel retainer bracket, which is mounted to the lock baseplate. Using a level, hold the bar and channel assembly horizontally against the door. Slide the end cap bracket into the end of the channel and, using the bracket as a template, mark and drill the two .157-inch-diameter mounting holes on the door. If the channel is too long, cut the channel and channel insert to the proper length and deburr edges.

9. Attach the pushbar to the lock at the clapper arm hinge bracket using the 1/2-inch screw and No. 10 internal tooth lockwasher provided.

10. Mount the end cap bracket to the door with two No. 10 screws provided, and tighten the lock securely to the door.

11. Attach the end cap to the end cap bracket using the 1/2-inch oval head screw provided.

12. For single doors only: Remove the keeper cover, roller, and pin. Install the keeper base on the door with two screws supplied. Reinstall the pin, roller, and cover with two screws.

    For double doors only: Install the rub plate for a 1 3/4-inch-wide door, from inside the door. Also install the 732 keeper with the two No. 10 screws. Do not tighten the screws fully because the keeper will require adjustment as mentioned in the next step.

13. For single doors only: Close the door, project the deadbolt, and adjust the keeper so that the door is tightly latched. Retract the deadbolt, hold the keeper, release the latch, and open the door. Open the keeper cover and tighten the screws. Drill a .157-inch-diameter hole, as shown on template, for the holding screw, and fasten the keeper with a No. 10 screw. Reinstall pin, roller, and cover with two screws.

    For double doors only: Close the door, project the bolt, and adjust the plastic slide on the 732 keeper so the door is tightly latched. Tighten the screws.

## Installing the control box

The control box for Model 715 is installed as follows:

1. Remove the control box cover.

2. Select a location for the control box on the hinge side of the door and mount it to the wall using the three 3/4-inch self-tapping screws.

## Wiring

Model 715 is wired in the following way:

1. A four-conductor No. 22-AWG cable is needed to connect the control box to the lock. Two ways of bringing electric current from the hinge side of the door frame to the door are as follows:

   a. Use the Armored Door Loop Model 271 by disconnecting one of the 271 end boxes and inserting the loose end of the armored cable into the 1/2-inch hole on the control box. Secure it with the retaining clip.

   b. Use a continuous conductor hinge with flying leads.

2. Connect one end of the 4-conductor cable to the control box terminal strip P2 (see FIG. 13-11) and the other end of the cable to the lock terminal strip P3.

The terminal strips are marked as follows:

1) SEN —sence
2) EM —electromagnet
3) +9VDC
4) —ground

Do not cross wires.

3. Connect one end of an approved twin lead 18-2 cable to the terminal strip at P1-1 and P1-2 (see FIG. 13-11), and connect the other end of the cable to the 12VAC 20VA transformer provided. Do not plug in the transformer at this time.

4. Connect one end of another approved twin lead 22-2 cable to the terminal strip at P1-3 and P1-4, connect the other end to the normally closed alarm relay contacts of one of the following: An approved supervised automatic fire detection system or an approved supervised automatic sprinkler system.

5. Install the control box cover.

## Operating Model 715

Before installing the lock cover, do the following:

1. For continuous alarm, leave the black jumper plug on the terminal strip, as installed. For 2-minute auto-alarm shutdown, remove the black jumper plug from the terminal strip.

2. Connect the 9-volt battery to its connector. A short beep will sound, ensuring that the 715 is powered and ready.

3. Plug in the transformer into a continuous 115 ac volt source. Note the red pilot lamp on the control box is lit.

4. Open the door and install the lock cover with the four screws that are supplied.

## Testing Model 715

To test a Model 715, do the following:

1. Lock and unlock the door using the cylinder key. Notice the small beep when the deadbolt is retracted, indicating a disarmed condition. The door can be opened by pushing the push bar, after a 15-second wait, without causing an alarm.

2. Lock the door again and push the push bar. Immediately the alarm will pulse loudly and one of the following will occur: The 715 latch will impede the opening of the door for approximately 15 seconds, then remain unimpeded until it is manually reset with the key. Or, if continuous alarm was selected above, the piezo sounder will remain on for 2 minutes, then reset.

3. In the event of a power failure the impeding latch will be disabled, but the 9-volt battery will provide standby power for the alarm circuit. This can be tested by disconnecting the transformer from the 115-volt source and pushing the pushbar. The door should now open immediately and the alarm should sound.

4. In the event of a fire panel alarm, the impeding latch will also be disabled. This can be tested by disconnecting the wire going to the control box P1-3, and pushing the pushbar. The door should open immediately and the alarm should sound.

## Battery test

Push and hold the test button on the control box and push the pushbar to alarm the unit. If the piezo sounder sounds weak or doesn't operate at all, replace the 9-volt battery.

*Chapter* **14**

# Electricity for locksmiths

$E$lectric strikes, electronic exit devices, and electromagnetic locks are just a few of the many electrical items locksmiths regularly install and service. More items are constantly being added to that list. Today it's difficult for a locksmith to stay competitive without working with electricity.

Not long ago, few locksmiths bothered to learn about electricity. Most just worked with mechanical locking devices, and left the installation and servicing of electrically controlled devices to electricians and alarm installers. This situation has changed within the last few years for two reasons: First, hardware and department stores were fiercely competing with locksmiths. Frequently such stores were so large they could afford to cut keys and install and sell locks at much lower prices than locksmiths could. Locksmiths had to offer new services to compete with those stores. Second, many locksmiths sell and install electrically controlled security devices because manufacturers now make such devices easy to install.

An extensive understanding of electricity isn't needed for installing most electric security devices. Usually the major difference between installing a mechanical locking device and an electric one is that the latter requires the running of wires. The instruction manual that comes with the device explains how to mount the various parts of the device and how to hook up the wires to those parts. The wires are color coded, which makes it easy to make proper connections.

Running the wires can be the most difficult part of installing an electrically controlled device. You should keep in mind, however, that this is usually more a matter of aesthetics than function. Even if the wires were just laying in a heap in the middle of a room, the electrically controlled device would operate if the wires were connected properly. However, most people prefer the wires to be less noticeable. Some wires must also be hidden to prevent people from tampering with them.

To install electrically controlled security devices, you need to know three things: the basic principles of electricity, how to run wires properly, and how to safely work with electricity.

## BASIC PRINCIPLES

For an electrically controlled device to operate, electric current must flow in a *complete circuit* (also called a *continuous circuit*). Basically, a complete circuit works in this way: A power source (such as a battery) must contain the amount of *voltage* needed to force a sufficient amount of current to flow through a wire, called a *conductor*, to an electrically controlled device, called a *load*, and then to flow through another conductor to return to the power source. That cycle continues until the power source runs down or the current flow is interrupted (FIG. 14-1).

In addition to wires, terminals, screws and strips of various types of metals are also used as conductors. Usually a combination of several types of conductors are used with an electrically controlled device. Regardless of which conductors are used, they must be connected to one another in such a way as to allow current to flow in an unbroken path.

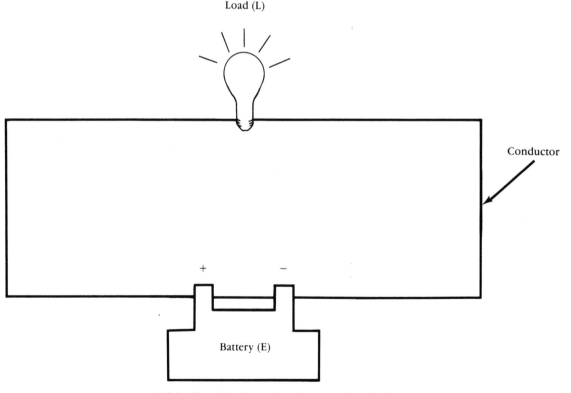

Load (L)

Conductor

+          −

Battery (E)

**14-1**  Electricity flows through a complete circuit.

Whenever you turn an electrically controlled device off, you are simply disrupting the current flow. For example, when you push a button or flip a switch to turn a light off, the button or switch moves a conductor—usually a small piece of wire or metal—away from a connecting conductor to prevent current from flowing (FIG. 14-2).

## Types of current

The two types of current are ac (*alternating current*) and dc (*direct current*). Alternating current flows through wires within the walls of your house; direct current flows from batteries. There are two practical differences between the two: ac can be more dangerous to work with, and *polarity* doesn't have to be observed when using an ac power source.

Polarity refers to negative and positive terminals of a power source. When using a battery, it's important to consider which terminal (negative or positive) each conductor must be connected to. When using ac power from a wall socket, either conductor can be connected to either terminal. For example: If you inserted an ac-powered device into a wall socket with the plug upside down, the device would still operate. However, if you insert batteries upside down in a dc-powered device, it would not operate.

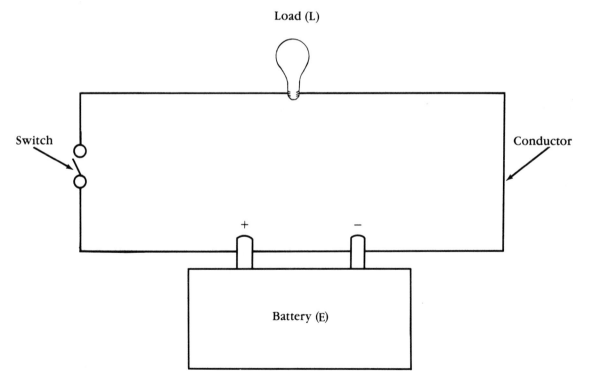

Load (L)

Switch

Conductor

+  —

Battery (E)

**14-2** Switches are used to turn the flow of electricity on and off.

## Resistance

Any opposition to electrical current flow is called *resistance*. There is some resistance in every circuit, because every load and every conductor contains resistance. Different types of material provide varying amounts of resistance. Rubber offers resistance to current flow, while copper conducts current very well. This is why rubber is often used to insulate electrical devices, and why copper is often used to make conductors.

## Controlling electricity

Electricity is controlled in two ways: by limiting its force and by directing its flow. The force must be controlled because too much could damage electrically controlled devices and too little wouldn't power the device. Force is controlled by using a power source that has the proper amount of voltage; by using the proper type, length, and thickness of conductors; and by using various circuit components.

## Circuit components

The various components within an electrically controlled device are the *circuit components*. Each serves a specific function within the device. Examples of such components include resistors, capacitors, transformers, etc. Each component and various types of each component are identified by a letter or a schematic symbol. This makes it easier to illustrate on a schematic diagram the components of a device and the relationships of those components.

## Resistors

*Resistors* are circuit components designed to reduce the amount of current flowing to other components within a circuit. Because different components in a circuit require different amounts of current, resistors are used in virtually every electrically controlled device. They are available in both fixed and variable types. Fixed resistors provide a certain level of resistance. Variable resistors, sometimes called *potentiometers* or *rheostats*, can provide more than one level of resistance.

## Capacitors

A *capacitor* stores electricity. It is made of two plates that are separated by insulation, such as air or glass. The insulation prevents the plates from making a direct electrical connection, which would cause the current to flow rather than be stored. Some current does flow through a capacitor, but some is also stored. Both fixed and variable types of capacitors are available.

## Transformers

Most of the electronic devices locksmiths install use *transformers*. Transformers allow you to use an ac power source from one circuit, such as is found in your wall socket, to power a device on another circuit. Depending on the type of transformer used (step-up or step-down), it will increase or decrease the current flow to make the level of current flow suitable for the device. Most of the transformers used by locksmiths are step-down transformers. Because they decrease the level of current flow, such transformers make it safe to install electrically controlled devices.

### Ohm's law

Voltage, the force that powers current, is measured in *volts*. Current is measured in *amperes* (also called *amps*). *Resistance* is measured in *ohms*.

The relationship between the basic elements of an electric circuit is expressed by *Ohm's law*. The law states that a resistance of 1 ohm passes a current of 1 ampere in response to an applied potential of 1 volt. Mathematically, Ohm's law is expressed as: $E = I \times R$, with E as voltage, I as current, and R as resistance.

Ohm's law expresses a basic principle of electricity that is very important to understand. When installing an electrically operated device, Ohm's law makes it possible to figure out how much voltage, current, and resistance is needed to properly power the device.

## RUNNING WIRES

Where wires can be run depends on the locations of the various parts of the locking system and the physical characteristics of the facility in which the system is installed. When deciding where to run wires, consider where they would be most unobtrusive and tamper resistant. This usually requires running the wires over a drop ceiling, under carpet near a wall, or along or behind baseboards.

When running a wire from one room to another, you might need to drill a small hole through a wall, preferably above a drop ceiling, or at another place that can't be seen. To run a wire from one floor to a floor on another level, you might be able to use an exiting hole such as one used for an air vent or water pipe.

Wire can be protected by running it through a conduit. Electrical metallic tubing (EMT) is commonly used for that purpose. Many local building codes require EMT to be used.

## SAFETY

Working with electricity can be very dangerous, unless certain safety factors are strictly adhered to. Before doing any electrical work, shut off the current coming into the building or automobile. In a building the current can be shut off at the service entrance panel by turning off the main circuit breaker or by pulling the main fuse. In an automobile, you would remove the battery cables from the battery. Be sure to wrap electrical tape around the ends of the cable.

All the tools you use should be properly insulated. Wear rubber soled shoes and stand on a rubber mat to resist electrical shock. Because water is a very good conductor of electricity, make sure your hands are dry before working with electricity.

Finally, always use properly insulated wires. Never use wires that are encased in frayed or torn insulation.

*Chapter* **15**

# Electromagnetic locks

$T$he electromagnetic lock was invented in the United States in 1970 and has gained considerable popularity. Today there are major manufacturers of the product throughout the United States, Canada, and Europe.

Electromagnetic locks are used primarily to secure emergency exit doors. When connected to a fire alarm system, the lock's power source is automatically disconnected when the fire alarm is activated. That allows the door to open freely so people can quickly exit.

Although the principle of operation of an electromagnetic lock is very different from that of a conventional mechanical lock, the former has proven to be a cost-effective, high-security locking device. Unlike a mechanical lock, an electromagnetic lock doesn't rely on the release of a bolt or a latch for security, but instead relies on electricity and magnetism.

A standard electromagnetic lock consists of two components: a rectangular electromagnet, and a rectangular ferrous metal strike plate. The electromagnet is installed onto a door's header; the strike plate is installed on the door in a position that allows it to meet the electromagnet when the door closes (FIG. 15-1). When the door is closed, and the electromagnet is adequately powered (usually by 12 to 24 dc volts at 3 to 8 watts), the door is secured. Some models provide over three times more locking force than typical deadbolt locks provide.

One of the biggest fears people have about electromagnetic locks is power failure. What happens if the power goes out, or if a burglar cuts the wire connecting the power source to the lock? Standby batteries are often installed with the lock to provide continued power in such cases. And the lock can't be tampered with from outside the door, because it is installed entirely inside the door. No part of the lock (or power supply wires) are exposed from outside the door.

**15-1** The strike plate of an electromagnetic lock is usually mounted on the top of a door in alignment with the electromagnet.

Securitron

Another important security feature of electromagnetic locks is they are fail-safe. That is, when no power is going to the electromagnet, the door will not be locked. That is why the lock meets the safety requirements of many North American building codes.

There are two major disadvantages with electromagnetic locks. The locks often cost from four to ten times more than typical high-security mechanical locks. And many people think electromagnetic locks are much less attractive than mechanical locks.

The electromagnetic lock we will discuss in this chapter is the Model 62 Magnalock, manufactured by Securitron. This lock can secure any door with a force of 1200 pounds. It is UL listed, and contains no moving parts. The electromagnet is housed in a compact stainless steel case, and operates on 3 watts of dc power. Internal electronics awarded two patents ensure compatibility with all standard card-reader access control systems (FIG. 15-2).

## INSTALLING A MODEL 62 MAGNALOCK

The Model 62 Magnalock has several features to make installation easy. The lock includes blind nuts to allow the installer to mount it to the frame without having to tap the mounting holes. The mounting holes are recessed into the lock and covered by tamper-resistant finish caps. The wire can be hidden in the frame of the door or run through conduit fitting from the lock.

**15-2** The Model 62 Magnalock.

To further help the installer, Securitron offers an installation kit that includes various tools and templates that make it easy for anyone to install a Model 62 Magnalock. Figure 15-3 shows an installation kit.

**15-3** The Magnalock installation kit makes it easy to install Magnalocks.

First survey the physical area in which the unit is to be placed, and determine the best method of mounting it. In this initial planning two considerations come into play: The mounting method must be strong enough so that the full holding power of the Magnalock can be effective, and the Magnalock and wiring must be protected from damage by intruders or vandals. Often an accessory bracket is necessary, either furnished by Securitron or made by the installer.

When installing the lock on an outswinging door, mount the lock under the door frame header in the corner farthest from the hinges (FIG. 15-4). Most commonly, it is positioned horizontally, but vertical positioning should also be considered. In some cases the horizontal header on an aluminum frame glass door is not as strong as the vertical extrusion; vertical mounting might be preferred in such an instance.

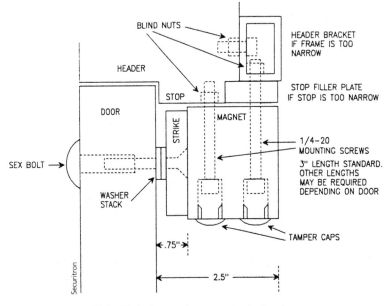

**15-4**  Typical mounting on outswinging door.

This type of installation places the Magnalock so the door swings away from it. This configuration is necessary for all facility doors, otherwise, the Magnalock would be on the outside of the building. For interior doors, the Magnalock should still be mounted in the same manner, unless security planning anticipates a physical assault on the Magnalock from that side of the door.

## Mounting the strike

Mount the strike before the magnet on the upper corner of the door. Position the top of the strike about $1/10$ inch below the point where the door meets the door stop, to permit free closing even if there is shifting of the door. Left/right positioning of the strike is dictated by the desired position of the magnet. The strike must be centered on the magnetic poles (three bars) and the magnet may be moved an inch or so in from the frame, so that the magnet mounting holes will not have to be drilled awkwardly in the corner.

When the strike position has been chosen, drill three holes in the door, following the template. Install the roll pins (metal for standard, plastic for senstat) in the strike, using a hammer. If metal roll pins have been supplied, be careful not to strike them too hard with the hammer, because it is possible to dent the strike surface of the pins by overdriving. This can degrade strike flatness and therefore holding force.

The strike is secured by the centrally located strike mounting screw. Place a washer stack consisting of two flexible washers between the strike and the door, with the strike mounting screw passing through the two washers to provide flexibility. Do not place the washers around the roll pins. The roll pins should float in their holes and not bind. Their only purpose is to prevent the strike from rotating.

Securing the strike to the door is crucial, because it is the only point of attack for an intruder from the outside. Accordingly, Securitron provides a $1^1/4$-inch-diameter sex bolt, machined from case hardened tool steel, to terminate the strike mounting screw. Commercial sex bolts made of brass or aluminum can be defeated with a chisel and should not be used.

Secure the strike to the door using the supplied sex bolt. It is sometimes difficult to align the strike mounting screw with the sex bolt. The following technique is recommended: Start the sex bolt in its $1/2$-inch hole, but thread the strike mounting screw into it before hammering the sex bolt down flush. When the strike mounting screw is started, hammer the sex bolt down and then screw the strike mounting screw in the rest of the way. This makes alignment much easier.

When the strike is mounted, make sure it flexes freely around the washer stack. This flexing allows the Magnalock to pull the strike into perfect alignment for maximum holding force, regardless of inaccuracies of alignment between the strike and the Magnalock that can occur during initial installation or during subsequent warping of the door or settling of the frame. The most common error made during installation of a Magnalock is mounting the strike in a rigid position.

## Mounting the magnet

The magnet mounts in the door frame header with four socket cap machine screws for metal frames or wood screws for wood frames. In mounting the Magnalock, these conditions must be followed:

- The frame header must present a flat surface for the magnet to mount to.
- The frame area selected must be structurally strong enough to yield a properly secure installation.
- The magnet face must be parallel to the strike plate.

- The magnetic poles (three metal bars on the Magnalock) must be centered on the strike.

- The magnet must make solid contact with the strike but still allow the door to close properly.

- The direction of the door opening must pull the strike directly away from the magnet rather than sliding it away. Electromagnets hold only weakly in the shear direction of pull.

Taking these points in order, the first is satisfied if the frame presents a flat surface wide enough for the magnet. If not, the use of stop filler plates and header brackets available from Securitron can usually solve the problem. A $2^1/2$-inch space is required from the door to the rear of the magnet for proper mounting.

The issue of the frame strength must be considered in selecting vertical or horizontal mounting. On aluminum headers the horizontal extrusion is often weak and can be snapped off so vertical mounting would be preferred. It is also possible to reinforce the header by adding a steel plate. Avoid mounting the magnet to a wobbly or weak support or the intrinsic security of the lock will be diminished.

Once a flat surface has been prepared for the magnet, it must be positioned so that its face is parallel to the strike plate and door. The magnetic poles must be centered on the strike plate, and the door must close properly, with the magnet making firm contact to the strike plate. When the magnet has been experimentally positioned so that these criteria have been met, it is ready for mounting.

Drill four holes for the mounting screws, and a $1/2$-inch-diameter wire hole should be drilled for electrical hookup. For proper strength, the mounting machine screws must be secured by nuts. To facilitate this, special blind finishing nuts with star washers are supplied with each Magnalock. The nuts are used as follows:

Drill a $3/8$-inch (9.5 mm) hole according to the template for each finishing nut. Press the nut up into the hole and lightly seat with a hammer tap so its knurl engages. Run the mounting screws through the magnet and into the finishing nuts. The star washers are placed between the magnet case and the blind nuts to provide friction, allowing the nuts to self collapse. If the blind nut collapsing tool furnished in the installation kit is used, the star washers are not needed. Don't forget to use the gold flat washers on the magnet mounting screw heads; they prevent the narrow screw heads from digging into the resin body of the magnet.

If the mounting screws don't line up with the blind nuts installed in the header, it is possible to enlarge certain of the mounting holes in the magnet to achieve alignment. Note that all Magnalock versions have a cable exiting on one end of the magnet. Never enlarge the two mounting holes on the end where the cable exits. Wiring runs through the resin in that part of the magnet and drilling out the mounting holes on that end can easily destroy the product—and void the warranty.

Either or both of the mounting holes on the opposite side of the magnet from cable exit may be enlarged, although you shouldn't go beyond

$9/32$ inch. If the holes still don't line up, you can mount only three of the four screws and still achieve a strong installation. Installers doing multiple jobs should obtain Securitron's installation tool kit, which includes a drill guide that yields excellent mounting hole alignment.

As the screws are tightened, you will feel an initial resistance—which is the nuts collapsing—and then a second stronger resistance as the screws seat. Do not over-torque. The collapsed nuts coupled with machine screws provide extremely strong mounting. The purpose of the star washers is to dig into the case of the magnet and into the base of the nuts. This helps the nuts collapse and prevents them from spinning. The nuts will work on any thickness metal frame, including concrete-filled headers. Once the nuts have collapsed, the star washers can be removed for a slightly neater installation.

### Installation on inswinging door with Z bracket

In cases where the Magnalock must be mounted on the inswinging side of the door to protect it from physical assault, the Magnalock body is mounted flush on the wall above the door frame. A Z bracket is affixed to the door, which positions the strike plate in front of the Magnalock. Securitron's F Series Magnalocks are used, because they have mounting holes through the face of the magnet and wire exit to the rear (FIG. 15-5).

Finishing nuts are supplied because often the lower two screws will be mounted in the metal door frame header. The top two screws might require appropriate anchors—when used in drywall, for instance. Securitron offers a Z bracket to allow for mounting on a wide range of commercial door/frame configurations.

### Aluminum frame glass door mounting

This is probably the most common door type that utilizes the Magnalock. Certain mounting problems can arise, depending on the exact configuration of the door and frame. Often the header is not wide enough to support the depth of the magnet. This can mean that none of the mounting screws can be run into the header, or that only two of the four will fit. Another aspect of the mounting screw problem is that two of the mounting screws might line up with the end of the header extrusion. Also, the wires might exit beyond the end of the header so that they will be exposed and vulnerable to tampering.

Most of these problems are solved by the use of Securitron's universal header bracket (part UHB-62). This extends the depth of the header either 1 or $1^1/2$ inches, depending on which way it is oriented. This usually allows mounting of all four screws, and because the bracket is a hollow extrusion, the wire is run inside the bracket and therefore is hidden.

Even with use of the bracket, it is possible that one set of mounting screws may end up lining up with the end of the header. Note that some adjustment of the magnet mounting position is possible. Instead of the two rubber washers supplied with the strike, one or three may be used. If the door is secured only by the Magnalock (there is no mechanical deadbolt), the door closed position may be altered to allow all four mounting

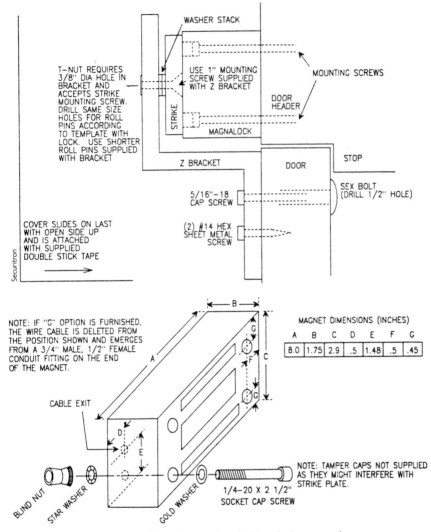

**15-5** Magnet dimensions and typical inswinging mounting.

screws to be used. Finally, note that an installation of this type of door is acceptable if only two mounting screws are used.

Because the screws run into steel nuts, the fastening technique is very strong. In fact, when the design decision was made to employ four mounting screws, one reason was to allow at least a pair to be used in the event of alignment problems with the header. It is best to use all four screws, but on this type of door—which is inherently not high security because the glass can be shattered for forced entry—firmly mounting two screws is acceptable.

Another problem that can arise with aluminum frame glass doors is that, in certain cases, the height of the aluminum rail at the top of the door is not sufficient to mount the strike and sex bolt. If the sex bolt is installed in the lowest area of the top rail, the top edge of the strike will protrude above the rail. To solve this problem, Securitron offers the offset strike. The holes in the strike plate are offset 1/4 inch from the center of the strike; this allows successful mounting on a narrow top rail. Approximately 10 percent of the holding force is lost because of the skewed position of the strike mounting screw. However, this is not significant on aluminum frame glass doors, because they are not high-security barriers. The offset strike can be supplied with the lock or ordered separately.

### Solid glass door mounting

The Magnalock is useful for securing 100 percent glass doors that have no aluminum rail. The magnet is suspended in normal fashion from the header. The difficulty in the installation is mounting the strike plate on the glass, because glass cannot be drilled. This is accomplished by using Securitron's Model GDB (glass door bracket) and Model AKG (adhesive kit for glass). The bracket is affixed to the glass surface by a special adhesive and the strike screws into the bracket conventionally. The adhesive provides a bond stronger than the holding force of the Magnalock, and is permanent (FIG. 15-6).

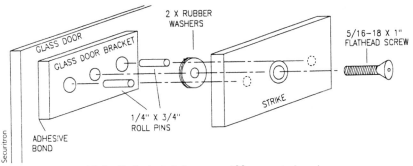

**15-6**   Strike installation on a 100 percent glass door.

Some doors that appear to be glass are laminated with plastic. If the Magnalock with glass door bracket is used on such a door, it might fail to work.

In some cases the header of a glass door is vertical glass. The magnet can be mounted on such a header by using a 3-×-3-inch aluminum angle bracket, available from Securitron. The bracket is glued to the vertical glass header with Securitron's adhesive kit for glass and the magnet is screwed to the bracket.

### Double door mounting

It is very common to use Magnalocks on double doors. Several mounting possibilities exist. In some cases one of the door leaves is pinned so only one leaf is used. It is secured by a single Magnalock in standard fashion. If both leaves are to be active, two Magnalocks can be used. For the most attractive installation, the leaves should be abutted, but if obstructions exist in the header that interfere with mounting, the magnets can be separated.

Another possibility is to use Securitron's split strike. In this method, a single Magnalock is mounted in the center of the header and a half-size strike is mounted on each leaf. This reduces the holding force to under 600 pounds for each leaf and lowers the door security to a "traffic control" level. It must be strictly understood that a strong kick will open this door, whereas if two complete Magnalocks are used, the lock strength typically exceeds that of the door. The split strike is available either as part of a complete Magnalock, or is supplied separately as a replacement for the standard strike. (Note: Certain electronic considerations also apply when Magnalocks are used on double doors.)

## Wood frame mounting

With a wooden frame, long wood screws may be used to mount the Magnalock. If, however, the frame is not solid enough to secure the Magnalock adequately, a wood frame bracket available from Securitron may be used (part WF-62). The Magnalock mounts to the bracket with machine screws and the bracket permits the wood screws to penetrate more deeply into the header (FIG. 15-7).

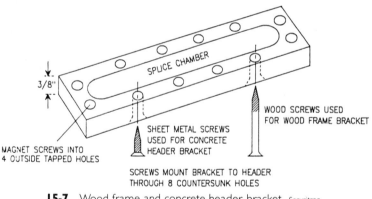

SPLICE CHAMBER

3/8"

WOOD SCREWS USED FOR WOOD FRAME BRACKET

SHEET METAL SCREWS USED FOR CONCRETE HEADER BRACKET

MAGNET SCREWS INTO 4 OUTSIDE TAPPED HOLES

SCREWS MOUNT BRACKET TO HEADER THROUGH 8 COUNTERSUNK HOLES

**15-7** Wood frame and concrete header bracket. Securitron

## Steel header filled with concrete mounting

The blind nuts function normally despite the presence of concrete, but a problem can occur in pulling the hook up wires. To help, Securitron offers the concrete header bracket (part CHB-62) which permits a range of techniques (see FIG. 15-7). The center of the bracket forms a splice chamber, which makes it less difficult to pull the wires back into the header. Alternately, the wires may be pulled through the edge of the bracket by drilling a hole, if it's impractical to drill the concrete.

Another technique for concrete headers is the use of Securitron's G version Magnalock. It incorporates a 1/2-inch female/3/4-inch male universal threaded conduit fitting. The conduit fitting is placed on the end of the magnet body, and the problem of pulling wires into concrete is bypassed because the wires can be run in pipe in a surface-mounted configuration. The mounting holes on G locks are counterbored from both sides to make the lock non-handed. To collapse the blind nuts (which otherwise would fall into the counterbore), flat washers are supplied to close the magnet surface. After the nuts have been collapsed, the flat washers should be discarded.

## Mounting the Magnalock on exterior gates

A popular application for the Magnalock is to secure motorized or manual exterior gates. The Magnalock provides several benefits in this application. Gates tend not to be precisely fitted, so electric bolts suffer from alignment failures. The Magnalock is designed to be self-aligning and tolerates considerable inconsistency in the gate-closed position, as regards upward/downward alignment, side alignment and twisting. The Magnalock is also fully sealed and waterproof so it is generally unaffected by tough environments.

Because of the wide variety of gates in existence, each installation has to be considered special. Usually bracketry must be made up on site. The concept is to mount the magnet on a fixed post and the strike plate to the swinging or sliding member of the gate. Both components should be positioned so that the strike plate slaps against the magnet face on closure.

Usually, the GF version of the Magnalock is preferred for gate installation. "G" refers to a conduit fitting mounted on the magnet end and "F" refers to mounting holes through the face (see FIG. 15-5). The magnet typically screws onto a back plate fashioned on site and the back plate is welded onto the fixed post.

A back plate or Securitron's Z bracket should also be provided for the strike plate. The strike plate cannot be directly welded to the gate because it will not be able to flex and self-align. It must be screwed onto a surface, with the washer stack used to provide flexibility. If Securitron's Z bracket is used, it typically bolts to the gate, rather than being welded, because it is aluminum.

In the case of very tall and large gates, a levering problem can exist. An intruder might be able to flex the gate enough to take up the slack in the strike mounting screw and then lever off the strike plate. If the installer or user determines that this might happen, a single Magnalock will not provide adequate security. Two must be used, typically at the top and bottom of the gate.

Figure 15-8 shows preferred special techniques for Magnalock mounting on three types of gates. The first drawing shows a single swinging gate. The general technique follows the principles discussed above, but the use of Securitron's Z bracket, which creates a neat installation, is also shown. Note that in some cases, the post that mounts the magnet is hollow. It is possible to use the F version (without conduit fitting) and pull the wires through the post, which might yield a neater and more secure installation.

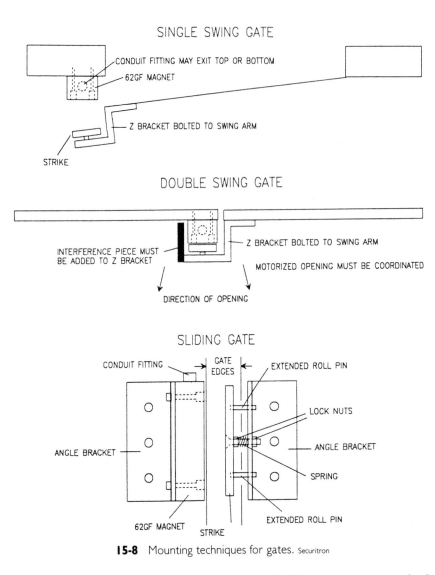

SINGLE SWING GATE

CONDUIT FITTING MAY EXIT TOP OR BOTTOM

62GF MAGNET

Z BRACKET BOLTED TO SWING ARM

STRIKE

DOUBLE SWING GATE

INTERFERENCE PIECE MUST
BE ADDED TO Z BRACKET

Z BRACKET BOLTED TO SWING ARM

MOTORIZED OPENING MUST BE COORDINATED

DIRECTION OF OPENING

SLIDING GATE

CONDUIT FITTING

GATE
EDGES

EXTENDED ROLL PIN

LOCK NUTS

ANGLE BRACKET

ANGLE BRACKET

SPRING

62GF MAGNET

STRIKE

EXTENDED ROLL PIN

**15-8**  Mounting techniques for gates. Securitron

The second drawing in FIG. 15-8 shows a double swinging gate, which presents a unique problem. The Magnalock is mounted in the same general way as on a single swinging gate, but because both arms move, an intruder pushing on the gate exerts a shearing force on the Magnalock. Electromagnets are not at all strong in this type of attack. Therefore, as the drawing shows, Securitron's Z bracket should be used with an interference piece. This blocks the shearing effect, while the strength of the magnet blocks one arm moving when the other is stationary. For this technique to work, the motorized operator must be coordinated: One arm must move first to clear the interference piece before the other arm starts moving. Gate operators can normally accomplish this.

The final drawing on FIG. 15-8 shows a special mounting technique for sliding. We recommend the GF type magnet and two 3-inch angle brackets (available from Securitron) for a neat installation. A special strike mounting technique that improves the reliability of the installation is shown in the drawing. The problem is that if the strike is mounted normally to the angle bracket and the gate is a powerful one which slams shut, the magnet might be impacted to the point where its mounting screws loosen or the bracket bends.

The strike mounting technique shown creates a shock absorber effect through the use of lock nuts at the rear of the strike and the rear of the bracket, together with a spring. Drill a through hole in the angle bracket mounting the strike and use extra-long roll pins. When the gate closes, the strike moves in against the spring and creates a shock absorbing action.

### Tamper-proofing the Magnalock

In situations where vandalism is more probable, the Magnalock should be protected from tampering. The magnet is inherently tamper-proof because it is totally sealed. However, the magnet mounting screws are vulnerable because the magnet can be dismounted if the screws are loosened. The Allen holes on the screws can be filled with a potting compound such as silicon.

Alternately, the entire hole in the magnet where the screw heads fit could be filled. Butyrate caps are supplied to close the mounting holes. These provide tamper proofing, because they can't be removed by hand, but can be pried off with a tool. The strike plate mounting screw is covered by the strike when the magnet is energized. If tampering is anticipated when the door is open, the screw socket head may be filled.

For added safety, Securitron inventories special tamper-proof screws for both magnet and strike mounting. These screws are identical Allen head types, but a special key is needed to install and remove the screws. It is highly unlikely that a vandal would have access to this type of key. Securitron supplies the tamper-proof screw sets with keys both in the form of a manual Allen wrench and in a bit key for use with a drill.

### General electrical characteristics

The Magnalock constitutes an electric load that draws 3 watts of power. Owing to patented internal circuitry, the Magnalock does not show the normal characteristics of an electromagnetic or other inductive load. Magnalock resistance also cannot be read with an ohmmeter because of the internal circuitry. Inductive kickback is suppressed, so arcing across control switch contacts need not be a concern.

This suppression also protects nearby access control or other computer equipment from possible interference and for the same purpose microwave radiation is also suppressed internally. The circuitry performs the additional functions of canceling residual magnetism (''stickiness'' on release) and accelerating field collapse so that the Magnalock releases instantly when power is removed from it.

## Standard lock

For operation, dc voltage (24 or 12 volts, depending on the model selected), must be provided to the lock. The red wire receives +12 volts or +24 volts, and the black wire, 0 volts (negative). If the lock is connected with reverse polarity, it will not function. The voltage source may be regulated, filtered, or pulsating dc (using transformer plus bridge rectifier). Half-wave pulsating dc generated by a transformer and single diode will not properly operate the Magnalock. An exact voltage level is not necessary. Less than standard voltage will proportionately reduce the holding force, but will cause no harm. Over-voltage up to 30 percent is acceptable.

The current draw is 125 mA for 24-volt versions or 250 mA for 12-volt (mA is an abbreviation for milliamperes). It is good practice to use power supplies with one-third extra capacity beyond the current requirements of the load. This greatly reduces the possibility of heat-induced power supply failure, and also allows for future expansion. Power supply cost is a small fraction of the job cost and should not be skimped on.

Switches may be wired as necessary between the Magnalock and power source. Internal circuitry eliminates inductive kickback, so neither electromechanical switches nor solid state devices will be damaged by arcing when the Magnalock is shut off.

Securitron recommends switching the Magnalock on the dc side of its power supply. If switched on the ac side, the power supply capacitor (in the case of a filtered supply) will discharge through the magnet, slowing its release.

## Wire gauge sizing

If the power supply is distant from the lock, voltage will be lost (dropped) in the connecting wires, and the Magnalock will not receive full voltage. The 24-volt Magnalock version is a much better choice for long wire runs, because it gives four times the resistance of the 12-volt version. Note that the correct calculation of wire sizing is a very important issue because the installer is responsible to ensure that adequate voltage is supplied to any load. In multiple device installations, the calculation can become quite complex.

The general practice of wire sizing in a dc circuit is to avoid causing voltage drops in connecting wires, which reduces the voltage available to operate the device. Because Magnalocks are low-power devices, they can be operated long distances from their power source. For any job that includes long wire runs, the installer must be able to calculate the correct gauge of wire to avoid excessive voltage drops.

To calculate the correct gauge, add the resistance of the Magnalock to the resistance in the power wires, then divide the wire resistance by the total resistance. This yields the fraction of voltage drop in the wires. For example, a single 24-volt Magnalock has a resistance of 192 ohms. If the wires completing the circuit between the Magnalock and its power source have a resistance of 10 ohms, the total resistance is 202 ohms. Dividing 10 ohms (the wire resistance) by 202 (the total resistance) yields roughly $1/20$ or 5 percent. If the input voltage is 24 volts, 5 percent of this voltage will be dropped in the wires (1.2 volts), leaving 22.8 volts to operate the Magnalock. This will cause a small reduction in holding force, but is generally acceptable.

To calculate the wire resistance, you need to know the distance from the power supply to the Magnalock and the gauge (thickness) of the wire. Magnalock resistances are 192 ohms for the 24-volt (dc) version and 48 ohms for the 12-volt version.

Suppose a single 24-volt Magnalock is 1200 feet from its power supply and you're using 20-gauge wire. The total length of the power wires is 2400 feet. Remember that you combine the wire lengths from the power supply to the lock and back to the power supply to get the total circuit wire length. The wire resistance then becomes $2.4 \times 10.1$ ohms, which is 24.2 ohms. Adding this to the Magnalock resistance of 192 ohms yields a total resistance of 216.2 ohms. 24.2 divided by 216.2 yields an 11 percent drop in the wires, which Securitron would consider excessive.

This problem can be dealt with in two ways: Either utilize 16-gauge wire, which would reduce the drop to a more acceptable 5-percent range, or provide extra voltage at the power supply. For instance, Securitron 24-volt power supplies are adjustable from 24 to 28 volts. You can therefore easily set the power supply to output 11 percent over-voltage, which will then deliver 24 volts at the lock. As mentioned, the Magnalock will accept up to 30 percent over-voltage without ill effects.

Note that the 12-volt dc Magnalock has one-fourth the resistance of the 24-volt version (48 ohms versus 192 ohms). This means that wire voltage drops are four times more significant in a 12-volt system than in a 24-volt system. In any job that has wire runs long enough to be of concern, always use 24 volts.

Note also that it is common to mount two Magnalocks on a double door and operate them as one lock (only two power wires). In this case, the resistance of the pair of locks is half the resistance of a single lock (96 ohms for 24 volts, 24 ohms for 12 volts).

If a common power wire is used in a loop structure, for instance, the many locks powered by the single loop will have an increasingly low combined resistance, so that the loop wire resistance will become more significant to the point that the locks don't receive enough voltage.

To find the combined resistance of multiple locks powered by a common wire, divide the resistance of one lock by the number of locks. For example, eight 24-volt Magnalocks would have a combined resistance of 192 divided by 8, which is only 24 ohms. Another method is to calculate the current in amps and divide that into the circuit voltage. Because each 24-volt Magnalock draws $1/8$ of an amp, eight would draw 1 amp. Dividing this into the same 24-volt input voltage yields a 24-ohm combined resistance.

In general, you have to be cautious about using common wires for loads in long-distance situations, unless you're very confident about your ability to calculate the correct configuration. Bear in mind, however, that whenever you are uncertain about the voltage drop in wiring, you can meter the voltage at the lock while it's connected and you will be able to see if it's receiving adequate voltage. If the lock is not connected when you make this measurement, the result will be false because the circuit will not see any lock resistance to compare with wire resistance. You will read the full input voltage. Also, you cannot measure Magnalock resistance with an ohmmeter. It has semiconductors inside that render ohmmeter readings meaningless.

## Senstat Magnalocks S and C

Securitron's optional patented Senstat electronics package provides lock status sensing. In many electrically controlled door security systems, status sensing is provided by a magnetic switch on the door. This indicates the door is closed, but not necessarily secured. Securitron's Senstat monitors the lock rather than the door and therefore provides a higher level of security.

**S Senstat Magnalock**   S versions provide a voltage signal on a third wire equal to the input voltage that is on when the door is secure. This is accomplished by conducting the input voltage through the strike, which ensures a flush fit of the strike to the magnetic core. Input power is also monitored so that if the lock is not powered, it will not report secure, even if the strike and magnet are flush.

The output signal is typically used to drive a load such as a relay coil, light, or buzzer, which will indicate lock status. The load must not draw more than 1 amp of current. Securitron monitoring control panels are designed to work with S Senstat Magnalocks. When using S Magnalocks on a metal door, it's important to insulate the strike from the door, to prevent low-voltage leakage into the door frame. The strike is furnished with insulating hardware (FIG. 15-9).

**C Senstat Magnalock**   C versions provide an isolated contact closure on a third and fourth wire that is closed when the lock is secure. The contact closure is between the strike and magnet face, so a flush fit is required for the lock to report secure. Although the C version does not directly monitor input power to the lock, it will generally not report secure if the strike merely lays on the magnet without power. If the C version is used on a metal door, the strike should be insulated from the door by utilizing the insulating hardware supplied with it. The C version is normally used because of its direct compatibility with popular access control and alarm monitoring circuits.

## Double door procedure for status reporting

Often two Magnalocks are mounted on a double door and are turned on and off together, having no separate control. It is possible to receive a separate Senstat status signal from each door, or you can combine the outputs so that if both locks are secure, the double door is secure—and if either lock is not secure, the double door is not secure.

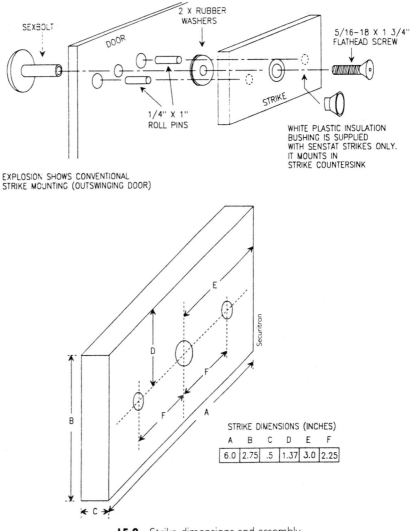

**15-9** Strike dimensions and assembly.

If the desired status output from the double door is of the dry contact type (C Senstat), the connection connects the C outputs in series. Connect one of the C status wires (white or green) from each lock together and take your output from the two wires (one from each lock). When both locks are secure, both Senstat circuits will be closed, and the double door will report secure as if it was a single lock.

If the desired status output from the double door is of the voltage type (S Senstat), you should use an S Magnalock together with a C Magnalock. The voltage output of the S Magnalock is passed through the C contacts and both locks must be secure for a single voltage output to be present from the pair (FIG. 15-10).

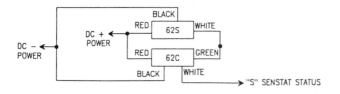

"S" AND "C" LOCKS ARE INTERWIRED AS SHOWN TO PROVIDE STATUS MONITORING AS IF ONE "S" LOCK WAS USED. THE DOUBLE DOOR REPORTS SECURE ONLY IF BOTH LOCKS ARE SECURE.

**15-10**   Double door wiring with S and C lock. Securitron

## Magnalock L version with indicator light

Magnalocks equipped with indicator lights have an LED mounting socket permanently installed in the magnet body. The indicator may be easily replaced in the event of vandalism or failure, because both the LED lamp and lens screw into the permanently mounted socket. On standard or C Senstat Magnalocks, the amber colored light indicates the magnet is powered. On S Senstat Magnalocks, the light is driven by the Senstat voltage output, and a green color indicates the door is secure. A vandal who removes the light and attempts to tamper with the socket will not be able to harm the function of the Magnalock, because the socket is short-circuit protected.

If a light is desired on an F type Magnalock, it must be mounted on the end of the lock because the rear of the lock is the mounting surface. This position interferes with Securitron's Z bracket cover, so Securitron does not recommend ordering lights with F Magnalocks—although it can be supplied on a special-order basis.

## Emergency release

Magnalocks are often wired into a system such that they can be simultaneously released in an emergency, either manually from one switch or automatically—often from the fire alarm system. It is the user's responsibility to accomplish this hookup correctly according to these instructions and good electrical practices. In general, Securitron recommends that a switch or relay be used to perform a series break of all dc power, which is the simple and sure way to make sure the doors release. Securitron power supplies have terminals for interconnection of such emergency release switches.

## TROUBLESHOOTING A MAGNALOCK

In this section are a list of problems that can occur with the Magnalock, and instructions for fixing them.

### No magnetic attraction between magnet and strike plate

Make sure the lock is being correctly powered with dc voltage. This includes connecting the power wires with correct polarity. Positive must go to red and negative to black. If the Magnalock is wired in reverse polarity, it will not be damaged, but it will not operate. If the unit continues to appear dead, it must be electrically checked with an ammeter. It must be powered with the correct input voltage and checked to see if it draws the specified current (125 mA at 24 dc volts or 250 mA at 12 dc volts.

The Magnalock cannot be checked with an ohmmeter, because it has semiconductors that render an ohmmeter reading meaningless. If the unit draws the correct current, it is putting out the correct magnetic field and the problem must be in the mounting of the strike.

### Reduced holding force

If this is the problem, usually the door could be kicked in without much effort. Check the strike and magnet face to see if some small obstruction is interfering with a flat fit. Even a small air gap can greatly reduce the holding force. If the strike and magnet are flat and clean, the cause is nearly always improper mounting of the strike: It is mounted too stiffly. The strike must be allowed to float around the rubber washer stack, which must be on the strike center mounting screw. The magnet then pulls it into flat alignment.

To correct the problem, try loosening the strike mounting screw to see if the lock then holds properly. It is also possible that the input voltage is too low. Be sure that you are not operating a 24-volt lock on 12 volts. If you are using 12 volts and suspect that you might have a 24-volt lock, the only way to check it is to use an ammeter. If it is a 24-volt lock, it will draw 60 mA (milliamperes) at 12 volts instead of the correct 250 mA.

### Lock buzzes or hums and does not hold

The lock will buzz if it's operated on ac or half-wave rectified dc power (transformer plus single diode). Make sure you are using full-wave rectified dc (transformer plus bridge), or filtered or regulated dc power. Another possible cause is that the strike and magnet are not flat together, owing to some obstruction or dent that must be corrected. This can cause a fluttering noise.

### The Senstat output does not report secure

Because of the simplicity of Securitron's patented Senstat design, this is almost always a case of the lock status sensor doing its job. It is not reporting secure because a small obstruction or a stiffly mounted strike is causing the Magnalock to hold at reduced force.

Correct the problem by cleaning the surfaces of the magnet and strike or establishing proper play in the strike mounting. If this doesn't work, you can verify function of the Senstat feature as follows:

Note that there are two thin vertical lines on the magnet face that can be said to separate the core into three sections, from left to right. The Senstat output is created by the strike establishing electrical contact between the leftmost and rightmost core segments. With the lock powered, use a pair of scissors and press the points respectively into the leftmost and rightmost core segments. The Senstat output should then report secure. This shows that the problem lies in the strike not making flat contact with the magnet face. If the scissors technique doesn't cause the lock to report secure, check to see if there is a broken Senstat wire. If this is not the case, the lock must be returned to the factory for replacement.

## Lock does not release

When the power is removed from it, the Magnalock must release. If internal circuitry, which eliminates residual magnetism, were to fail completely, the lock would only exhibit "stickiness" at a rough level of 5 pounds. Therefore if the lock will not release, the problem is either mechanical bonding via vandalism meaning glue has been applied between the strike, or a failure to completely release power.

Failure to completely release power is generally a wiring integrity problem. What happens is that an upstream switch removes power from the wires going to the Magnalock, but through an installation error, the wires have their insulation abraded between the switch and lock. Partial or full power can leak in from another Magnalock or other dc device with similarly abraded wiring. This is most likely to occur at the point where the wire cable leaves the lock case and enters the door frame. Another potential problem area is an improper splice on wiring in conduit.

Either a metal door frame or the metal conduit is capable of leaking power between multiple devices with abraded wires, thereby bypassing switches. A good way to check this electrically (as opposed to visually removing and inspecting the wires) is to use a meter. Check for leakage between the power supply positive or negative and the door frame and conduit. Magnalocks should be powered by isolated dc voltage, without any earth ground reference to positive or negative.

## Rusted lock

Both the Magnalock core and strike plate are cadmium plated and sealed following a military specification. Cadmium provides the highest degree of rustproofing possible on ferrous metal. Because of this plating and the sealed nature of the magnet, the Magnalock is weatherproof and may be used outdoors. If rusting appears, the most common cause is improper cleaning. If steel wool is used, for instance, it can strip off the relatively soft cadmium. Once the plating has been removed, it cannot be restored in the field, so the lock will have to be periodically cleaned and coated with oil or other rust inhibitor. A rusty Magnalock will still function, but at reduced holding force. If the product is installed in a heavily corrosive atmosphere, such as near the ocean, it will eventually rust, even if non-abrasive cleaners are used. The only solution in this case is continued periodic removal of the rust.

## Electronic noise interference with access control system

Electric locks, being inductive devices, return voltage spikes on their power wires and also emit microwave radiation when switched. This can interfere with access control electronics, causing malfunctions—particularly with those units employing high-speed dynamic RAM. Access control contractors often employ installation techniques designed to isolate the access control electronics from the electric lock. These include separate circuits for the lock, shielded wiring, and other techniques. The Magnalock, however, is heavily used in access control installations and includes internal electronics that suppress both inductive kickback and radiation.

The Magnalock has been extensively tested and accepted by numerous access control manufacturers and has been used in thousands of installations without incident. An apparent noise problem is therefore usually not caused by the Magnalock. The access control equipment might be faulty or installed improperly. Check with the manufacturer for proper installation procedures.

There is a problem that can arise with the Magnalock: If the Senstat version is being used, the strike plate (which passes current) must be isolated from a metal door and frame. Securitron supplies insulating hardware to accomplish this, but the hardware might not have been used or the strike might be scraping against the header. Check for full isolation between the strike and the door frame, when the door is secure, using an ohmmeter. The presence of lock voltage potential in the door frame can interfere with the ground reference of access control system data communication and therefore cause a problem.

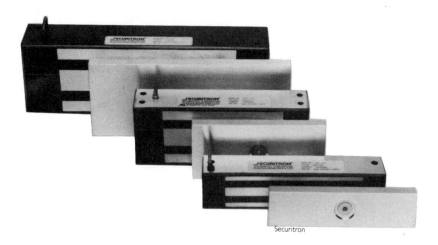

Securitron

**15-11** Securitron offers a "family" of Magnalocks.

## OTHER MAGNALOCKS

In addition to the Model 62, as discussed in this chapter, Securitron manufactures two other Magnalock models (FIG. 15-11). Model 92 is the company's strongest model; it secures a door with a holding force of 3000 pounds. It is for extremely high-security applications. Model 32 Magnalock provides for 600 pounds of holding force (FIG. 15-12). It is especially useful for internal traffic control doors and sliding doors, where high security is not a requirement.

Model 32 comes in a convenient size (8 × 1.5 × 1.75 inches), and is the smallest of the three models. Called the "Mini-Mag," Model 32 has many of the same features as Model 62, including vandal- and weather-resistant stainless steel casing, mil-spec cadmium plating for rust resistance, and low power consumption (300 mA at 12 dc volts or 150 mA at 24 ac volts). Pairs of Model 32's can be mounted at the top and bottom of a door to yield a total holding force of 1200 pounds.

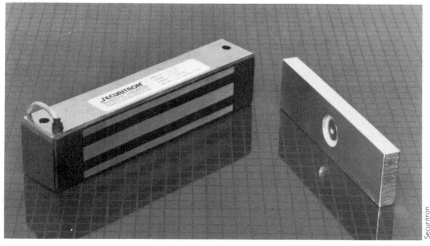

Securitron

**15-12**   The Mini-Mag Model 32 Magnalock has 600 pounds of holding force.

*Chapter* **16**

# Electric strikes

$A$n electric strike is architectural hardware designed to make frequently used doors more secure and convenient to use (FIG. 16-1). These strikes are usually mounted in the frame of a door and use electricity to either hold or release a latch. A switch, which can be located close to or far away from the strike, can be used to activate and deactivate a strike.

There are several popular manufacturers of electric strikes. This chapter reviews some of the models made by Adams Rite Manufacturing Company.

## STRIKE SELECTION

With two exceptions, all Adams Rite electric strikes have flat faces. The exceptions are the 7801 and the 7831 models, which have radiused faces to match the nose shape of paired narrow stile glass doors.

The basic Adams Rite size conforms to American National Standards Institute strike preparation: $1^1/4 \times 4^7/8$ inches. However, two other sizes are offered to fill or cover existing jambs or opposing stile from previous installations of: M.S. deadlock strike (7830 and 7831) or discontinued Series 002 electric strikes (7810).

Strikes are available with round corners for installation in narrow stile aluminum, where preparation is usually done by router, and with square corners for punched hollow metal ANSI (American National Standards Institute) preparation or wood mortise.

Strikes with vertically mounted solenoids are designed to slip into hollow metal stile sections as shallow as 1.6 inches. Horizontal solenoids for wood jambs require an easily bored $3^5/8$-to-$4^1/2$-inch-deep mortise, depending on the model.

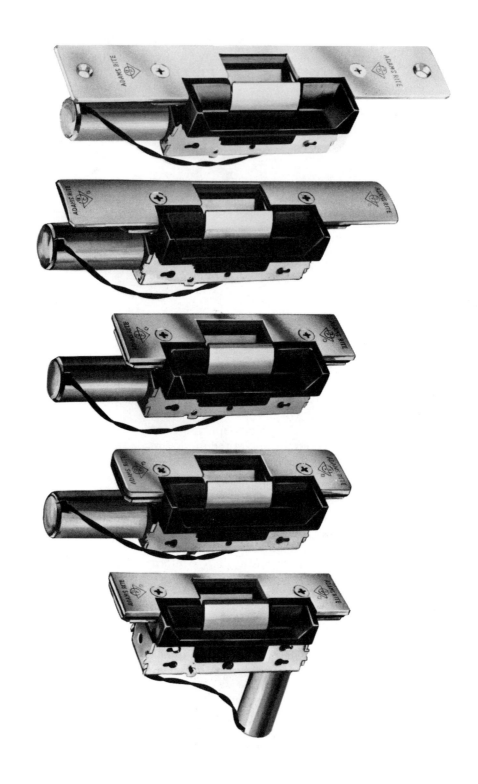

**16-1** Electric strikes provide control over frequently used doors. Adams Rite Manufacturing Co.

If the jamb was previously fitted with an 002 electric strike, the new 7810 unit will fit with minor alterations. In the case where a hollow jamb was originally prepared to receive the bolt from an Adams Rite M.S. dead-lock and a 4710 latch is to be substituted, the 7830 (flat jamb) or 7831 (radius inactive door) will cover the old strike cutout.

The standard lip on all basic Adams Rite electric strikes accommodates 1³/₄-inch-thick doors that close flush with the jamb. If the door/jamb relationship is different, a long clip can be added if it is so specified.

Adams Rite electric strikes 7800 through 7831 mate with all Adams Rite Series 4500 and 4700 latches. The 7840, 7840 ANSI, and 7870 are designed for mortise latches of the makes shown. All standard operation strikes are unhanded, and so can be installed for either right- or left-hand doors.

## ELECTRICAL CONSIDERATIONS

The first electrical factor to determine is whether the operation is to be intermittent or continuous. If the door is normally locked and released only momentarily from time to time, it is intermittent. If it is rare that the strike is activated (unlocked) for long periods, the duty is continuous. A seldom-used requirement is continuous/reverse action, also called "fail safe," in which the strike is locked only when its current is switched on.

For a normal intermittent application, specify an electric strike using 24 ac volts. This gives enough power for almost any entrance, even one with a wind-load situation. Yet this low-voltage range is below that requiring UL or Building Code supervision. Reliable transformers are available in this voltage.

The buzzing sound inherent in ac-activated strikes is usually not considered offensive in intermittent use. In fact it acts as the "go" signal to a person waiting to enter. However, if silence is desired, specify a strike using 24 dc volts. This dc current operates the strike silently. A buzzer or pilot light can be wired in if a signal is required.

Continuous duty is required when the strike will be energized for more than 60 seconds at a time. Most continuous-duty applications can be supplied through the same 115/24-volt ac transformer used for intermittent jobs. The factory will automatically add those components necessary to achieve continuous performance at the voltage specified.

For long periods of unlocking, a Fail-Safe Reverse Action strike can be obtained. This might be required to provide the same service as a continuous-duty strike, but will preserve current because it is on dc power. It could also be used to provide a "fail-safe" unlocked door in case of power failure.

If a visual or other signal is required to tell the operator what the electric strike is doing, a monitoring strike is needed. This feature is specified by adding the proper dash number to the strike's catalog number. Two sensor/switches are added: One is activated by the latchbolt's penetration of the strike and the other by the solenoid plunger that blocks the strike's release.

Low voltage for electric strike operation is obtained by the use of a transformer, which steps down the normal 115-volt ac power to 12, 16, 24 or other lower voltage. For this operation, three items must be specified: Input voltage (usually 115 volts); output voltage (12, 16, or 24 volts); and capacity of the transformer, measured in volt-amps.

Skimping on the capacity of the transformer to save a few dollars will not provide adequate power for the door release. Adams Rite electric strikes for intermittent duty models draw less than 1 amp (regardless of duty), and use a 4602 current limiter, which stores electrical energy for high-use periods.

The wire must carry the electrical power from the transformer through the actuating switch (or switches) to the door release. It must be large enough to minimize frictional line losses and deliver most of the output from the transformer to the door release. For example, a small-diameter garden hose won't provide a full flow of water from the nozzle, particularly if it's a long run. Neither will an undersized wire carry the full current.

When insufficient electrical power is suspected in a weak door release, this simple check can be made: Measure the voltage at the door release while the unit is activated. If the voltage is below that specified on the hardware schedule, the problem is in the circuit—probably an under-capacity transformer if the current length is short. A long run might indicate both a transformer and wire problem.

## ELECTRICAL CONSIDERATIONS: Questions and answers

1. How do you define intermittent or continuous duty? When there are only short periods of time that the switch will be closed and the coil energized, the duty is classified as continuous.

2. What is the electrical voltage source? The majority of electric strikes are applied where 115-ac volt, 60-cycle alternating current is the power source. Unless there are specific customer demands to do otherwise, it is recommended that the 115-volt, 60-cycle source be stepped down to a 24-volt, 60-cycle source using the Adams Rite 4605 transformer. Why 24 volts? There are two reasons: First, at 24 volts there is very little safety problem. Second, popular low-voltage replacement transformers are readily available at any electrical supply house. (Caution: Make sure any replacement transformer has minimum rating of 20 volts.)

3. Should the strike be audible or silent? First understand what causes the strike to be audible or silent. Alternating current changes direction 60 times per second. The noise you hear in the audible unit is that very brief period of time in which the solenoid plunger is released from the pole piece as the current builds back up to a peak. This is why a continuous-duty alternating current system is not recommended. The solenoid simply beats itself to death. By contrast, dc current flows in one direction and when the coil is energized the plunger remains seated until the circuit is

broken. Silent operation can be achieved in an ac circuit by specifying an Adams Rite 4603 rectifier be installed between the transformer secondary and the solenoid coil, which then sees only dc current.

4. What about current draw? The Adams Rite Electric Strike combines standard components that provide the maximum mechanical force necessary to do the job. To the overwhelming majority of intermittent-duty applications, the current draw for this heavy-duty performance poses no problem. But there are applications where, because of sensitive components somewhere else in the electrical system, a low maximum amount of current flow can be tolerated. In this case, specific hardware is used.

   Specially wound coils can be used to compromise the mechanical force range through which the strike gives peak performance. This means a solenoid is substituted that meets the current draw requirements but doesn't have the strong plunger "pull" of the high-current solenoid. In most instances, customer's requirements have been satisfactorily met with standard available parts without the added expense of specially wound coils.

5. What is "surge" current? Surge current is the momentary high draw necessary to start the plunger moving from its rest position. Considerably less current is required to "hold" the plunger, in just the same way it requires more energy to start a freight train moving than it does to keep it moving.

6. What is a current limiter? A current limiter is simply a combination of known electrical components wired in such a way as to provide the extra surge of current without damaging effects on other equipment in the circuit. One such combination uses: capacitors to store electrical energy in exactly the same way an air reservoir stores pneumatic energy, diodes, resistors of correct value, and switches. This combination stores electrical energy and upon demand (closing a switch), provides the solenoid coil extra momentary charge. It does so outside the circuit so that the circuit sees only the current required to hold the coil in position.

## TROUBLESHOOTING

Accurate checking of an electrical circuit requires the proper tool. Purchase a good 200,000-volt-per-ohm *volt-ohm-milliammeter* (VOM). Simpson Model 261, one of several on the market, costs about $130. Read the instructions and make some practice runs on simple low-voltage circuits.

### How to check voltage

1. Zero the pointer.
2. Be sure power is turned off to the circuit being measured.

3. Set the function switch to the correct voltage to be measured (+dc or ac).

4. Plug black test lead into the common (−) jack. Plug the red test lead into the (+) jack.

5. Set the range selector to the proper voltage scale. Caution: It is important that the selector be positioned to the nearest scale above the voltage to be measured.

6. Connect black test lead to the negative side of the circuit and the red lead to the positive side. This is applicable to dc circuit only. Turn power on to circuit to be tested. If the pointer on the VOM moves to the left, the polarity is wrong. Turn switch function to − dc and turn power back on. Pointer should now swing to right for proper reading on the dc scale.

7. Be sure to turn circuit off before disconnecting VOM.

Note: For an ac circuit, connections are the same, except you don't have to worry about polarity.

## How to check dc current

1. Zero the pointer.

2. Be sure power is turned off to the circuit being measured.

3. Connect black test lead to the − 10-amp jack and the red test lead to the +10-amp jack.

4. Set range selector to 10 amps.

5. Open the circuit to be measured—by disconnecting the wire that goes to one side of the solenoid, for example. Connect the meter in series: Hook the black lead to one of the disconnected wires and the red lead to the other wire.

6. Turn the power on to the circuit and observe the meter. If the pointer moves to the left, reverse the leads in the − 10-amp and +10-amp jacks.

7. Turn power back on to the circuit and read amperage on dc scale.

8. Turn power off before disconnecting VOM.

## How to check line drop

Measure line drop by comparing voltage readings at the source (the transformer's secondary or output side) with the reading at the strike connection.

## How to locate shorts

Again, the VOM is the most reliable instrument for detecting a short. This is accomplished by setting up the VOM to measure resistance.

1. Set the range switch at position R×1.

2. Set function switch at +dc.

3. Connect the black test lead to the common negative jack and the red test lead to the positive jack.

4. Zero the pointer by shorting test leads together.

5. Connect the other ends of the test lead across the resistance to be measured. In the case of a solenoid, connect one end of test lead to one coil terminal, the other to the other terminal.

6. Watch meter. If there is no movement of the pointer, the resistance being measured is *open*. If the pointer moves to the peg on the right hand side of the scale, the resistance being measured is shorted *closed*. If you get a reading in between these two extremes, there is probably no problem with the solenoid.

If strike will not activate after installation, proceed as follows:

1. Check fuse or circuit breaker supplying system.

2. Check to make sure all wiring connections are securely made. When wire nuts are used, take care to be sure both wires are twisted together.

3. Check the solenoid coil rated voltage (as shown on coil label) to make sure it corresponds to the output side of the transformer within ±10 percent.

4. Using VOM, check the voltage at the secondary (output) side of the transformer.

5. Using VOM, check the voltage at the solenoid to make sure there are no broken wires, bad rectifiers, or bad connections.

6. Check coil for short.

If strike won't activate after use, proceed as follows:

1. Check fuse or circuit breaker supplying system.

2. Make sure you have a transformer—one shot of 115 ac volts and the coil is ruined.

3. Make sure the rated voltage of the transformer and the rated voltage of the coil correspond with ±10 percent.

4. Check coil for short.

## Overheating

If transformer overheats, proceed as follows:

1. Make sure the rated voltage of the transformer and the rated voltage of the coil correspond within ±10 percent.

2. Make sure the volt-amp rating is adequate. We recommend 40 to 20 volts as the minimum.

If rectifier overheats, proceed as follows:

1. The rectifier might be wired wrong, which means the overheating is a temporary situation because it burns out.

2. There might also be too many solenoids being supplied by a single rectifier, and more current is being pulled through than the diodes can handle.

If solenoid overheats, proceed as follows:

First define overheating in this case: The coils used by Adams Rite are rated to have a temperature rise rating of 65 degrees Centigrade (149 degrees Fahrenheit) above ambient. However, all 7800 Series continuous-duty units should run 200 degrees F. or less in a 72-degree environment.

The vast majority of these intermittent-duty units never see the kind of use that brings the coil to maximum rating. If a coil gets extremely hot on very short pulses at two- or three-second intervals, either the wrong coil or the wrong transformer output is being used. The same is basically true for continuous-duty coils. If the coil temperature exceeds the ratings, it has to be because the coil voltage or the transformer are improperly coordinated.

Set the meter up as if testing for a short and obtain the exact resistance.

Figures 16-2 through 16-13 are installation instructions for several models of Adams Rite electric strikes.

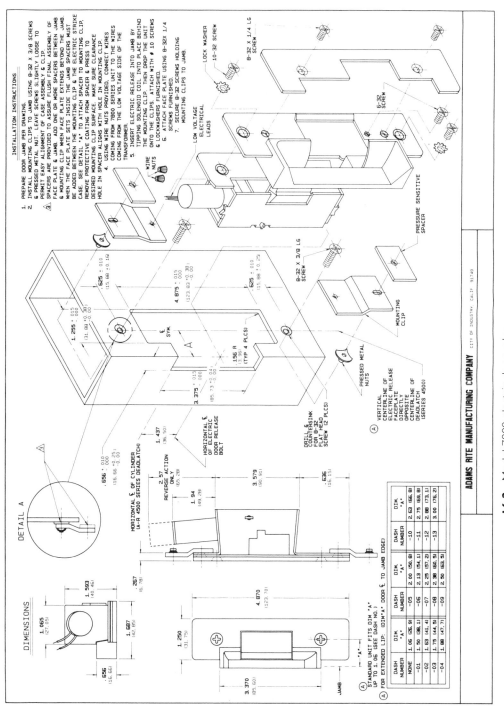

## INSTALLATION INSTRUCTIONS

1. PREPARE DOOR JAMB PER DRAWING.
2. INSTALL MOUNTING CLIPS TO JAMB USING 8-32 X 3/8 SCREWS & PRESSED METAL NUT. LEAVE SCREWS SLIGHTLY LOOSE TO PERMIT EASY ALIGNMENT OF CASE ASSEMBLY & CLIP.
3. SPACERS ARE PROVIDED TO ASSURE FLUSH FINAL ASSEMBLY OF FACE PLATE & JAMB. ADD ONE OR MORE SPACERS BETWEEN JAMB & MOUNTING CLIP WHEN FACE PLATE EXTENDS BEYOND THE JAMB. WHEN THE FACE PLATE SETS INSIDE THE JAMB SPACERS MUST BE ADDED BETWEEN THE MOUNTING CLIP & THE ELECTRIC STRIKE CASE. SEE DETAIL "A" TO ATTACH SPACER TO MOUNTING CLIP. REMOVE PROTECTIVE COATING FROM SPACER & PRESS TO DESIRED MOUNTING CLIP SURFACE. MAKE SURE CLEARANCE HOLE IN SPACER ALIGNS WITH HOLE IN MOUNTING CLIP.
4. USING WIRE NUTS PROVIDED, CONNECT WIRES COMING FROM 7800 SERIES UNIT TO THE WIRES COMING FROM THE LOW VOLTAGE SIDE OF THE TRANSFORMER.
5. INSERT ELECTRIC RELEASE INTO JAMB BY TIPPING SOLENOID COIL INTO PLACE BEHIND THE MOUNTING CLIP THEN DROP THE UNIT ONTO THE CLIPS. ATTACH WITH # 10 SCREWS & LOCKWASHERS FURNISHED.
6. ATTACH FACE PLATE USING 8-32X 1/4 SCREWS FURNISHED.
7. SECURE 8-32 SCREWS HOLDING MOUNTING CLIPS TO JAMB.

DETAIL A

DIMENSIONS

**ADAMS RITE MANUFACTURING COMPANY** CITY OF INDUSTRY, CALIF 91749

**16-2** Model 7800 electric door release. Adams Rite Manufacturing Co.

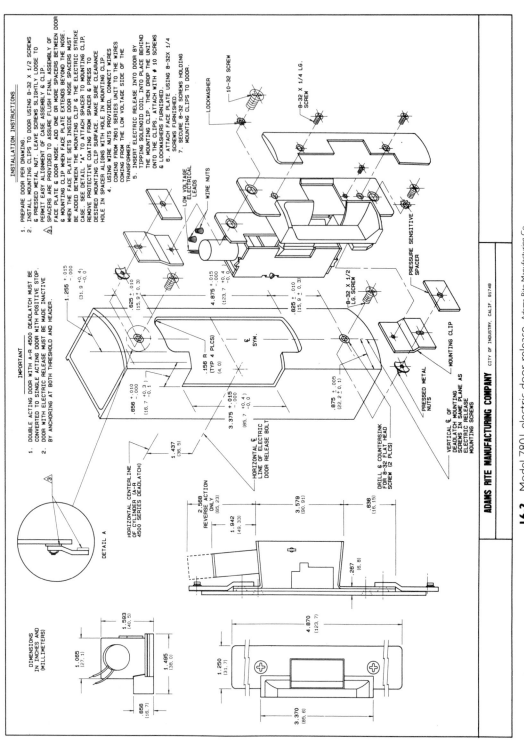

**INSTALLATION INSTRUCTIONS**

1. PREPARE DOOR PER DRAWING.
2. INSTALL MOUNTING CLIPS TO DOOR USING 8-32 X 1/2 SCREWS & PRESSED METAL NUT. LEAVE SCREWS SLIGHTLY LOOSE TO PERMIT EASY ALIGNMENT OF CASE ASSEMBLY & CLIP.

⚠ SPACERS ARE PROVIDED TO ASSURE FLUSH FINAL ASSEMBLY OF FACE PLATE. ADD ONE OR MORE SPACERS BETWEEN DOOR & MOUNTING CLIP WHEN FACE PLATE EXTENDS BEYOND THE NOSE. WHEN THE FACE PLATE SETS INSIDE DOOR NOSE SPACERS MUST BE ADDED BETWEEN THE MOUNTING CLIP & THE ELECTRIC STRIKE CASE. SEE DETAIL "A" TO ATTACH SPACER TO MOUNTING CLIP, REMOVE PROTECTIVE COATING FROM SPACER & PRESS TO DESIRED MOUNTING CLIP SURFACE. MAKE SURE CLEARANCE HOLE IN SPACER ALIGNS WITH HOLE IN MOUNTING CLIP.

4. USING WIRE NUTS PROVIDED, CONNECT WIRES COMING FROM 7801 SERIES UNIT TO THE WIRES COMING FROM THE LOW VOLTAGE SIDE OF THE TRANSFORMER.

5. INSERT ELECTRIC RELEASE INTO DOOR BY TIPPING SOLENOID COIL INTO PLACE BEHIND THE MOUNTING CLIP. THEN DROP THE UNIT ONTO THE CLIPS. ATTACH WITH # 10 SCREWS & LOCKWASHERS FURNISHED.
6. ATTACH FACE PLATE USING 8-32X 1/4 SCREWS FURNISHED.
7. SECURE 8-32 SCREWS HOLDING MOUNTING CLIPS TO DOOR.

**IMPORTANT**

1. DOUBLE ACTING DOOR WITH A-R 4500 DEADLATCH MUST BE CONVERTED TO SINGLE ACTING DOOR WITH POSITIVE STOP.
2. DOOR WITH ELECTRIC RELEASE MUST BE MADE INACTIVE BY ANCHORING AT BOTH THRESHOLD AND HEADER.

DIMENSIONS IN INCHES AND (MILLIMETERS)

LOW VOLTAGE ELECTRICAL LEADS

WIRE NUTS

10-32 SCREW

LOCKWASHER

8-32 X 1/4 LG. SCREW

PRESSURE SENSITIVE SPACER

MOUNTING CLIP

8-32 X 1/2 LG. SCREW

PRESSED METAL NUTS

1.255 +.015 −.000
(31.9 +0.4 −0.0)

.625 ±.010
(15.9 ± 0.3)

4.875 +.015 −.000
(123.7 +0.4 −0.0)

.625 ±.010
(15.9 ± 0.3)

.156 R (TYP 4 PLCS)
(4.0)

SYM

.650 +.010 −.000
(16.7 +0.3 −0.1)

.875 ±.005
(22.2 ± 0.1)

3.375 +.015 −.000
(85.7 +0.4 −0.0)

VERTICAL ℄ OF DEADLATCH MOUNTING SCREWS IN SAME PLANE AS ELECTRIC RELEASE MOUNTING SCREWS

HORIZONTAL CENTERLINE OF CYLINDER (1A-R 4500 SERIES DEADLATCH)

HORIZONTAL ℄ LINE OF ELECTRIC DOOR RELEASE BOLT

1.437 (36.5)

DRILL & COUNTERSINK FOR 8-32 FLAT HEAD SCREW (2 PLCS)

DETAIL A

REVERSE ACTION ONLY

2.568 (65.23)

1.942 (49.33)

3.579 (90.91)

.636 (16.15)

.267 (6.8)

1.593 (40.5)

1.065 (27.1)

1.495 (38.0)

.656 (16.7)

4.870 (123.7)

1.250 (31.7)

3.370 (85.6)

**Adams Rite MANUFACTURING COMPANY** CITY OF INDUSTRY, CALIF. 91749

**16-3** Model 7801 electric door release. Adams Rite Manufacturing Co.

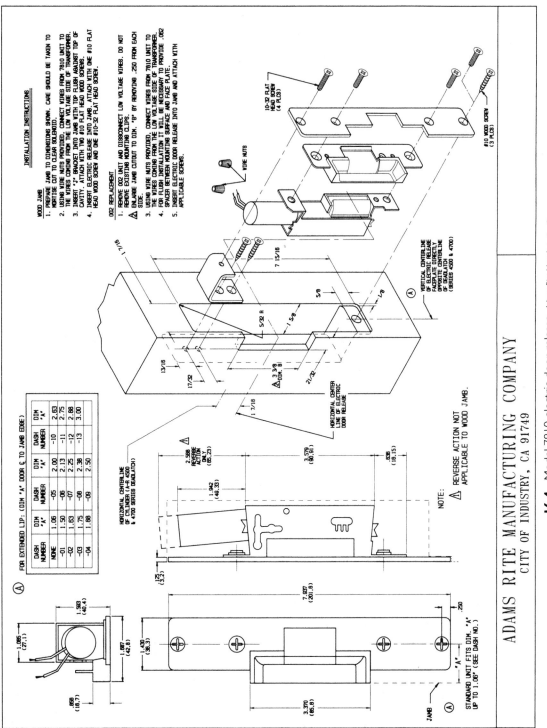

## ADAMS RITE MANUFACTURING COMPANY

CITY OF INDUSTRY, CA 91749

### INSTALLATION INSTRUCTIONS

**WOOD JAMB**

1. PREPARE JAMB TO DIMENSIONS SHOWN. CARE SHOULD BE TAKEN TO MORTISE CUT TO CLEAR SOLENOID.
2. USING WIRE NUTS PROVIDED, CONNECT WIRES FROM 7810 UNIT TO THE WIRES COMING FROM THE LOW VOLTAGE SIDE OF TRANSFORMER.
3. INSERT "J" BRACKET INTO JAMB WITH TOP FLUSH AGAINST TOP OF CAVITY. ATTACH WITH TWO #10 FLAT HEAD WOOD SCREWS.
4. INSERT ELECTRIC RELEASE INTO JAMB. ATTACH WITH ONE #10 FLAT HEAD WOOD SCREW AND ONE #10-32 FLAT HEAD SCREW.

**002 REPLACEMENT**

1. REMOVE 002 UNIT AND DISCONNECT LOW VOLTAGE WIRES. DO NOT REMOVE EXISTING MOUNTING CLIPS.
2. ENLARGE JAMB CUTOUT TO DIM. "B" BY REMOVING .250 FROM EACH SIDE.
3. USING WIRE NUTS PROVIDED, CONNECT WIRES FROM 7810 UNIT TO THE WIRES COMING FROM THE LOW VOLTAGE SIDE OF TRANSFORMER.
4. FOR FLUSH INSTALLATION IT WILL BE NECESSARY TO PROVIDE .062 SPACER BETWEEN MOUNTING SURFACE AND FACE PLATE.
5. INSERT ELECTRIC DOOR RELEASE INTO JAMB AND ATTACH WITH APPLICABLE SCREWS.

10-32 FLAT HEAD SCREW (4 PLCS)

#10 WOOD SCREW (3 PLCS)

WIRE NUTS

1 7/8

7 15/16

5/8

1 3/8

5/32 R

13/16

17/32

3 3/8 DIM. B

21/32

1 7/8

VERTICAL CENTERLINE OF ELECTRIC RELEASE FACEPLATE DIRECTLY OPPOSITE CENTERLINE OF DEADLATCH (SERIES 4500 & 4700)

HORIZONTAL CENTER LINE OF ELECTRIC DOOR RELEASE

### FOR EXTENDED LIP: (DIM "A" DOOR ₵ TO JAMB EDGE)

| DASH NUMBER | DIM "A" | DASH NUMBER | DIM "A" | DASH NUMBER | DIM "A" |
|---|---|---|---|---|---|
| NONE | 1.06 | -05 | 2.00 | -10 | 2.63 |
| -01 | 1.50 | -06 | 2.13 | -11 | 2.75 |
| -02 | 1.63 | -07 | 2.25 | -12 | 2.88 |
| -03 | 1.75 | -08 | 2.38 | -13 | 3.00 |
| -04 | 1.88 | -09 | 2.50 | | |

HORIZONTAL CENTERLINE OF CYLINDER (4-R 4500 & 4700 SERIES DEADLATCH)

2.568 REVERSE ACTION ONLY (65.23)

1.942 (49.33)

3.579 (90.91)

.636 (16.15)

.125 (3.2)

NOTE:

⚠ REVERSE ACTION NOT APPLICABLE TO WOOD JAMB

1.593 (40.4)

1.065 (27.1)

1.687 (42.8)

.658 (16.7)

7.937 (201.6)

1.430 (36.3)

.250

3.370 (85.6)

JAMB

STANDARD UNIT FITS DIM. "A" UP TO 1.06" (SEE DASH NO.)

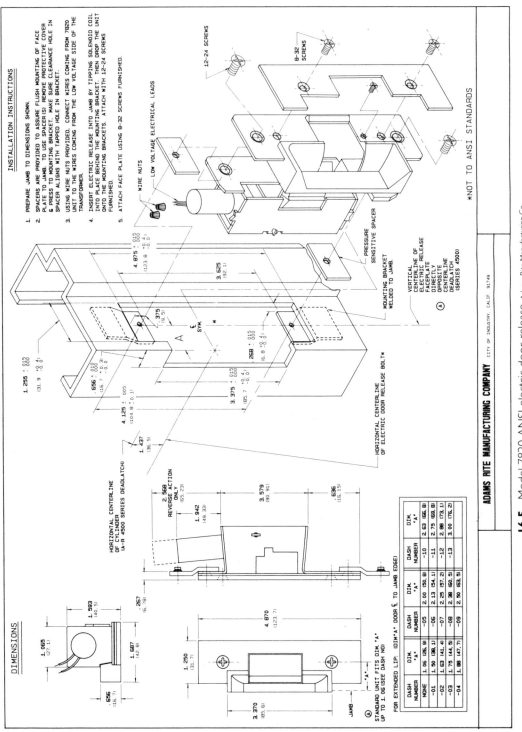

**16-5** Model 7820 ANSI electric door release. Adams Rite Manufacturing Co.

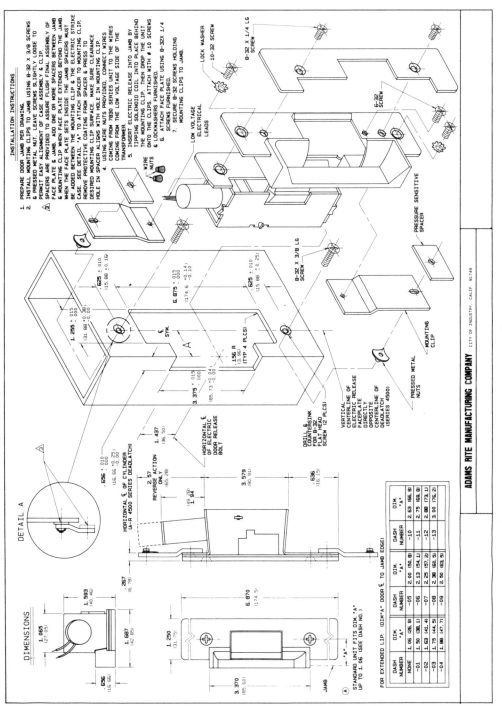

**ADAMS RITE MANUFACTURING COMPANY** CITY OF INDUSTRY, CALIF. 91749

**16-6** Model 7830 electric door release. Adams Rite Manufacturing Co.

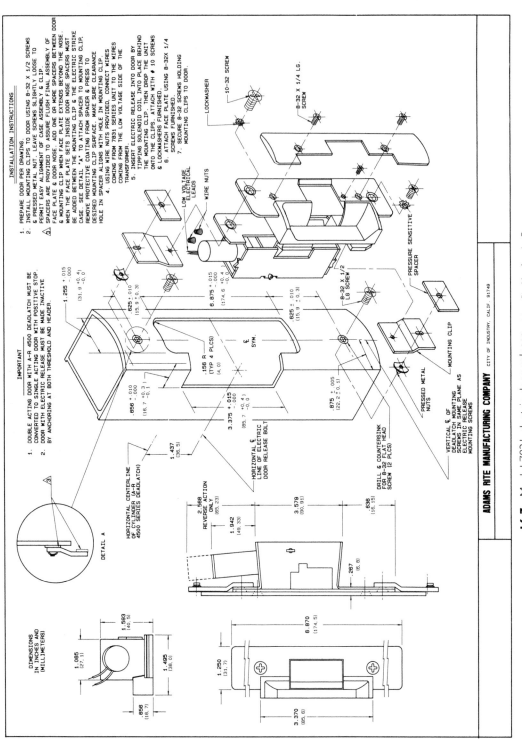

**INSTALLATION INSTRUCTIONS**

1. PREPARE DOOR PER DRAWING.
2. INSTALL MOUNTING CLIPS TO DOOR USING 8-32 X 1/2 SCREWS & PRESSED METAL NUT. LEAVE SCREWS SLIGHTLY LOOSE TO PERMIT EASY ALIGNMENT OF CASE ASSEMBLY & CLIP.
3. SPACERS ARE PROVIDED TO ASSURE FLUSH FINAL ASSEMBLY OF FACE PLATE & DOOR NOSE. ADD ONE OR MORE SPACERS BETWEEN DOOR & MOUNTING CLIP WHEN FACE PLATE EXTENDS BEYOND THE NOSE. WHEN THE FACE PLATE SETS INSIDE DOOR NOSE SPACERS MUST BE ADDED BETWEEN THE MOUNTING CLIP & THE ELECTRIC STRIKE CASE. SEE DETAIL "A" TO ATTACH SPACER TO MOUNTING CLIP. REMOVE PROTECTIVE COATING FROM SPACER & PRESS TO DESIRED MOUNTING CLIP SURFACE. MAKE SURE CLEARANCE HOLE IN SPACER ALIGNS WITH HOLE IN MOUNTING CLIP.
4. USING WIRE NUTS PROVIDED, CONNECT WIRES COMING FROM 7831 SERIES UNIT TO THE WIRES COMING FROM THE LOW VOLTAGE SIDE OF THE TRANSFORMER.
5. INSERT ELECTRIC RELEASE INTO DOOR BY TIPPING SOLENOID COIL INTO PLACE BEHIND THE MOUNTING CLIP. THEN DROP THE UNIT ONTO THE CLIPS. ATTACH WITH # 10 SCREWS & LOCKWASHERS FURNISHED.
6. ATTACH FACE PLATE USING 8-32X 1/4 SCREWS FURNISHED.
7. SECURE 8-32 SCREWS HOLDING MOUNTING CLIPS TO DOOR.

**IMPORTANT**

1. DOUBLE ACTING DOOR WITH A-R 4500 DEADLATCH MUST BE CONVERTED TO SINGLE ACTING DOOR WITH POSITIVE STOP. DOOR WITH ELECTRIC RELEASE MUST BE MADE INACTIVE BY ANCHORING AT BOTH THRESHOLD AND HEADER.

10-32 SCREW

LOCKWASHER

8-32 X 1/4 LG. SCREW

LOW VOLTAGE ELECTRICAL LEADS

WIRE NUTS

PRESSURE SENSITIVE SPACER

8-32 X 1/2 LG SCREW

MOUNTING CLIP

PRESSED METAL NUTS

1.255 +.015 -.000
(31.9 +0.4 -0.0)

6.875 +.015 -.000
(174.6 +0.4 -0.0)

.625 ±.010
(15.9 ±0.3)

.625 ±.010
(15.9 ±0.3)

€ SYM.

.156 R (TYP 4 PLCS)
(4.0)

.656 +.010 -.000
(16.7 +0.3 -0.1)

3.375 +.015 -.000
(85.7 +0.4 -0.0)

.875 ±.005
(22.2 ±0.1)

VERTICAL € OF DEADLATCH MOUNTING SCREWS IN SAME PLANE AS ELECTRIC RELEASE MOUNTING SCREWS

DRILL & COUNTERSINK FOR 8-32 FLAT HEAD SCREW (2 PLCS)

HORIZONTAL € LINE OF ELECTRIC DOOR RELEASE BOLT

1.437
(36.5)

HORIZONTAL CENTERLINE (A-R 4500 SERIES DEADLATCH)

DETAIL A

DIMENSIONS IN INCHES AND (MILLIMETERS)

1.593
(40.5)

1.065
(27.1)

1.495
(38.0)

.656
(16.7)

2.568
(65.23)

REVERSE ACTION ONLY

1.942
(49.33)

3.579
(90.91)

.636
(16.15)

.267
(6.8)

6.870
(174.5)

1.250
(31.7)

3.370
(85.6)

**ADAMS RITE MANUFACTURING COMPANY** CITY OF INDUSTRY, CALIF. 91749

**16-7** Model 7831 electric door release. Adams Rite Manufacturing Co.

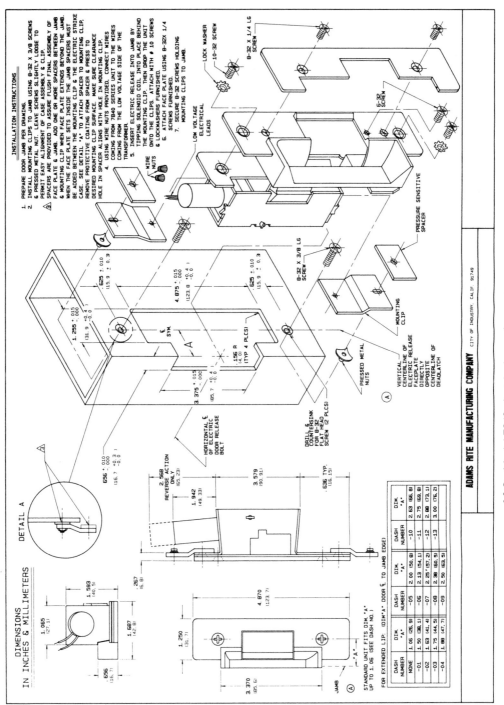

## INSTALLATION INSTRUCTIONS

1. PREPARE DOOR JAMB PER DRAWING.
2. INSTALL MOUNTING CLIPS TO JAMB USING 8-32 X 3/8 SCREWS & PRESSED METAL NUT. LEAVE SCREWS SLIGHTLY LOOSE TO PERMIT EASY ALIGNMENT OF CASE ASSEMBLY & CLIP.
3. SPACERS ARE PROVIDED TO ASSURE FLUSH FINAL ASSEMBLY OF FACE PLATE & JAMB. ADD ONE OR MORE SPACERS BETWEEN JAMB & MOUNTING CLIP WHEN FACE PLATE EXTENDS BEYOND THE JAMB. WHEN THE FACE PLATE SETS INSIDE THE JAMB SPACERS MUST BE ADDED BETWEEN THE MOUNTING CLIP & THE ELECTRIC STRIKE CASE. SEE DETAIL "A" TO ATTACH SPACER TO MOUNTING CLIP. REMOVE PROTECTIVE COATING FROM SPACER & PRESS TO DESIRED MOUNTING CLIP SURFACE. MAKE SURE CLEARANCE HOLE IN SPACER ALIGNS WITH HOLE IN MOUNTING CLIP.
4. USING WIRE NUTS NUTS PROVIDED, CONNECT WIRES COMING FROM 7840 SERIES UNIT TO THE WIRES COMING FROM THE LOW VOLTAGE SIDE OF THE TRANSFORMER.
5. INSERT ELECTRIC RELEASE INTO JAMB BY TIPPING SOLENOID COIL INTO PLACE BEHIND THE MOUNTING CLIPS. THEN DROP THE UNIT ONTO THE CLIPS. ATTACH WITH # 10 SCREWS & LOCKWASHERS FURNISHED.
6. ATTACH FACE PLATE USING 8-32X 1/4 SCREWS FURNISHED.
7. SECURE 8-32 SCREWS HOLDING MOUNTING CLIPS TO JAMB.

DETAIL A

DIMENSIONS
IN INCHES & MILLIMETERS

STANDARD UNIT FITS DIM "A"
UP TO 1.06 (SEE DASH NO.)

FOR EXTENDED LIP. (DIM"A" DOOR ₵ TO JAMB EDGE)

| DASH NUMBER | DIM "A" | DASH NUMBER | DIM "A" | DASH NUMBER | DIM "A" | DASH NUMBER | DIM "A" |
|---|---|---|---|---|---|---|---|
| NONE | 1.06 (26.9) | -05 | 2.00 (50.8) | -10 | 2.63 (66.8) |
| -01 | 1.50 (38.1) | -06 | 2.13 (54.1) | -11 | 2.75 (69.8) |
| -02 | 1.63 (41.4) | -07 | 2.25 (57.2) | -12 | 2.88 (73.1) |
| -03 | 1.75 (44.5) | -08 | 2.38 (60.5) | -13 | 3.00 (76.2) |
| -04 | 1.88 (47.7) | -09 | 2.50 (63.5) | | |

**ADAMS RITE MANUFACTURING COMPANY** CITY OF INDUSTRY CALIF 91749

**16-8** Model 7840 electric door release. Adams Rite Manufacturing Co.

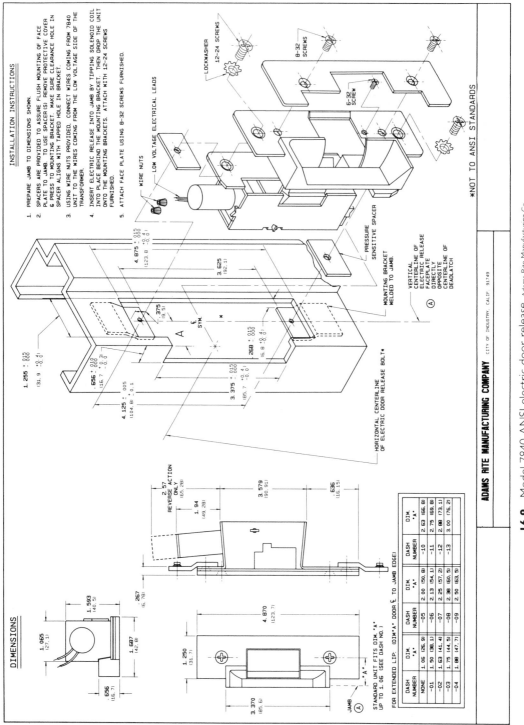

INSTALLATION INSTRUCTIONS

1. PREPARE JAMB TO DIMENSIONS SHOWN.
2. SPACERS ARE PROVIDED TO ASSURE FLUSH MOUNTING OF FACE PLATE TO JAMB. TO USE SPACER(S) REMOVE PROTECTIVE COVER & PRESS TO MOUNTING BRACKET. MAKE SURE CLEARANCE HOLE IN SPACER ALIGNS WITH TAPPED HOLE IN BRACKET.
3. USING WIRE NUTS PROVIDED, CONNECT WIRES COMING FROM 7840 UNIT TO THE WIRES COMING FROM THE LOW VOLTAGE SIDE OF THE TRANSFORMER.
4. INSERT ELECTRIC RELEASE INTO JAMB BY TIPPING SOLENOID COIL INTO PLACE BEHIND THE MOUNTING BRACKET. THEN DROP THE UNIT ONTO THE MOUNTING BRACKETS. ATTACH WITH 12-24 SCREWS FURNISHED.
5. ATTACH FACE PLATE USING 8-32 SCREWS FURNISHED.

\*NOT TO ANSI STANDARDS

DIMENSIONS

| DASH NUMBER | DIM. "A" | DASH NUMBER | DIM. "A" | DASH NUMBER | DIM. "A" |
|---|---|---|---|---|---|
| NONE | 1.06 (26.9) | -05 | 2.00 (50.8) | -10 | 2.63 (66.8) |
| -01 | 1.50 (38.1) | -06 | 2.13 (54.1) | -11 | 2.75 (69.8) |
| -02 | 1.63 (41.4) | -07 | 2.25 (57.2) | -12 | 2.88 (73.1) |
| -03 | 1.75 (44.5) | -08 | 2.38 (60.5) | -13 | 3.00 (76.2) |
| -04 | 1.88 (47.7) | -09 | 2.50 (63.5) | | |

STANDARD UNIT FITS DIM. "A" UP TO 1.06 (SEE DASH NO.)

FOR EXTENDED LIP: (DIM"A" DOOR ℄ TO JAMB EDGE)

---

**ADAMS RITE MANUFACTURING COMPANY** CITY OF INDUSTRY, CALIF. 91749

**16-9** Model 7840 ANSI electric door release. Adams Rite Manufacturing Co.

FOR EXTENDED LIP: (DIM "A" DOOR ℄ TO JAMB EDGE)

Ⓐ

| DASH NUMBER | DIM "A" | DASH NUMBER | DIM "A" | DASH NUMBER | DIM "A" |
|---|---|---|---|---|---|
| NONE | 1.06 (26.9) | -05 | 2.00 (50.8) | -10 | 2.63 (66.8) |
| -01 | 1.50 (38.1) | -06 | 2.13 (54.1) | -11 | 2.75 (69.8) |
| -02 | 1.63 (41.4) | -07 | 2.25 (57.2) | -12 | 2.88 (73.1) |
| -03 | 1.75 (44.5) | -08 | 2.38 (60.5) | -13 | 3.00 (76.2) |
| -04 | 1.88 (47.7) | -09 | 2.50 (63.5) | | |

INSTALLATION INSTRUCTIONS & NOTES

1. PREPARE DOOR JAMB PER DRAWING.
2. USING WIRE NUTS PROVIDED, CONNECT WIRES COMING FROM 7850 SERIES UNIT TO THE WIRES COMING FROM THE LOW VOLTAGE SIDE OF TRANSFORMER.
3. INSERT ELECTRIC RELEASE INTO JAMB & ATTACH WITH # 10 WOOD SCREWS.
4. ATTACH FACE PLATE USING 8-32 X 1/4. SCREWS FURNISHED.

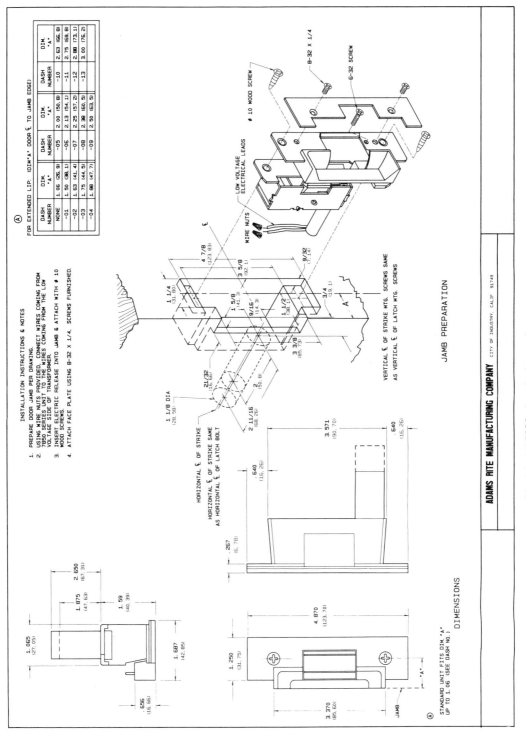

8-32 X 1/4

6-32 SCREW

# 10 WOOD SCREW

LOW VOLTAGE ELECTRICAL LEADS

WIRE NUTS

JAMB PREPARATION

VERTICAL ℄ OF STRIKE MTG. SCREWS SAME AS VERTICAL ℄ OF LATCH MTG. SCREWS

HORIZONTAL ℄ OF STRIKE SAME AS HORIZONTAL ℄ OF LATCH BOLT

DIMENSIONS

STANDARD UNIT FITS DIM "A" UP TO 1.06 (SEE DASH NO.)

ADAMS RITE MANUFACTURING COMPANY   CITY OF INDUSTRY, CALIF  91749

16-10  Model 7850 electric door release. Adams Rite Manufacturing Co.

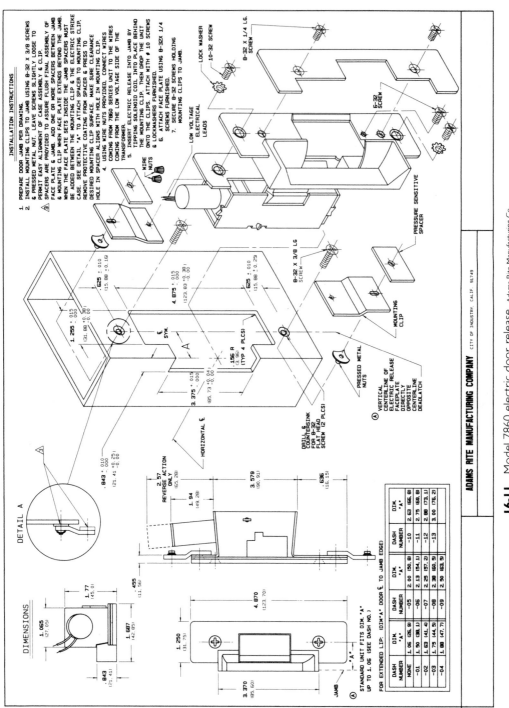

**INSTALLATION INSTRUCTIONS**

1. PREPARE DOOR JAMB PER DRAWING.
2. INSTALL MOUNTING CLIPS TO JAMB USING 8-32 x 3/8 SCREWS & PRESSED METAL NUT. LEAVE SCREWS SLIGHTLY LOOSE TO PERMIT EASY ALIGNMENT OF CASE ASSEMBLY & CLIP.
3. SPACERS ARE PROVIDED TO ASSURE FLUSH FINAL ASSEMBLY OF FACE PLATE & JAMB. ADD ONE OR MORE SPACERS BETWEEN JAMB & MOUNTING CLIP WHEN FACE PLATE EXTENDS BEYOND THE JAMB. WHEN THE FACE PLATE SETS INSIDE THE JAMB SPACERS MUST BE ADDED BETWEEN THE MOUNTING CLIP & THE ELECTRIC STRIKE CASE. SEE DETAIL 'A' TO ATTACH SPACER TO MOUNTING CLIP, REMOVE PROTECTIVE COATING FROM SPACER & PRESS TO DESIRED MOUNTING CLIP SURFACE. MAKE SURE CLEARANCE HOLE IN SPACER ALIGNS WITH HOLE IN MOUNTING CLIP.
4. USING WIRE NUTS PROVIDED, CONNECT WIRES COMING FROM 7860 SERIES UNIT TO THE WIRES COMING FROM THE LOW VOLTAGE SIDE OF THE TRANSFORMER.
5. INSERT ELECTRIC RELEASE INTO JAMB BY TIPPING SOLENOID COIL INTO PLACE BEHIND THE MOUNTING CLIP. THEN DROP THE UNIT ONTO THE CLIPS. ATTACH WITH # 10 SCREWS & LOCKWASHERS FURNISHED.
6. ATTACH FACE PLATE USING 8-32x 1/4 SCREWS FURNISHED.
7. SECURE 8-32 SCREWS HOLDING MOUNTING CLIPS TO JAMB.

DETAIL A

DIMENSIONS

---

**ADAMS RITE MANUFACTURING COMPANY** CITY OF INDUSTRY, CALIF. 91749

**16-11** Model 7860 electric door release. Adams Rite Manufacturing Co.

| DASH NUMBER | DIM. 'A' | DASH NUMBER | DIM. 'A' |
|---|---|---|---|
| NONE | 1.06 (26.9) | -05 | 2.00 (50.8) |
| -01 | 1.50 (38.1) | -06 | 2.13 (54.1) |
| -02 | 1.63 (41.4) | -07 | 2.25 (57.2) |
| -03 | 1.75 (44.5) | -08 | 2.38 (60.5) |
| -04 | 1.88 (47.7) | -09 | 2.50 (63.5) |
| | | -10 | 2.63 (66.8) |
| | | -11 | 2.75 (69.8) |
| | | -12 | 2.88 (73.1) |
| | | -13 | 3.00 (76.2) |

STANDARD UNIT FITS DIM. 'A'
UP TO 1.06 (SEE DASH NO.)
FOR EXTENDED LIP, (DIM'A' DOOR ℄ TO JAMB EDGE)

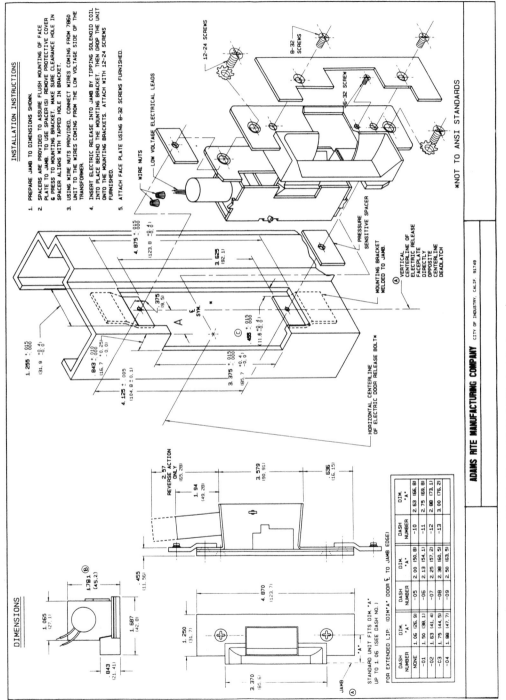

INSTALLATION INSTRUCTIONS

1. PREPARE JAMB TO DIMENSIONS SHOWN.

2. SPACERS ARE PROVIDED TO ASSURE FLUSH MOUNTING OF FACE PLATE TO JAMB. TO USE SPACER(S), REMOVE PROTECTIVE COVER & PRESS TO MOUNTING BRACKET. MAKE SURE CLEARANCE HOLE IN SPACER ALIGNS WITH TAPPED HOLE IN BRACKET.

3. USING WIRE NUTS PROVIDED, CONNECT WIRES COMING FROM 7860 UNIT TO THE WIRES COMING FROM THE LOW VOLTAGE SIDE OF THE TRANSFORMER.

4. INSERT ELECTRIC RELEASE INTO JAMB BY TIPPING SOLENOID COIL INTO PLACE BEHIND THE MOUNTING BRACKET. THEN DROP THE UNIT ONTO THE MOUNTING BRACKETS. ATTACH WITH 12-24 SCREWS FURNISHED.

5. ATTACH FACE PLATE USING 8-32 SCREWS FURNISHED.

*NOT TO ANSI STANDARDS

DIMENSIONS

STANDARD UNIT FITS DIM "A"
UP TO 1.06 (SEE DASH NO.)

FOR EXTENDED LIP: (DIM "A" DOOR ℄ TO JAMB EDGE)

| DASH NUMBER | DIM "A" | DASH NUMBER | DIM "A" | DASH NUMBER | DIM "A" |
|---|---|---|---|---|---|
| NONE | 1.06 (26.9) | -05 | 2.00 (50.8) | -10 | 2.63 (66.8) |
| -01 | 1.50 (38.1) | -06 | 2.13 (54.1) | -11 | 2.75 (69.8) |
| -02 | 1.63 (41.4) | -07 | 2.25 (57.2) | -12 | 2.88 (73.1) |
| -03 | 1.75 (44.5) | -08 | 2.38 (60.5) | -13 | 3.00 (76.2) |
| -04 | 1.88 (47.7) | -09 | 2.50 (63.5) | | |

ADAMS RITE MANUFACTURING COMPANY  CITY OF INDUSTRY, CALIF. 91749

**16-12**  Model 7860 ANSI electric door release. Adams Rite Manufacturing Co.

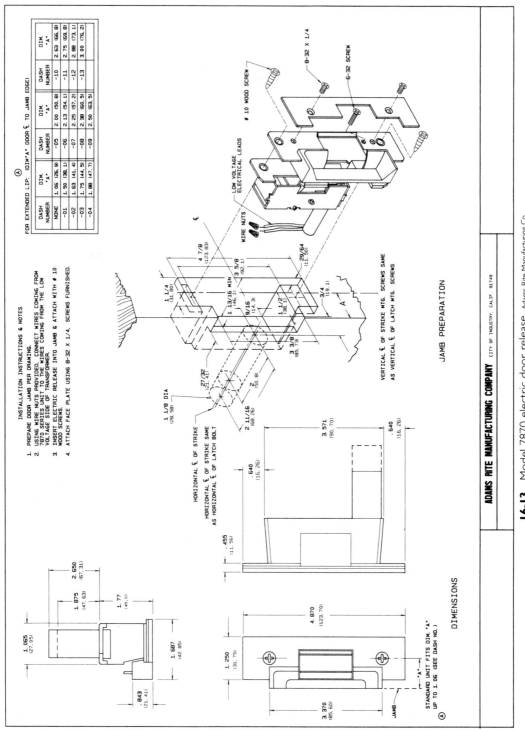

INSTALLATION INSTRUCTIONS & NOTES

1. PREPARE DOOR JAMB PER DRAWING.
2. USING WIRE NUTS PROVIDED, CONNECT WIRES COMING FROM THE 7870 SERIES UNIT TO THE WIRES COMING FROM THE LOW VOLTAGE SIDE OF TRANSFORMER.
3. INSERT ELECTRIC RELEASE INTO JAMB & ATTACH WITH # 10 WOOD SCREWS.
4. ATTACH FACE PLATE USING 8-32 X 1/4. SCREWS FURNISHED.

Ⓐ FOR EXTENDED LIP. (DIM"A" DOOR ℄ TO JAMB EDGE)

| DASH NUMBER | DIM. "A" | DASH NUMBER | DIM. "A" | DASH NUMBER | DIM. "A" |
|---|---|---|---|---|---|
| NONE | 1.06 (26.9) | -05 | 2.00 (50.8) | -10 | 2.63 (66.8) |
| -01 | 1.50 (38.1) | -06 | 2.13 (54.1) | -11 | 2.75 (69.8) |
| -02 | 1.63 (41.4) | -07 | 2.25 (57.2) | -12 | 2.88 (73.1) |
| -03 | 1.75 (44.5) | -08 | 2.38 (60.5) | -13 | 3.00 (76.2) |
| -04 | 1.88 (47.7) | -09 | 2.50 (63.5) | | |

JAMB PREPARATION

8-32 X 1/4
6-32 SCREW
# 10 WOOD SCREW
LOW VOLTAGE ELECTRICAL LEADS
WIRE NUTS

4 7/8 (123.83)
3 5/8 (92.1)
29/64 (11.50)
1 1/4 (31.80)
1 13/16 MIN (46.1)
3/4 (19.1)
9/16 (14.3)
1 1/2 (38.1)
A
3 3/8 (85.73)
27/32 (21.41)
2 (50.8)
2 11/16 (68.26)
1 1/8 DIA (28.58)

HORIZONTAL ℄ OF STRIKE
HORIZONTAL ℄ OF STRIKE SAME AS HORIZONTAL ℄ OF LATCH BOLT

VERTICAL ℄ OF STRIKE MTG. SCREWS SAME AS VERTICAL ℄ OF LATCH MTG. SCREWS

DIMENSIONS

2.650 (67.31)
1.875 (47.63)
1.77 (45.0)
1.065 (27.05)
1.687 (42.85)
.843 (21.41)

3.571 (90.70)
.540 (16.26)
.540 (16.26)
.455 (11.56)

4.870 (123.70)
1.250 (31.75)
3.370 (85.60)
JAMB

Ⓐ STANDARD UNIT FITS DIM. "A" UP TO 1.06 (SEE DASH NO.)

ADAMS RITE MANUFACTURING COMPANY  CITY OF INDUSTRY, CALIF. 91749

**16-13** Model 7870 electric door release. Adams Rite Manufacturing Co.

*Chapter* **17**

# Key duplicating machines

$A$ key duplicating or key cutting machine cuts keys by tracing their patterns onto key blanks. It consists of three basic parts: a pair of vises—one to hold a key, and one to hold a blank—that move in unison; a key guide or stylus that traces the profile of a key; and a cutter wheel that cuts a key blank to the shape of a key.

The pattern key is placed in the vise near the stylus. The key blank is placed in the vise near the cutter wheel. When the key and blank are properly positioned, the machine is turned on and the cutter wheel makes cuts in the blank that are identical to those of the key.

Some machines can duplicate only one type of key; others are designed to duplicate a wide variety of keys. This chapter explains how to operate and service several popular key duplicating machines.

## HALF-TIME KEY MACHINE

Manufactured by Precision Products, Inc., the Half-Time key machine is designed to cut two flat keys at a time (FIG. 17-1). L-shaped vise handles allow fast seating of all flat keys. The rotating stylus changes to four guide widths: .055, .062, .072, and .088 inch. The machine is operated as follows:

1. Place the original key to be duplicated into the right side vise, securely against the key stop.

2. Select the proper key blank for duplication. Secure a pair of these keys into the left side vise, against the key stop.

Note: The Half-Time is designed so that the vise jaws need be lightly tightened. Do not overtighten. To release keys, only a quarter turn of the handle is required.

Locking spring loaded tracer

4-way cam

Safety light

Carbide cutter wheel

Precision Products, Inc.

Parallel-operating
key vises

**17-1** The Half Time key machine can cut two flat keys simultaneously.

Caution: If the vise spring-loaded handle is forced, the internal stud may be stripped. To place the handle in a convenient position, lift up and rotate it around the post to a comfortable position and let it rest.

3. Turn the Half-Time on.

4. Push the carriage forward to lift the original key to the tip of the tracing stylus. Move the carriage left or right to locate the stylus into the first bitting to be duplicated.

5. Push the carriage forward until the stylus collar stops against the stylus block.

The cutter on the Half-Time is not a side-milling slotter. All carriage movements must be straight in and out when the stylus is engaged with the original key. Any lateral movement will cause damage to the solid carbide cutter.

6. Lower the carriage until the stylus is clear of the original key. Move the carriage left or right to locate the stylus into the next

bitting to be duplicated. Push the carriage forward until the collar stops against the stylus block.

7. Repeat Step 6 until all bits of the original key have been duplicated onto the pair of blanks.

8. Allow the carriage to rest at the bottom of its stroke.

9. Inspect the pair of new keys. If there is any material left in a bitting, locate that bit under the cutter and gently push forward until the collar stops against the block. If there is now excess material, the pair of new keys are completed.

10. Turn the machine off.

11. Remove the pair of keys.

12. Remove any burring on the side of the bottom key with a small file or deburring brush. The top key will not have any burring on it.

13. Repeat Steps 6 through 12 to duplicate as many new keys as needed.

To make adjustments in the depths of the bits, the stop collar has been indexed with increments of .001 inch. To make an adjustment in the depth of cuts, loosen the two set screws on the stylus collar. Rotate the collar clockwise to make the cuts shallower. Rotate the collar counterclockwise to deepen the cuts. To measure the amount of adjustment, rotate and count the increments on the collar against the index centered on the stylus block.

To check for proper cutter height against the stylus, secure the same type key blank in each vise. Slide the carriage forward until the stylus engages the key blank and is stopped against the block. Rotate the cutter slowly by hand. It should make light contact with the blank under it. If the cutter will not rotate, the stylus is too deep. If it turns freely, it is too shallow. Make corrections by rotating the stop collar on the stylus and then retighten the set screws.

## Maintenance

1. Keep the Half-Time free of excess dirt, dust, and chip debris.
2. Occasionally lubricate bearing surfaces with a very light amount of oil, then wipe off with a clean dry cloth. The motor is a sealed bearing type and requires no lubrication.
3. Inspect all screws and keep them snug.

## Troubleshooting

1. If the new keys don't operate the lock, check the depths of the cuts. Make corrections by adjusting the stop collar.

2. If the carriage movements feel binding or sluggish, check the hardened rods for cleanliness. Also, if the Half-Time has been bolted down to the work surface, release the pressure from each securing screw until the binding has been released. The securing screws must have balanced pressure.

3. If the vise jaws do not slide up and down smoothly, check under the top jaw against the beating post for cleanliness. Apply a very light coat of oil, then wipe off with a clean dry cloth.

4. When the cutter blade becomes dull, remove it with two open-end wrenches and have it sharpened. When it is replaced, make adjustments to zero—the stylus to the cutter. Note: The blade securing nut is reverse threaded.

## BORKEY 986

Distributed by DiMark International, the Borkey model 986 is a semi-automatic key machine designed to cut cylinder keys (FIG. 17-2). The machine is operated in the following way:

1. Place the machine on a strong, stable workstand or bench. Bolting isn't necessary. Check the current on your supply, which must be two-phase ac, and compare with that shown on the machine. Unplug the machine before undertaking any repairs, removing housing or moving the machine.

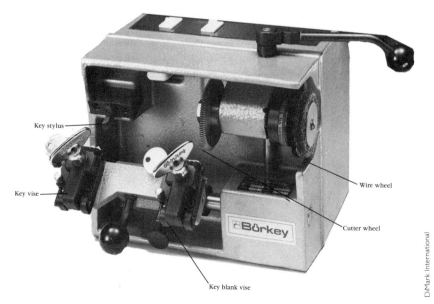

**17-2**  The Borkey 986 is a semi-automatic machine for duplicating cylinder and cross keys.

The hand lever (FIG. 17-3) for the moving of the slide carriage is on top of the machine body. Loosen the screw and the hand lever can be turned to any working position.

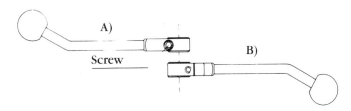

**17-3** Loosen the screw to turn the hand lever   DiMark International

2. Firmly fasten the pattern key in left-hand vise, with shoulder about 2 millimeters from vise side (FIG. 17-4). Use alignment gauge lever to line up accurately. Place the key blank in the same manner into the right-hand vise. Tighten down. Disengage the alignment gauge levers.

**17-4**   Engage the feeler on side of key near shoulder.
DiMark International

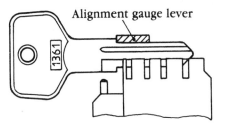

When cutting very small keys, place the delivered pins behind the pattern key and the blank. To prevent the key and blank from slipping during the duplicating process, insert small pins into the keyway grooves or place the bolts—mounted on left side of vises—below the bows.

3. Turn the machine on. You should then see a red light. Hold the carriage knob with your left hand. Loosen the security device on the right side below the slide carriage by pulling out the bolt. Gently engage the feeler on side of key near shoulder. Start cutting from left to right using the lever on right side of the machine.

   Note: Special springs ensure correct pressure against the cutter, obviating need to press by hand.

4. Return lever and slide carriage into start position, remove ready cut key and pass on wire brush to remove swarf.

## Cutting keys without shoulders

By using the flat bars (without cut-out sections) that are delivered with the accessories, shoulderless keys can be aligned at the tip of the key (FIG. 17-5). Depending on the length of the key, the flat bars are to be inserted in the left or right slots of the vises.

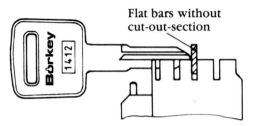

**17-5**   Flat bars are used to align shoulderless keys at the shoulder. DiMark International

## Cutting cross keys (100 mm)

Depending on the length of cross key shoulder, insert the flat bars with cut-out section (FIG. 17-6) on top—into the exterior or interior left slot of the vises. The edges of the cut-out section are used as stops (FIG. 17-7) and, depending on the position of the key shoulder, they are lifted or pressed down or upwards. After fastening the key and blank in the vises, the flat bars are pushed back.

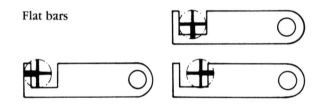

**17-6**   Insert the flat bars with cutout section. DiMark International

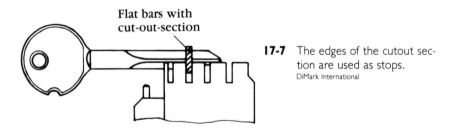

**17-7**   The edges of the cutout section are used as stops. DiMark International

It's important that both key and blank are always fastened in the vises with the same rip. Cross keys with ring or shoulder are to be aligned at the left outside of the vises (FIG. 17-8). The upper part of the vise must be protected against moving by placing the flat bars into the right slot of the vises.

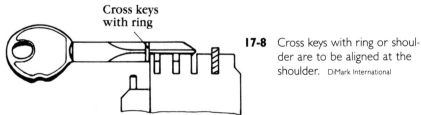

**17-8**   Cross keys with ring or shoulder are to be aligned at the shoulder. DiMark International

### Checking the cutting depth

Place check-up keys into the vises and tighten down. Bring slide carriage forward to feeler and cutter. Feeler (#2 in FIG. 17-9) and cutter must now equally touch the edges of the check-up keys.

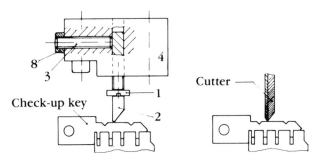

**17-9** Place check up keys into the vises. DiMark International

### Readjusting cutting depth

Loosen nut (#8 in FIG. 17-9) and threaded pin (#3 in FIG. 17-9) of the feeler. Adjust cutting depth by turning the adjustment screw (#1 in FIG. 17-9) to the right to withdraw the feeler, to the left to advance the feeler. Tighten threaded pin and nut again.

### Checking lateral distance

Place check-up keys into the vises. Push them against the left side of the vises and tighten down.

### Alignment gauge levers

Both gauge levers must now touch the shoulders of the check-up keys equally. To readjust, loosen screws (#6 of FIG. 17-10) of one of the gauge levers. Push gauge lever against shoulder of check-up key and tighten down screw (#6 of FIG. 17-10) again.

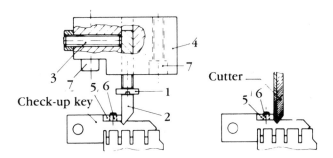

**17-10** Bring slide carriage forward to feeler and cutter. DiMark International

## Distance feeler and cutter

Bring slide carriage forward to feeler and cutter. Feeler and cutter must equally touch the cut of the check-up keys. To readjust, loosen screw (#7 of FIG. 17-10) of feeler base. Move feeler base until feeler and cutter equally touch the cut of the check-up keys. Tighten screw (#7 of FIG. 17-10) again.

To change the cutter, switch off machine. Unscrew cutter nut, turning to the right. Retain spindle by using 6 millimeter bar.

## BORKEY 954 REXA 3/CD

Distributed by DiMark International, the Borkey 954 Rexa 3/CD (FIG. 17-11) precisely duplicates a variety of key types.

The carriage is released by lowering slightly and pulling out on the trigger release below the vises and carriage. After cutting the key, it is not necessary to pull out the release to lock the carriage back down. It will automatically lock down.

When releasing the carriage, hold it firmly or the spring tension will cause the carriage to fly forward—which will damage the vises, feeler, and cutter. Always return the top shoulder alignment gauge to the rear before bringing the carriage assembly into the cutting position. If the cutter hits the alignment gauge lever, it might knock out a cutter tooth.

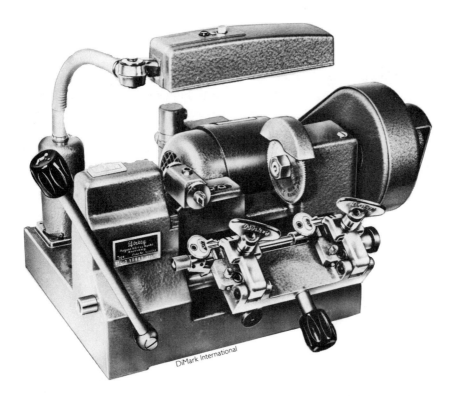

**17-11** The Borkey 954 Rexa 3/CD is simple to use and duplicates keys accurately.

## Cutting a key

1. Select either the front or rear sets of vises. This will depend on whether keys can be gripped flat on the bottom of the vises or can ride on their milling along the top of the front set of vises, or whether the keys should be gripped on a center milling. In the latter case, rear jaws will be used.

2. Place the original key in the left vise and the blank in the right vise. The key may be gauged by the top shoulder using the alignment gauge, by the bottom shoulder using the lower side of the vise, or by the tip using the slots in the vise and the flat tip gauges provided.

   In the case of double-sided keys, the milling or ridges of the key may rest on the top of the vise, but not at the bottom of the vise. The Rexa vises grip keys that have a bare minimum to grasp.

3. Pull down the carriage slightly, holding the carriage firmly as the trigger release is pulled out. Turn on the machine and, using the lever handle to move the carriage, begin cutting from the bow of the key to the tip. Do not cut over the tip. Reverse back to the bow.

4. Pull the carriage down to lock it in place, and turn the machine off. The entire process of cutting a key takes only a few seconds.

## Changing the cutter

1. Remove the cutter guard. Using the wrench provided, turn the cutter nut toward you in the direction of the arrow on the cutter. Stabilize the cutter axle by inserting the round rod provided into the hole which is drilled in the cutter axle to the left of the belt guard.

2. Remove the cutter nut and the cutter. When placing the new cutter on the machine, remember to note the direction indicated on the side of the cutter. The cutter will be ruined if placed on the shaft in the wrong direction.

## Maintenance

The Rexa should be maintained regularly to guarantee the long life for which it was designed. Maintenance includes cleaning the filings from the machine with a soft brush, and oiling certain areas with a 20-30 W motor oil—not with a cleaner like LPS or WD-40. The following are areas to oil regularly:

- On the top of the cutter shaft at the V grooves, next to the belt guard and the cutter.

- Along the main carriage axle shaft at all exposed points, moving the lever to the right and left to bathe the shaft in oil. (Then lightly wipe with a cloth to remove excess oil.)

- At the oil cup at the left side of the machine, near the main axle shaft.
- On the alignment gauge axle where it rides in the casting.
- Under the wing nuts where the pressure washers are located, and along the threaded posts on which the wing nuts tighten. (This keeps the vises gripping well.)

## FRAMON DBM-1 FLAT KEY MACHINE

Manufactured by Framon Mfg. Co., Inc., the DBM-1 key machine (FIG. 17-12) is designed to duplicate flat keys. The procedure for using the machine is as follows:

**17-12**   The DBM-1 is designed to accurately duplicate flat keys.

1. Set all keys from the tip for spacing.
2. Insert pattern key in right-hand vise, with tip of blank protruding slightly beyond left side of vise (FIG. 17-13). The reason for this position is to allow cutting of tip guide on blank, if blank tip can be cut without cutting vise.
3. Push guide shaft rearward and lock into this position by tightening locking knob (FIG. 17-14). This relieves spring pressure so tip setting is easier. Lift yoke and set tip of pattern key against right hand side of guide. While holding this position, insert blank and set tip of blank against right hand side of cutter (FIG. 17-15). This procedure assures proper spacing.

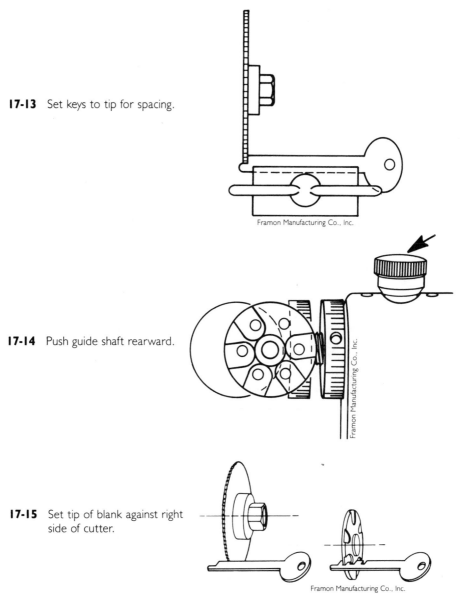

**17-13** Set keys to tip for spacing.

Framon Manufacturing Co., Inc.

**17-14** Push guide shaft rearward.

Framon Manufacturing Co., Inc.

**17-15** Set tip of blank against right side of cutter.

Framon Manufacturing Co., Inc.

4. Release guide by loosening locking knob, and key is ready to be cut.

5. Set cut in pattern key against guide and lift yoke into cutter to make cut (FIG. 17-16). Lower yoke and repeat for next cut. Follow this procedure until all cuts (including throat cut) are made and key is complete. All cuts should be made with a straight in motion. This will ensure clean square cuts.

You will notice there is no side play in guide assembly, so all cuts on duplicate key will be the same width as the pattern key.

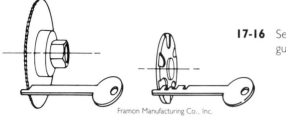

**17-16**  Set cut in pattern key against guide.

Framon Manufacturing Co., Inc.

## Adjustments

With tip setting for spacing there is no problem with improper spacing.

Check depth settings using two blanks that are the same and drawing blank against cutter guide. If depth adjustments are needed, simply loosen set screw on depth ring and adjust ring until cutter barely touches blank. To make cuts deeper, rotate depth ring and adjust ring counter-clockwise. To make cuts shallower, rotate ring clockwise. Tighten set screw after adjustment is made, but do not overtighten.

Another way to check depth is to make one cut on duplicate key and check depth cut on both pattern key and duplicate key. For example, if cut on duplicate key is .003 inch deeper than cut on pattern key, loosen set screw on depth ring, rotate ring .003 inch clockwise, and tighten set screw. Note: Calibrations on depth ring are in increments of .001 inch.

Check cutter guide setting. The guide must be set to same width of cutter used. The DBM-1 is supplied with one .045 width cutter. This is the best width for general work. Cutter widths of .035, .055, .066, and .088 inch will be available if needed. Cutter width of .100 (LeFebure) can be obtained by using an .045- and an .055-inch at the same time. All of these cutters are solid carbide.

To set guide, simply loosen cap screw. Rotate guide to cutter width and tighten cap screw. The screw in guide shaft will align guide. No adjustment is required when changing guide settings.

## Maintenance

Yoke rod, guide shaft, and vise studs should be lubricated using very fine oil sparingly. Do not use motor oil. Be sure to wipe off all excess oil. To lubricate guide shaft, unscrew locking knob and put one or two drops of oil in the opening, then replace knob.

Other than these parts, cleanliness is the best maintenance.

## ILCO UNICAN MODEL .023 KEY MACHINE

Manufactured by Ilco Unican Corporation, the Model .023 key machine (FIG. 17-17) is designed to duplicate a variety of cylinder keys.

### Setup procedure

1. Place the machine on a level sturdy surface before cutting keys. Unlatch cover and lift to a completely open position. Remove the hook from the eye on left side of the carriage (FIG. 17-18).

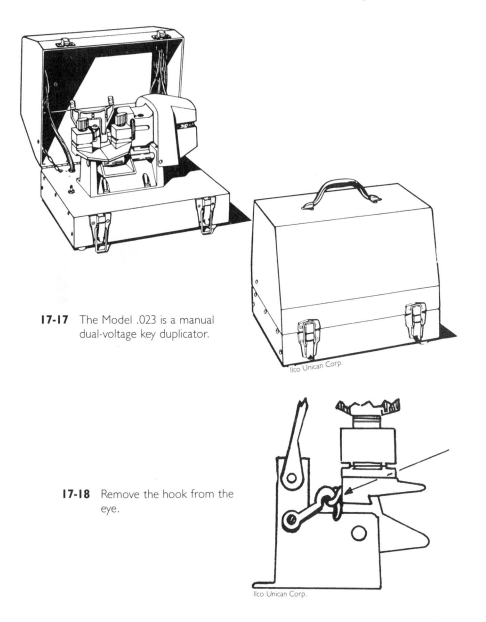

**17-17** The Model .023 is a manual dual-voltage key duplicator.

Ilco Unican Corp.

**17-18** Remove the hook from the eye.

Ilco Unican Corp.

2. Insert appropriate plug into available power supply. Machine operates on 110 volts (220-volt optional) or 12 dc volts, with no other attachments or converter required. The machine is equipped with both cords for ac or dc application and will work equally as well from either power source. It automatically converts from one to the other without touching any switches. If both plugs should be installed in separate receptacles at the same time, the power pack will draw current from the strongest source only, and will not cause damage to the machine.

The motor is protected against overload by a circuit breaker type starting switch. To reset this switch, simply depress the start button after a few seconds delay.

## Transportation procedure

1. Disengage electrical cord from power source and wrap cord on metal angles provided in cover.
2. Slide carriage to extreme left-hand position and engage hook and eye. Note: For transportation, hook and eye must be secure to prevent possible damage to the machine.
3. Close cover and engage both latches.

## General instructions

1. Do not make adjustments on this machine. It has been adjusted and thoroughly tested at the factory.
2. Never work on this or any machine without disconnecting the power cord.
3. Do not attempt to cut keys until you have read the operating instructions and are sure that you understand the mechanism and its operation.
4. Wear eye protection when operating the machine.
5. Keep the carriage spindle clean of chips and lightly lubricate it with a thin film of 3-in-1 or light machine oil. Keep cutter shaft bearings well lubricated using the oil cup.
6. Do not use pliers to tighten wing nuts. They are designed to give adequate pressure when tightened by hand.
7. Take proper care of the cylinder key duplicating cutter. Cutters can be dulled or broken by improper handling. They are shipped in perfect condition and will last a long time if handled with reasonable care. Accurate duplication requires proper seating of key and blanks in four-way jaws. Keep jaws and machine free from chips. Brush off regularly. The accumulation of dirt and chips shortens the life of the machine and reduces its accuracy.
8. Periodically remove the top jaws of each vise to check the springs and to clean the jaw out thoroughly. After doing so, apply a few drops of light machine oil or a touch of grease on the four vertical surfaces that guide the top jaw.
9. When duplicating a key, avoid an irregular jerk in the movement of the carriage. Acquire a smooth steady motion, using both hands on the carriage, holding it behind the key clamps. Apply the same degree of pressure each time a key is duplicated. Excessive pressure may cause over-cutting; too light pressure will result in under-cutting. Practice on a few keys to learn to apply a uniform pressure to the carriage.

10. The motor speed varies with the force of the cutter. This is normal for any battery-operated motor. Too much pressure on the cutter will cause the motor to draw excessive current and trip the circuit breaker.

11. After removing key from clamp, remove any burrs that remain on the edge of the cut surface by drawing lightly through the deburring brush.

### Selecting jaw position

The .023 is equipped with four-way jaws that do not require adapters (FIG. 17-19). The jaw positions are labeled as follows: Standard, Wide, A, and W.

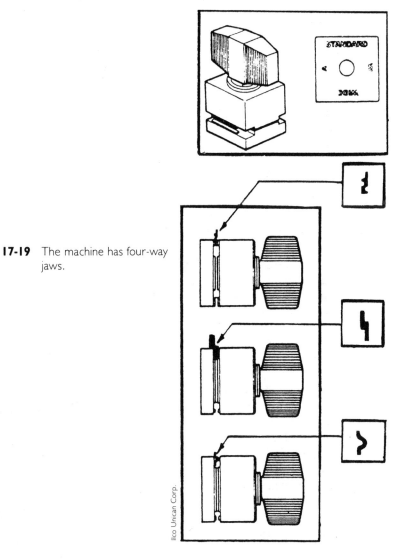

**17-19**   The machine has four-way jaws.

Ilco Unican Corp.

- Standard width is used on most commercial keys and house keys. Examples include keys made by Dominion, Schlage, and Weiser.
- Wide width is used on wide blade large keys, and for some European locks that are cut on both sides—where the center milling of the key can rest on the top of the jaw. Examples include keys made by Ford and Chicago.
- A width jaw is for duplicating Schlage "A" double-sided keys, some foreign automobile keys, and other keys that require a firm grip.
- W width jaw is for duplicating Schlage "W" double-sided keys. The center of the key is held on ridge in jaw.

From experience, you'll find many other uses for the various jaw positions.

### Changing jaw positions

1. Turn the wing nut at least seven complete turns counterclockwise (FIG. 17-20).

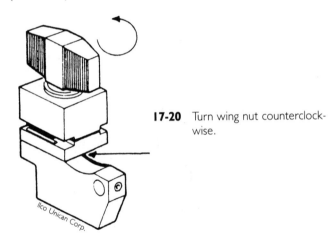

**17-20**  Turn wing nut counterclockwise.

2. Place fingers in the recessed roles provided, then lift upward and rotate until desired position is reached. Be sure to raise lower jaw enough to clear its seat.

### Duplicating regular cylinder keys

1. Select the proper key blank.
2. Make sure the machine's jaws are in the required position.
3. Place the blank in the right-hand jaw (FIG. 17-21). Using the wing nut, lightly clamp blank in place. Swing the shoulder guide over the rest on the blank. With the wing nut loose enough, slide the blank using your thumb. Push on the head of the blank until the shoulder abuts the guide. Use your index finger to press down tip

**17-21** Place the key blank in the right-hand jaw.

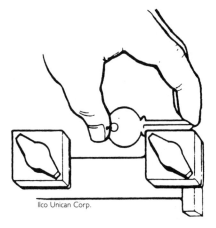

Ilco Unican Corp.

of blank, making sure the blank is seated in the jaw. Tighten the wing nut securely. Swing the shoulder back out of the way.

4. Insert the pattern key in the left hand jaw in the same manner.

### Duplicating irregular keys

The majority of the keys presented for duplication are easily cut. The four-way jaws simplify the job of holding irregularly shaped keys such as Ford, Schlage double-sided, etc. (FIG. 17-22). The following instructions will help you with the duplication of most irregular cylinder keys.

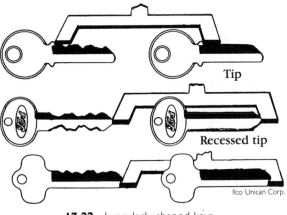

**17-22** Irregularly shaped keys.

### Duplicating Ford double-sided, Best, Falcon, and other keys without shoulders:

1. Rotate jaws to Standard or Wide, as required by the size of the key.
2. Place the key blank in right-hand jaw. The blank should be located in the jaw so the tip of the blank is parallel with back portion of the jaw.

3. Move carriage to extreme left until the tip of the key abuts the key gauge. Maintaining this position, place pattern key in left vise and slide key until it abuts the opposite end of key gauge.

4. Tighten keys securely and return key gauge to upright position.

5. Turn machine on and proceed as outlined in duplication of regular keys.

6. If key is cut on two sides, turn over and duplicate reverse side. In the case of Ford keys and most foreign double-sided keys, only the blank needs to be turned over, because the cuts are the same on both sides.

   Note: The aforementioned procedures can also be used in duplicating broken keys that do not have a shoulder to gauge from.

### Duplicating Chicago double-sided keys:

1. Set jaws to Wide or other position that would provide a secure grip. Some of the less popular Chicago keys (41N, 41FD, etc.) can be secured firmly in the A or W jaw position.

2. Proceed as for regular keys by using shoulder guide to gauge shoulders, although the jaws provide a secure grip.

### Duplicating Chicago double-sided keys (Wide)

1. Set jaws to Wide or other position needed to provide a secure grip.

2. Proceed as for regular keys by using shoulder guide to gauge shoulders.

3. Duplicate as for regular keys. Turn over to duplicate reverse side.

4. Remove finished key and deburr.

### Duplicating Schlage double-sided keys

1. Rotate jaw to position A or W as required.

2. When inserting key in jaws, the shoulder of the key should abut the jaw (FIG. 17-23), because the shoulder guide cannot be used in this case.

3. After cutting one side, turn the original key and blank over to cut the other side.

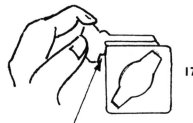

**17-23**    Abut shoulder of key against jaw. Ilco Unican Corp.

Shoulder against jaw

## Maintenance

Clean out vises continually. They must be kept free of cuttings, dirt, or any foreign matter at all times. Keep a small paint brush handy for this purpose. Occasionally, lubricate the wing nut studs and the four inner surfaces of the vise clamps that align the top and bottom portions with grease. Wipe carriage spindle with 3-in-1 or light oil to clean and lubricate.

A few drops of medium lubricating oil once a month is sufficient for the oil cup in the cutter shaft. One or two drops of 3-in-1 or light machine oil once a month is sufficient for the oil cups on the motor.

Belt tension is automatically maintained by the freely suspended motor. Under normal conditions it needs no attention. To replace belt, simply lift up motor and the belt will be free to remove from the pulleys. After replacing belt, make sure that the motor mounting bracket is free to move on its pivot.

Cutter replacement is necessary when the blank is being worn rather than cut away. The copying dog requires frequent adjustment when inaccurate cutting occurs. New cutters are available for prompt supply. It is wise to keep one or two extra cutters on hand at all times. When the last one is put on the machine, order a replacement.

The copying dog is set and adjusted perfectly before the machine leaves the factory and no further adjustments should be necessary. However, as a cutter becomes worn or is replaced by a new cutter, it will be necessary to reset the copying dog.

To replace the copying dog, loosen top screw and place a blank key into each of the jaws. Use two identical blanks. Turn the cutter by hand in cutting direction. If the rotary cutter bites into the blank, turn the adjusting screw in front to advance the copying dog and move the carriage away from the cutter. If the milling cutter does not touch the blank, turn the adjusting screw in the opposite direction. Properly adjusted, the milling cutter should just graze the key blank. After adjusting, tighten the set screws. The spacing adjustment is fixed and requires no attention.

## Power supply

The solid state power supply, located within the machine base, normally doesn't need servicing. If, however, an electrical failure should occur, or the machine power supply is greater than what is required, a self-resetting circuit breaker that is located in the power supply will trip the ac circuit to protect the power transformer. The breaker will self-reset after a few seconds and start the motor unless the start switch is turned off immediately. Do not touch any rotating parts while the current is tripped.

Suspected defects in the power supply should be inspected only by qualified servicepeople. Machines under warranty should be returned to factory for examination and corrective action.

## ILCO UNICAN 018 KEY MACHINE

Manufactured by Ilco Unican Corporation, the Model 018 Lever-Operated key machine is a precision crafted semi-automatic machine designed to duplicate a wide variety of cylinder keys.

The machine comes with an accessory pack that consists of test keys, Allen wrenches, straight wires, a Ford shoulder adapter, and a screwdriver (FIG. 17-24). The test keys are used for adjusting cutter to stylus. Allen wrenches are used for making adjustments on machine and performing periodic maintenance. Straight wires raise narrow keys requiring deep cuts in vise, and keep special keys from tilting in vise. The Ford shoulder adapter allows use of gooseneck gauge on Ford keys.

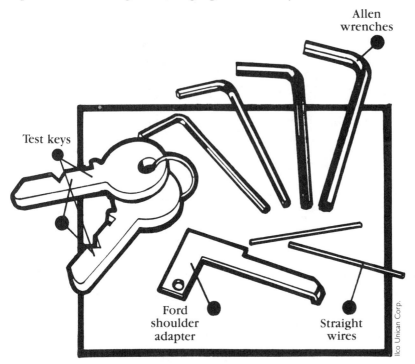

**17-24**    The .018 Lever Operated key machine comes with an accessory pack.

Some features of the 018 include the following: a manual switch to activate the cutting wheel and the wire brush, reversible vises for accommodating various kinds of keys and blanks, and a gooseneck gauge for aligning the key with the blank. It is also equipped with a safety switch that must be disengaged for the machine to operate.

## Operating instructions

To duplicate keys with the 018, proceed as follows:

1. Select the proper key blank.

2. Set the reversible vises to the proper positions. The vise assemblies rotate to accommodate various types of keys. To rotate a vise, loosen the wing nut three or four turns from the closed position. Using the thumb and forefinger, grasp the bottom section of the

**17-25** Set the reversible vises.

LIFT . . .
ROTATE . . .
RESEAT

Ilco Unican Corp.

vise. Lift this section up and free of the carriage. Rotate the vise assembly one-half turn and reseat lower section into groove of carriage (FIG. 17-25).

Note the arrow on the top of each vise. Both arrows should always be pointing in the same direction, either toward or away from the machine. Do not cycle machine without key and blank in vise jaws.

3. Position the key and blank in vises. The upper vise holds the blank and the lower vise holds the pattern key. A key or blank is always placed in the vise with the bow to the left.

To position single-sided keys: Raise protective shield to allow access to wing nuts for loading keys. Loosen the wing nut on the lower vise. Insert key to be duplicated with the bow to the left and cuts facing the stylus. Leave approximately 1/4 inch between the shoulder of the key and the left side of the vise. Tighten the wing nut just enough to hold the key in place. Repeat this procedure on the key blank in the upper vise. The thin edge of the blank should be facing the cutter (FIG. 17-26).

To position a double-sided key: Clamp the center edge of the blade against the face of the vise. Insert the key with the bow to the left, leaving a 1/4-inch margin between bow and left side of the vise. Tighten the wing nut just enough to hold the key in place. Repeat this procedure on the blank in the upper vise (FIG. 17-27).

To position a double-sided corrugated key: Clamp the key in the vise in the center groove. Position the key with the bow to the left, leaving a 1/4-inch margin between shoulder and left side of vise. Tighten the wing nut just enough to hold the key in place. Repeat this procedure on the key blank in the upper vise.

To position other keys: Narrow keys with deep cuts such as General Motors sometimes require the straight wire to be placed under the key and blank. This raises the key in the vise enabling the cutter to reach full depths.

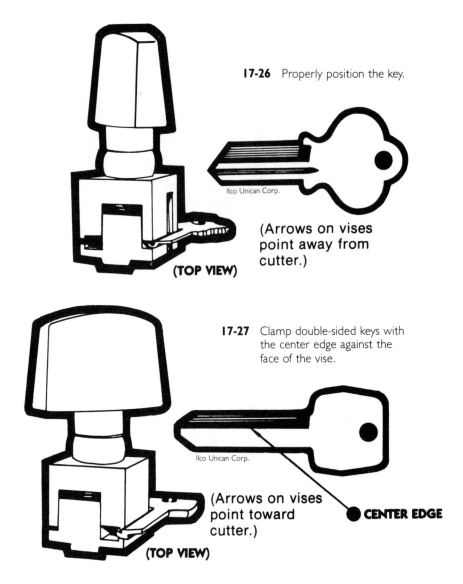

**17-26**  Properly position the key.

Ilco Unican Corp.

**(Arrows on vises point away from cutter.)**

**(TOP VIEW)**

**17-27**  Clamp double-sided keys with the center edge against the face of the vise.

Ilco Unican Corp.

**(Arrows on vises point toward cutter.)**

**(TOP VIEW)**

**CENTER EDGE**

Some keys with rounded or wide millings are difficult to clamp firmly in the vises. Such keys have a tendency to tilt and roll during the cutting cycle. To get a good gripping surface, place a straight wire into the milling closest to the bottom of the key and blank and insert both into position. Straight wires must be used on both key and blank.

4. Lock the gooseneck gauge in place. Pull the carriage toward you with your right hand using the lower wing nut. At the same time, pull the gooseneck gauge forward with your left hand until the pin on the carriage locks in the groove on the gooseneck gauge (FIG. 17-28).

**17-28** Pull the carriage toward you.

5. Align key and blank in vises. The key and blank are aligned by using the gooseneck gauge as a common point of reference. The object is to have the shoulder of both the key and blank butted firmly against the gooseneck gauge. When they are in this position, they are aligned. Aligning key and blank in vises is critical. Take time to be sure they are aligned properly.

   To align a key with a top shoulder, loosen the wing nut on the bottom vise assembly and slide the pattern key to the right until the shoulder touches the gooseneck gauge. Holding the blade firmly in place (FIG. 17-29), tighten the wing nut. Repeat this procedure on the blank.

**17-29** Align keys with shoulders to the gooseneck gauge.

To align Ford keys from the tip, place adapter over pattern key with bent ear over the tip of the key. Hold the bow of the key and the adapter between the thumb and middle finger on the edge of both the key and adapter.

6. Loosen the wing nut and push the pattern key and adapter to the right. This will abut the adapter to the gooseneck. Keep the bottom edge of the key blade and adapter pressed firmly against the face of the vise. Tighten wing nut and remove the adapter. Repeat this procedure with the key blank in the top vise.

When aligning keys or blanks in the vises, the index finger should force the blade of the key or blank down into the vise. The bow of the key may be held with the thumb and middle finger. This method of holding the key or blank keeps the edge of the blade flush against the inside of the vise and keeps the shoulder of the key butted against the gooseneck gauge (FIG. 17-30).

Ilco Unican Corp.

**17-30**  Use your index finger to force the blade of the key or blank into the vise.

When aligning a double-sided key, the center thickness of the key is forced down against the face of the vise in the same manner.

7. Return the carriage and gooseneck gauge to the starting position. Place your right hand on the carriage at the wing nut. Your left hand should be on the gooseneck knob. Pull the carriage toward you slightly to relieve the tension on the gooseneck gauge and push the gooseneck gauge down. As you lower the gooseneck gauge, make certain it is in contact with safety switch. The gooseneck gauge must be in the down position or the machine will not operate.

8. Begin the cutting cycle by activating the starter switch. Pull down on operating lever slowly, until the lever bottoms out. Push the lever upwards to starting position. When the lever is returned to starting position, the cutting cycle has been completed. Shut switch off and remove key from vise.

If you are cutting a double-sided key, remove the blank and flip it over. Go back to Step 3 and begin positioning the key blank with the uncut edge facing the cutter. Repeat Steps 3 – 7.

9. The duplicate key will have small burrs on the blade. The burrs should be removed. Turn manual switch on and hold the key gently against the face of the spinning brush. Turn manual switch off and the job is complete (FIG. 17-31).

**17-31** Hold the key against the face of the spinning brush.

## Adjusting the 018

After considerable usage or when replacing a cutter, the machine might require adjustment. There are two basic adjustments: spacing and depth. Both spacing and depth refer to the alignment of the stylus and the cutter. Proper spacing assures that cuts copied from the pattern key will be the proper distance from the shoulder. Proper depth adjustment assures that cuts made on the blank will match the depth of those on the pattern key. Depth adjustment should be correct before attempting to adjust spacing.

Caution: Before making any adjustments, unplug the machine.

To adjust for depth of cuts, proceed as follows:

1. Remove the hood by releasing the screws on both sides of the machine.

2. Insert and align two identical, uncut key blanks in the vises.

3. Pull the plunger knob out and rotate it one-quarter turn in either direction to lock the plunger away from the cam follower plate. Remove the carriage tension spring to free the carriage. Move the carriage to the right so that the stylus is centered on the blade of the blank.

4. If the cutter and stylus are both touching the blanks, the machine is properly adjusted for depth of cut. The key should be touching the cutter only slightly. To check, turn the pulley that rotates the cutter a few turns by hand.

   The cutter should only nick the blank for only a small part of one full rotation. This occurs because no cutter is perfectly round. Make adjustments on the high point of the cutter.

5. If the stylus touches the key blank in the lower vise but does not touch the blank in the upper vise, the cuts will be too shallow

**17-32** The cuts will be too shallow if the stylus touches the blank in the lower vise, but the cutter doesn't touch the blank in the upper vise.

(FIG. 17-32). To make the proper adjustment, loosen the depth locking screw slightly by turning it counterclockwise. Turn the depth adjusting screw clockwise until the cutter, which should be rotated by hand, nicks the blank in the upper vise. Tighten the depth locking screw securely.

6. If the cutter touches the blank in the upper vise but the stylus does not touch the blank in the lower vise, the cuts will be too deep. Loosen the depth locking screw slightly. Turn the depth adjusting screw until you are able to rotate the cutter so that it barely nicks the blank. Check the adjustment at another point and tighten the depth locking screw securely.

## Adjusting for spacing of cuts

To adjust for spacing of cuts, proceed as follows:

1. Remove the hood by releasing the screws on both sides of the machine.

2. Locate the matching pair of test keys from the accessory pack supplied with the 018. Rotate each vise assembly to point arrow away from the cutter. Insert and align one test key in each vise.

3. Pull the plunger knob out and rotate it one-quarter turn in either direction to lock the plunger away from the cam follower plate. Remove the carriage tension spring to free the carriage (FIG. 17-33).

4. Slide the carriage to the right until the cutter and guide are centered in the grooves of their respective test keys. When the cutter is centered in the groove of the upper test key and the stylus is centered in the groove of the lower test key (FIG. 17-34), the machine is properly adjusted for spacing.

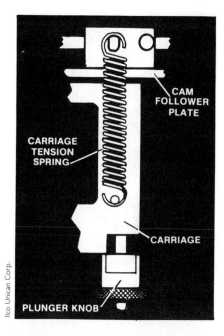

**17-33** Remove the carriage tension spring to free the carriage.

**17-34** The machine is properly adjusted for spacing when the cutter is centered in the groove of the upper test key and the stylus is centered in the groove of the lower test key.

Caution: Never rotate the cutter when adjusting the machine for spacing. Rotating the cutter will damage the test key in the upper vise.

5. If the cutter touches only the right side of the groove and the stylus touches only the left side of the groove, cuts in the keys you duplicate will be too far from the shoulder (FIG. 17-35). To make proper adjustment, loosen space locking screw slightly by turning

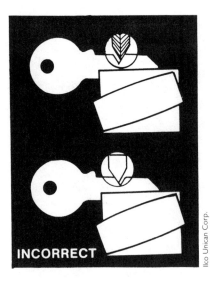

**17-35** Cuts will be too far from the shoulder if the cutter touches only the right side of the groove and the stylus touches only the left side of the groove.

Ilco Unican Corp.

it counterclockwise. Then turn the space adjusting screw counter-clockwise until cutter and guide are centered in the grooves. Tighten space locking screw securely (FIG. 17-36).

6. If the cutter touches only the left side of the groove and the stylus touches only the right side of the groove, cuts in the duplicate key will be too close to the shoulder. To make proper adjustment, loosen space locking screw slightly and turn space adjusting screw clockwise until cutter and guide are centered in the grooves. Tighten the space locking screw securely.

7. After the spacing has been checked or adjusted, remove the test keys from the vises. Unlock the plunger. Slide the carriage to the left until plunger drops into the cam follower plate. Return carriage tension spring to the proper post.

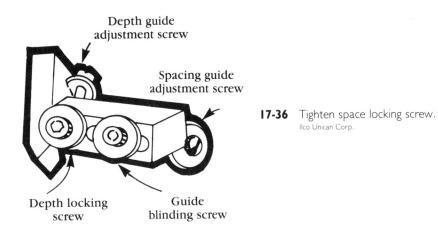

**Depth guide adjustment screw**

**Spacing guide adjustment screw**

**17-36** Tighten space locking screw.
Ilco Unican Corp.

**Depth locking screw**

**Guide blinding screw**

### Changing the cutter

To change the cutter for a 018 key machine, proceed as follows:

1. Remove the hood by releasing the screws on both sides of the machine.

2. Remove cutter shaft nut and cutter shaft spacer. When removing the cutter shaft nut, hold the cutter shaft with a $9/16$-inch open-end wrench in the flats provided.

3. Remove the worn cutter and replace with a new one. (Note: Be sure to install cutter with the arrow pointing in the same direction as rotation of the cutter shaft.)

4. Check for depth of cut. Adjust if necessary.

### Minimizing cutter shaft end play

To minimize cutter shaft end play, proceed as follows:

1. Remove the hood by releasing the screws on both sides of the machine and linkage screw.

2. Remove machine screw and washers at right end of cutter shaft. When removing screw and washer, hold cutter shaft secure with wrench in flats provided.

3. Remove deburring brush.

4. Loosen two Allen screws that secure machine pulley to main cutter shaft.

5. Tighten hex nut gradually, by hand, until all end play is eliminated. Then tighten hex nut an additional one-eighth turn.

6. Tighten two Allen screws in machine pulley.

7. Replace washer and screw. Hold cutter shaft with wrench in flats provided. Caution: Do not over-tighten hex nut.

# Chapter 18

# Key coding machines

*W*ith a key coding machine you can make keys to a lock without using a pattern key. Manufacturers often stamp code numbers on the locks and keys they make. Unless a lock has been rekeyed, you can use a key coding machine to make keys based on the locks' code number.

Find the code number from the customer, or by examining the lock or the key. Then look in a code book to find out which blank to use, where to space the cuts on the blank, and how deep to make each cut.

You could use the information obtained from a code book to make a key by hand using a file and a dial caliper, but a key coding machine can make the key faster. With some models, you can also duplicate keys.

## THEORY OF CODE KEY CUTTING

In cutting keys by code, you make the notches or cuts in the key blank by using the code number stamped on the original key or on the lock. You also need to use code charts, which are numeric listings of code numbers. Each code number has a corresponding number that is the key combination.

For example: A 1986 Chevrolet automobile uses code number 1V89. The V indicates the series of numbers for these cars. Refer to the V Series lists and locate 1V89. Adjacent to 1V89 is the number 322135, which represents the actual cuts in the key.

### Spacing and depth of cuts

There are two critical dimensions in code cutting: spacing of the cuts and depth of the cuts. Spacing is where the cuts are placed on the key and

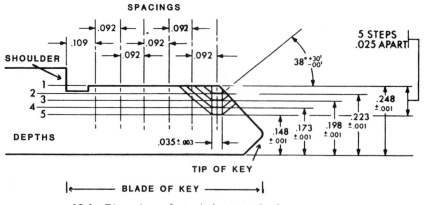

**18-1**   Dimensions of a typical automotive key. Ilco Unican Corp.

actually represents the distances between the centers of adjacent cuts (FIG. 18-1). The component that sets the spacing dimension is the spacing plate.

Note: Spacing is determined from left to right, also expressed as the bow or shoulder of the key (left) to the tip (right).

Depth refers to how deeply the cut is made into the blade of the key. Each depth has a definite dimension, which is measured from the bottom of the cut to the bottom of the key blade. To simplify the code numbers, each depth is identified by a single number. For example, in the V Series, there are five depths, identified as 1, 2, 3, 4, and 5. For key number 1V89, the cuts are 322135, which means that the key does not use depth 4.

In a key coding machine, the component that sets the depths is the depth knob (FIG. 18-2). When the knob is turned so that a 4 centers over the index mark, a 4 depth will be cut into a key. Turning the knob to 5 will produce a 5 depth, etc.

When reading the key cuts, pay attention to the order in which the cuts appear. In the example of key 1V89, the key cuts are 322135. The first number is 3, so a 3 depth must be cut in the first spacing, which is the spacing nearest the shoulder or bow of the key. The second number is 2,

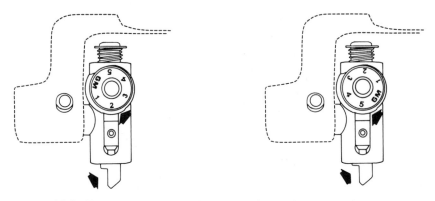

**18-2**   The depth knob is used to set the depth of a cut. Ilco Unican Corp.

so a 2 depth must be cut in the second spacing, from the shoulder or bow of the key. The last number in 322135 is a 5, and a 5 depth must be cut in the sixth spacing, which is the spacing closest to the tip of the key.

Some code charts are based upon the reverse sequence of cuts—that is, the spacings start at the tip of the key and run from right to left. With these charts, the first number of a key combination must be cut in the first spacing nearest the tip of the key. The following cuts then are made in succession from that point.

Suppose you cut a key from bow to tip and it doesn't operate the lock. Try another blank, this time reversing the direction of cutting. Use the same numbers but cut from tip to bow. The key might work. If not, the code charts do not apply to the lock, or someone has changed the combination of the lock, or the lock is malfunctioning.

## THE KD80 CODE CUTTING KEY MACHINE

Manufactured by Ilco Unican Corporation, the KD80 (FIG. 18-3) has the dual capability of cutting keys by code and duplicating keys. Because code cutting is primarily centered around car keys, the KD80 is available with optional kits to cut keys for vehicles manufactured by General Motors, Ford, and Chrysler. The model numbers for the various kits are as follows: basic machine—KD80; basic machine with Chrysler kit—AVCE-KD80; basic machine with Ford 5 and 10 code kit—011-00147; basic machine with GM kit—074-00010.

The KD80 is never used as is. It must be equipped with the components in a specific kit to permit code cutting of a specific brand of car key.

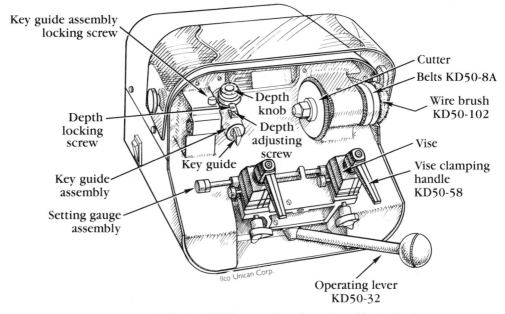

**18-3** The KD80 can cut keys by code and by duplication.

For example, the KD80 to cut General Motors keys would be part number 074-00010 and it would be provided with the components to specifically cut GM keys. These components are a key guide assembly, a spacing plate, a depth knob, and a cutter. When installed, these components will set the KD80 to cut keys by code for GM vehicles, using original factory dimensions. Keys cut by code will be like the original keys supplied with the lock.

## Setting up

After unpacking the KD80, set it on a sturdy work bench. It's not necessary to bolt down the unit because its 70-pound weight makes it rigid enough for normal use. Attach the lever handle and the safety shield. Then locate the parts identified in FIG. 10-3.

Plug the electric cord into an outlet and depress the on-off switch to make sure that the machine operates. Turn it off.

## Cutting a key

Before cutting a key with the KD80, be sure that the machine is provided with the proper components for cutting the brand of key you want. If making a GM key, you should have the GM depth knob installed on the GM key guide, the GM cutting wheel and key guide assembly on the machine, and the GM spacing plate. You should also have the key cuts written down on paper, in proper sequence.

The procedure for cutting the key is as follows:

1. Clamp the spacing plate into the left vise jaw in any random position. Then rotate the key gauge up and set the left finger in contact wit the shoulder of the spacing plate.
2. Clamp the key blank into the right vise. Make sure that the blade of the blank is resting flat against the key rest surface of the vise, and that the shoulder of the blank is touching the right finger of the setting gauge.
3. Determine the first depth cut of the key combination.
4. Turn the depth knob to set the number corresponding to the first depth cut at the index mark.
5. Depress the on-off switch to turn on the machine.
6. Move the carriage lever so that the first notch of the spacing plate is under the key guide. Line up the point of the key guide with the center line of the first space in the spacing plate.
7. Raise the carriage (lift the lever). Continue lifting until the center line of the first space touches the tip of the key guide. At this point, the cutter will be cutting the key blank. Move the carriage carefully from side to side to widen the cut. Don't let the guide move away from the bottom of the cut, or it might jump into the adjacent space in error. When the cut has been widened, lower the carriage lever.

8. Turn the depth knob to set the number corresponding to the second depth cut at the index mark.

9. Move the carriage lever so that the second notch of the spacing plate lines up with the key guide. Proceed exactly as you did with the first cut.

10. Raise the carriage and continue lifting the lever until the center line of the second space touches the tip of the key guide. Move the carriage from side to side to widen the cut. At this point, the cutter will be cutting the blank.

11. Continue the same procedure for each successive depth cut—that is, rotate the depth knob to the appropriate depth number and move the carriage to the appropriate spacing position.

12. After the last depth cut is made, remove the notched key from the right vise jaw.

13. Deburr the new key by running the underside of the key against the wire brush. Do not over-brush and do not turn the notches of the key into the brush. If you over-brush, you could take away more metal and reduce the accuracy to which the key was cut.

## Cutting other car keys

The KD80 code cutter can be "converted" to cut keys for other vehicles, such as Chrysler or Ford. Generally, the procedure is the same. The numbers of Chrysler and Ford keys also are code numbers that are arranged in a series. The series translate into actual key combinations, which represent the depths of the cuts. The cuts are made according to the depths and spacings.

Major differences occur, however, with the actual values of the depth and spacing dimensions. The GM spacing measurements are not the same as Ford. The Chrysler depth and spacing measurements aren't the same as GM and Ford. They're all different. As a result, different spacing plates, depth knobs, and cutters are needed for each brand of vehicle. The setting gauge is used to position keys with shoulders.

If you have a KD80 set to cut GM keys and you want to cut Chrysler keys, you must replace the GM key guide assembly and depth knob with the Chrysler key guide assembly and depth knob. You must also use the Chrysler spacing plate and install the Chrysler cutter. Once the proper parts are installed, you can cut the Chrysler key using the same procedure as described earlier.

## Clamping Ford double-sided keys

Ford keys are cut on both sides of the key blade and are called double-sided keys, because they first must be cut on one side, then turned over and cut on the other side. The key cuts are the same on both sides.

To clamp the Ford double-sided key in its vise jaw, lay the center ridge of the key on top of the jaw surface. Then tighten the vise jaw. After cutting, the vise jaw is loosened and the key blank is turned over for reclamping. Again, the center ridge of the key is rested on the top surface of the vise jaw, as for the first cuts.

Because Ford keys don't have a shoulder, it is not accurate to use the key gauge for aligning the keys in the vise jaws. On the Ford key, the tip of the key is the shoulder, so Ford keys are aligned by the tips.

On the KD80, there are two ways to align the tips of Ford keys: Use the service bar in one of the vise jaw slots to serve as a stop (use the same slot in both jaws), or use the special key gauge set on the right side of each vise jaw. Remove the service bar, or drop the key gauge before cutting.

### Cutting Ford five-cut keys

Ford vehicles from 1965 through 1984 used pin tumbler locks that had five pin tumblers and five depths. Key numbers were listed in the code charts running from FA000 to FA1863 (ignition) and FB000 to FB1863 (glove and trunk).

### Cutting Ford ten-cut keys

Selected 1985 and 1986 models in the Ford line use a key that has ignition and door cuts in one side of the key. The blank needed for the ten-cut system is slightly longer and wider than the previous five-cut blank. Even though the ten-cut blank looks like the five-cut blank, the two are different and cannot be used interchangeably.

A quick way to measure the key blanks is to line up the tips and then check the lengths of the blades. The ten-cut key contains ten cuts (FIG. 18-4).

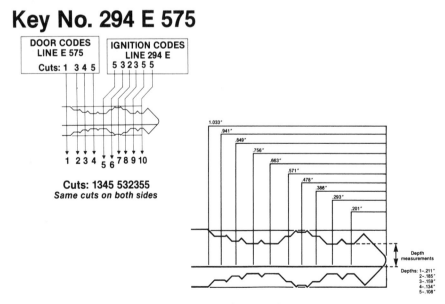

**18-4** Depths and spaces of a key code. Ilco Unican Corp.

The ten cuts operate either one of two disc tumbler cylinders—one for the door and the other for the ignition lock. The ignition cylinder uses six tumblers, in positions 5, 6, 7, 8, 9, and 10. These are the positions closest to the tip of the key.

The door lock also has six tumblers, but these are placed in positions 1, 2, 3, 4, 5, and 6—which are the positions closest to the bow of the key. Note that the tumblers in the fifth and sixth positions are the same in both locks.

The ten-cut system uses the new FC series of code charts, which runs from FC100 to 344 (for the ignition lock codes) and FC501 to 624 (for the door lock codes). The charts contain five columns, identified as A, B, C, D, and E. The letter designation carries over to the key combination, which will have three numbers, a letter, and then three more numbers. The cuts of the key are located by referring to the charts, using both the numbers and letter of the key combination.

To illustrate how key cuts are determined, assume there is a key with the number 294E575. Because E is in the combination, you must refer to the E column in the charts, for both numbers 294 and 575. After locating 294 in the charts, go across the columns to the E column and find the number 532355, which represents the combination of the ignition lock, or the actual key depths. After locating 575 in the charts, go across to column E to number 1345; this represents the combination of the door lock, or the actual key depths. Both groups of numbers together give a key combination of 1345532355 (FIG. 18-4).

Next, determine whether you are going to cut from bow to tip (left to right) or from tip to bow (right to left). This will depend upon how the code charts are set up. Factory charts are set up so that spacings run from tip to bow. With these charts, the first number of the ignition combination is cut in the first spacing (which is the spacing closest to the tip of the key). The other cuts are then made in sequence. The first number of the door combination is cut into the seventh spacing, and the remaining cuts are made in sequence.

According to the factory charts, therefore, the combination for key 294E575 reads as 5431553235. In essence, both numbers produce the same cuts in the key, but it depends on the direction in which the key is cut (either right to left or left to right).

Align the key in the same manner: Raise the small key gauge at the right side of the vise jaw and push the tip of the key against the key gauge.

The spacing plate and the depth control knob must be changed when switching from a five- to ten-code key. Do not use one set of components for the other style of key.

## Cutting Merkur keys

Keys for the Merkur can be cut on the KD80 machine using FM components—that is, the spacing plate, the key guide assembly, the cutter, and the depth knob all marked FM. The change in components is necessary because the dimensions of the Merkur lock are different from other Ford locks.

The Merkur key is double-sided, but it does not have a shoulder. As a result, alignment of the original key and the blank can be made with the setting gauge. The Merkur key blanks are imprinted "TX" on the blades.

When clamping the key or blank in the vise jaw, set it so that its bottom rests against the bottom of the vise. The key blank should not rest on the top surface of the vise jaw.

Because of the wide groove on each side of the Merkur key, the key may rotate or tilt towards the cutter. A tilted position of the key is not acceptable because the cuts will be inaccurate. To prevent the key from tilting, insert the 1.2 millimeter service pin into the back groove of the key (FIG. 18-5). The tip of the service pin should be about 1/8 inch away from the shoulder of the key.

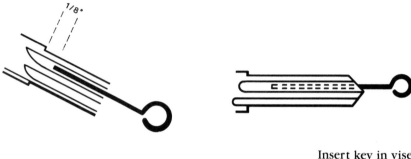

Tip of service pin should
be 1/8" from shoulder

Insert key in vise
with service pin
in back groove

**18-5**   Insert service pin into back groove of the key. Ilco Unican Corp.

With the pin inserted, clamp the key in the vise jaw, making sure that the bottom of the key rests against the bottom of the vise. The service pin should remain in the groove during the actual cutting, in both the original pattern key and the key blank.

When turning the keys over to make the second side cuts, take extra care to make sure that the keys do not tilt fore and aft. The bottom-most points of the cut side should rest against the bottom of the vise and the horizontal lines of the key should match the horizontal lines of the vise.

## Duplicating keys

To set the KD80 code cutter to duplicate a key, you simply have to turn the depth knob to its neutral setting. This neutral setting is at the two letters that identify the knob. On the GM depth knob, there are five numbers (1, 2, 3, 4, and 5) and two letters (GM) on its surface. When the knob is turned so that the GM is pointing to the index mark, the knob is at its neutral setting. In this position, the tip of the key guide is in the same plane as the tip of the cutter, having been preset at the factory.

The alignment of the key guide tip and the cutter is commonly referred to as the adjustment. Checking the adjustment is done in the conventional manner. Two key blanks are clamped in the left and right vise

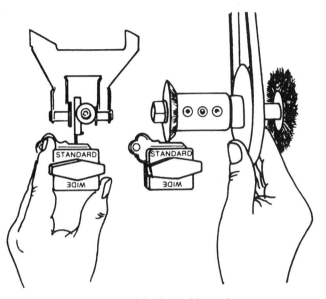

**18-6**   Check the adjustment of the key guide to the cutter. Ilco Unican Corp.

jaws. The carriage then is lifted so that the left key blank touches the tip of the key guide. At this point, the right key blank also should touch the cutter. If the right blank does not touch, the key guide must be moved in or out until both blanks touch (FIG. 18-6).

To adjust the key guide, loosen the depth locking screw (slotted screw) and turn the depth adjusting screw one-eighth turn or less, as needed. Then retighten the depth locking screw.

When the adjustment is correct and the depth knob is at the neutral setting, the KD80 machine is ready for duplicating a key. Clamp the pattern key in the left vise and the key blank in the right vise. Use the setting gauge to properly line up the shoulders of both keys. Swing the setting gauge away, turn the machine on, and move the carriage with the lever to cut the duplicate key. When moving the lever sideways, use a steady even motion. Avoid rapid or jerky movements because this could damage the cutter.

When cutting the Merkur double-sided key, use the key gauge to align the shoulder of the pattern key with the key blank. Keep the service pins inserted in the keys. When turning the keys over to make the second side cuts, make sure the keys do not tilt fore and aft. The bottom-most points of the cut side should rest against the bottom of the vise and the horizontal lines of the keys should match the horizontal lines of the vise.

## The cutter

Like all cutting instruments, cutters should be treated with care. Harsh and abusive treatment will ruin a good cutter quickly. Let the cutter do its work by applying a steady, moderate pressure and cutting only those materials it's designed to cut (brass, brass nickel-plated, or nickel silver keys).

In normal operation, the cutter should rotate downward (when looking at the machine from the operator's position). Do not alter the direction of the rotation of the cutter.

To replace the cutter, insert the stabilizing rod into the hole of the cutter shaft and use a wrench to unscrew the cutter nut (FIG. 18-7). The cutter nut has a left-hand thread and must be turned clockwise to loosen. Be sure that the nut is tight after the new cutter has been installed on the shaft.

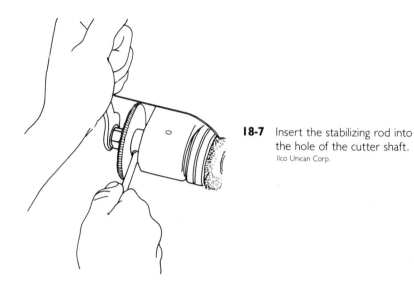

**18-7**  Insert the stabilizing rod into the hole of the cutter shaft.
Ilco Unican Corp.

## EXACTA CODE KEY CUTTER

Manufactured by Ilco Unican Corporation, the Exacta Code key cutter is a mechanical type of key coding machine called a key bitting punch. It stamps or punches cuts into a key, rather than grinding them. Because a key bitting punch is mechanical, it's very useful to take on outside service calls.

The Exacta requires a minimum of care; the only danger to the machine is dirt and the only maintenance necessary is keeping the machine clean. Use a soft brush to make sure the key insert slot, key insert, and punch die assembly in particular are kept free of chips and dust.

Empty the plastic chip box occasionally. Lubrication should be unnecessary because all moving parts are treated with a permanent lubrication.

## Setting up

The Exacta Code key cutter is to be set up in the following way (refer to the parts numbers in FIG. 18-8):

1. Select proper depth knob (P-10), spacing plate (P-11), key insert (P-12), and key blank from your code book.
2. Assemble depth knob by sliding on shaft over pin, keying it to shaft. Insert screw and tighten until knob is resting against shoulder.
3. Install spacing plate by placing dowel pin and swing into position where it will be held by detent.
4. Slide key insert into slot of carriage from right or left side of carriage, according to instruction in your code book.
5. In order to cut some keys, punch and die assembly (P-13) must be replaced. Select proper punch and die from code book. Remove lever (P-20) by unfastening and removing hand lever (P-23) and die screw (P-24). Pull out punch and die assembly. Slide proper punch and die assembly in position, making sure that the frontal flat portion of the die is lined parallel to carriage. Tighten screw (P-24). Replace lever and lever screw (P-23).

## Operating instructions

To operate the Exacta Code key cutter, proceed as follows:

1. Place spacing lever (P-15) into extreme left hole of spacing plate (P-11).
2. Set depth knob (P-10) to extreme C.W. position.
3. Insert key blank on carriage (P-18) from same side the key insert (P-12) was installed, or according to code book instruction.
4. Locate key blank with right-hand shoulder guide or one of the following methods, according to instructions in code book:
   - Left-hand shoulder guide.
   - Tip of key against stop on insert.
   - Shoulder of key against insert.
5. Lock key blank of carriage by tightening carriage lever. Be sure that blank is level and straight with insert (P-12). Do not over-tighten lever or carriage could be hard to move.
6. Return shoulder guide to rest position against side of carriage.
7. Set spacing lever (P-15) in position 1 on spacing plate.
8. Set depth knob to required depth.
9. Press lever (P-20) firmly to complete cut.

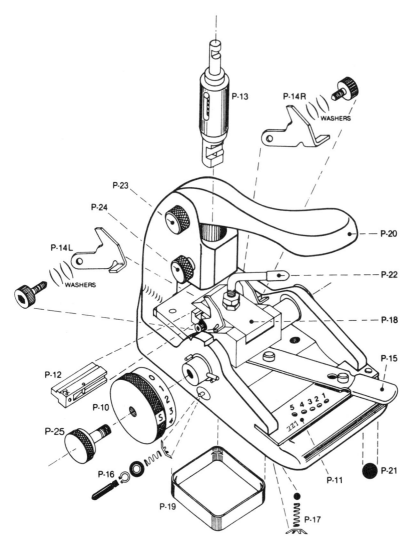

## SPARE PARTS LIST

P-10   Depth knob (knobs for different makes are identified by letters A, B, etc.)

P-11   Spacing plate (plates for different makes are identified by numbers 100, 101, etc.)

P-12   Key insert (inserts for different makes are identified by numbers 1, 2, etc.)

P-13   Punch and die assembly

P-14R  Shoulder guide assembly — Right side

P-14L  Shoulder guide assembly — Left side

P-15   Spacing lever assembly

P-16   Detent assembly for depth knob

P-17   Detent assembly for spacing plate

P-18   Carriage assembly

P-19   Plastic box

P-20   Lever

P-21   Felt pads (set of 4)

P-22   Carriage lever

P-23   Lever screw

P-24   Knob set screw

P-25   Depth knob screw

**18-8**   An exploded view of the Exacta key bitting punch. Ilco Unican Corp.

10. Return depth knob to extreme C.W. position before attempting to advance spacing lever to next position.

11. Set spacing lever in subsequent positions and repeat procedure for all cuts.

12. When all cuts have been made, return spacing lever to the extreme left position of spacing plate.

13. Loosen carriage lever and remove key.

14. For keys cut on two sides, repeat procedure on second side.

15. For keys inserted from right-hand side only: When key is inserted from left side, spacing lever (P-15) is moved to extreme right hole in spacing plate (P-11).

**18-9**   The Framon DC-300 is designed to cut automotive keys by code and by duplication.

## FRAMON DC-300 DUPLICATING CODE MACHINE

Manufactured by Framon Manufacturing Co., Inc., the DC-300 (FIG. 18-9) is designed primarily to cut automotive keys by code and to duplicate keys. The machine uses depth cams and spacing keys.

The basic machine at time of purchase includes one cam and five spacing keys. This furnished cam is the No. 1 cam and has depths for cutting the five-pin Ford, the ten-wafer Ford, American Motors, Chrysler, and General Motors keys. Each of the five spacing keys are numbered both for identification and with the number of spaces for each manufacturer. Also included is an adjusting Allen wrench and instruction/service manual.

The machine can be purchased with either 100-volt ac or 12-volt dc power.

## Cutter head

The DC-300 cutter head is mounted on precision-grade sealed bearings for accuracy and long life. The cutter is precision ground of M3 tool steel and is the precise configuration for most automotive keys.

## Yoke

The yoke has two vises. The left vise is used for the blank to be cut. The right vise is used to hold the spacing key required when cutting by code and for holding the pattern key when duplicating (FIGS. 18-10 and 18-11).

All spacing keys are used in the right-hand or guide-side vise, and keys to be cut by code or duplicated are used in the left-hand or cutter-side vise.

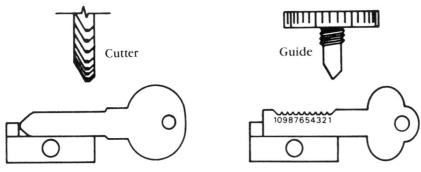

**18-10** Typical set-up to cut keys by code using spacing key and blank.
Framon Manufacturing Co., Inc.

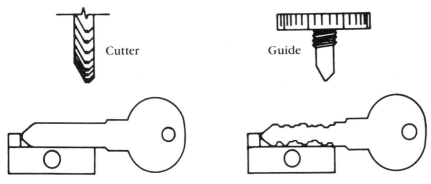

**18-11** Typical set-up to cut duplicate keys using a pattern key and blank.
Framon Manufacturing Co., Inc.

## Key blank information

Most automotive key blanks are made without bottom shoulders and should be inserted into guide-side vise with the tip of the key against the built-in stop in the vise. Key blanks that have a bottom shoulder must be inserted with the bottom shoulder against the right-hand side of the guide vise (FIG. 18-12). The spacing key will always be inserted with the tip of the key against the stop in the guide vise.

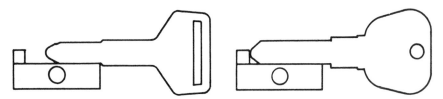

**18-12**  Insert keys that have a bottom shoulder with the bottom shoulder against the right hand side of the guide vise. Framon Manufacturing Co., Inc.

## Cams and cam post

As previously mentioned, the Framon DC-300 comes furnished with the No. 1 cam. This cam will handle all domestic automobile locks: Ford ten-wafer, Ford five-pin, American Motors, Chrysler, and General Motors.

Six other cams are available with depth increments from .0138 up to .040. Each cam is numbered on the rear surface. The No. 1 depth is common to all cams and this depth on any cam is used for duplicating keys.

To change cams, loosen cam lock knob at right rear of cam post, withdraw cam pin, and slide cam out. Slide new cam into slot, push cam pin into place, and tighten cam lock knob.

Note: The counterbore at the rear of the cam is set to hold a wavy washer. Be sure to replace this washer each time cams are changed (FIG. 18-13).

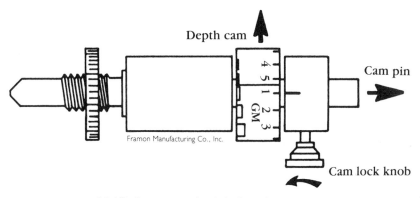

**18-13**  Loosen cam knob lock to change cams.

## Cam post and guide

The key guide in the cam post is spring-loaded to allow the guide to enter cuts in either the spacing key (when cutting by code) or in a pattern key (when duplicating). This system is used to allow straight cuts.

The adjusting ring on the guide shaft is what controls the depth of each cut. If keys are cut too high, the adjusting ring is rotated counterclockwise to change depth. If keys are cut too deep, the adjusting ring is rotated clockwise to change depth. To rotate adjusting ring, loosen the set screw in the edge of ring and tighten when adjustment is made. Do not overtighten set screw.

## Cutting keys by code

To cut a General Motors key, code 3V86, with cuts of 133545, proceed as follows:

1. Insert key blank in left-hand vise, with tip against tip stop in vise.
2. Insert spacing key No. 14 in right-hand vise, with tip against tip stop in right-hand vise (FIG. 18-14).

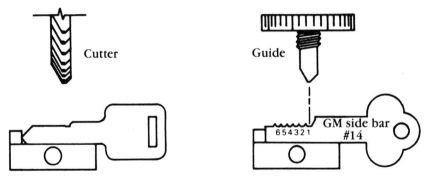

**18-14** Insert spacing key in right-hand vise. Framon Manufacturing Co., Inc.

3. Insert No. 1 cam in cam post and rotate cam so GM No. 1 cut is aligned with guide mark at top of cam post (FIG. 18-15).
4. Lift yoke so guide engages No. 1 cut on spacing key, then continue lifting yoke until cutter engages key and cut is made.

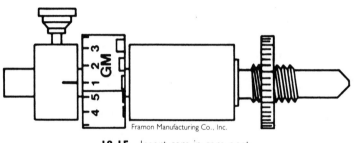

**18-15** Insert cam in cam post.

5. Rotate cam to No. 3 cut and, following same procedure, make the next two cuts.

6. Rotate cam to No. 5 cut and make cuts in No. 4 and No. 6 spacing positions.

7. Rotate cam to No. 4 cut and make cut in No. 5 position on spacing key. Key is now complete.

## Spacing keys

Each cut on spacing keys are numbered from bow to tip. This allows operator to make cuts of the same depth anywhere on any key without changing cam settings. Each spacing key is numbered according to the provided chart. Framon can furnish spacing keys for any code series needed; over 250 spacing keys are available.

## Duplicating keys

To duplicate a key with the DC-300, proceed as follows:

1. Insert key blank in left-hand vise, with tip against stop in vise.

2. Insert pattern key in right-hand vise with tip stop.

3. Rotate cam to No. 1 cut (No. 1 cut on any cam can be used for duplicating).

4. Set guide in each cut on pattern key. Make each cut by lifting yoke until it comes to a complete stop at each cut, and key is complete (FIG. 18-16).

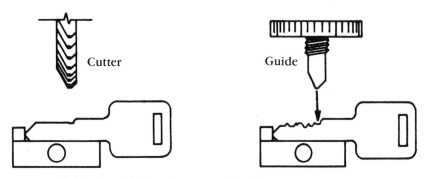

Cutter        Guide

**18-16**  Set guide in each cut on pattern key. Framon Manufacturing Co., Inc.

If blank and pattern keys have a bottom shoulder, keys should be inserted in vises with bottom shoulder against right-hand side of vise. Otherwise use tip stop.

## Lubrication

A small amount of fine oil can be used on the yoke slide rod. Wipe off all excess oil. If guide needs lubrication, remove cam, push guide rearward, remove snap ring at rear of guide shaft, and slide shaft out. Lubricate as needed and reverse procedure to replace.

The depth plunger under guide shaft is spring-loaded and will eject itself when guide shaft is removed. This rod can be lubricated. Replace all parts.

## THE BORKEY 989 TOP-CUT

Distributed by DiMark International, the Borkey 989 Top-Cut (FIG. 18-17) cuts straight and curved track as well as drill-type keys to pattern or to code.

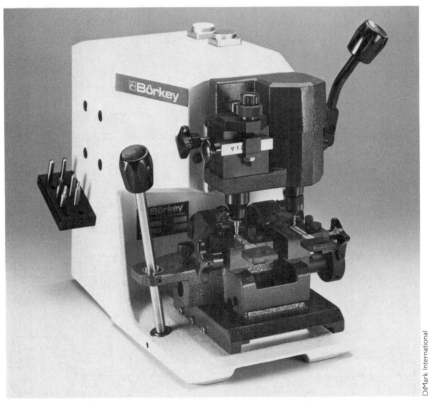

**18-17** The Borkey 989 Top-Cut is designed to cut a variety of key types by code and by duplication.

### Changing the cutter

To release the cutter, insert the pin punch provided into the hole over the cutter chuck. Hold pin punch (FIG. 18-18) securely and loosen lower nut, turning to the left with the wrench provided. The cutter will drop out of the bottom. Insert new cutter in chuck and push it right up to the top, as far as it will go. Tighten by turning the nut to the right. Be sure it is firmly tightened in the collet.

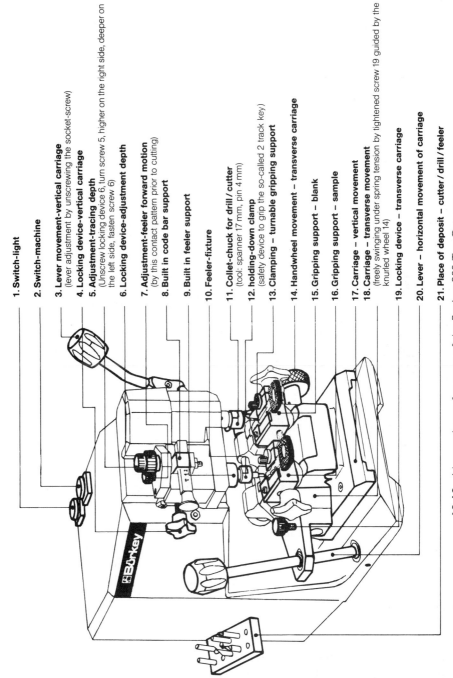

1. **Switch-light**

2. **Switch-machine**

3. **Lever movement-vertical carriage**
   (lever adjustment by unscrewing the socket-screw)

4. **Locking device-vertical carriage**

5. **Adjustment-tracing depth**
   (Unscrew locking device 6, turn screw 5, higher on the right side, deeper on the left side, fasten screw 6)

6. **Locking device-adjustment depth**

7. **Adjustment-feeler forward motion**
   (by this contact pattern prior to cutting)

8. **Built in code bar support**

9. **Built in feeler support**

10. **Feeler-fixture**

11. **Collet-chuck for drill / cutter**
    (tool: spanner 17 mm, pin 4 mm)

12. **holding-down clamp**
    (safety device to grip the so-called 2 track key)

13. **Clamping – turnable gripping support**

14. **Handwheel movement – transverse carriage**

15. **Gripping support – blank**

16. **Gripping support – sample**

17. **Carriage – vertical movement**

18. **Carriage – transverse movement**
    (freely swinging under spring tension by tightened screw 19 guided by the knurled wheel 14)

19. **Locking device – transverse carriage**

20. **Lever – horizontal movement of carriage**

21. **Place of deposit – cutter / drill / feeler**

**18-18**  Nomenclature for parts of the Borkey 989 Top-Cut. DiMark International

## Removing the guide

Use the same wrench to loosen the nut above the guide, turning the nut to the left about one-half turn. The guide will drop out of the bottom. Insert the new guide into the collet-chuck as far as it will go and tighten the nut firmly.

The locking nut that is above and on the right side of the casting has been preset at the factory, so it is not necessary to make adjustments to it. This controls a concentric adjustment and should not be adjusted by the operator, except with special instructions from factory representatives.

## Cutting drilled keys

Install the drilled key and guide, following instructions given above. Remove key disc holders from the top of both jaws and put them in a safe place. Put blank keys in both vises.

Referring to FIG. 18-19, loosen lower knob (6) and turn upper knob (5) counterclockwise two to three turns to release previous settings. Turn knob (7) clockwise two to three turns to raise the guide higher than the cutter. Lower cutter/guide assembly so the cutter lightly touches the blank. Securely tighten left thumb turn (4) so that the cutter still lightly touches the key. Turn knob (7) until the guide just touches the left key and then stop turning.

Hold knob (6) stationary and turn knob (5) clockwise so it is lightly finger tight. Turn knob (6) clockwise to lock in the calibration. Release cutter/guide assembly so it goes back up.

Turn on the machine to test depth. Cutter should just touch the right key blank when cutter/guide assembly is lowered. It should make a small mark, but no actual cut. If depth is too high or low, repeat complete instructions above.

Once the depth is correctly set, the guide can be put in the spring-loaded mode by turning knob (7) counterclockwise several turns. If spring-loaded mode is not desired, knob (7) should be adjusted at this time.

Check angle of jaws to be sure they are in the right position for the key you are cutting. They are normally set in the 0 position. To tilt jaws, the large thumb turn at the right rear must be loosened. Jaws tilt in unison. The detent will index at 0.

Proceed to cut key by sample or by depth keys. Using the large knurled knob on the right of the carriage, adjust the carriage assembly so that the cutter is near the cutting line on either the left or right side of the keys. Because the carriage is somewhat self-centering, it will find the center of the cut as each cut is placed on the key. You might want to pull all cuts to the left of center on one side, then turn the key over and do all the left cuts on the opposite side before adjusting the carriage to do all the right cuts.

To lock the left-right travel of the carriage so that shallow cuts can be centered in line, hold the guide and cutter in a deep cut on that side of the key and lock the carriage with the round black screw that is to the right of the main carriage handle.

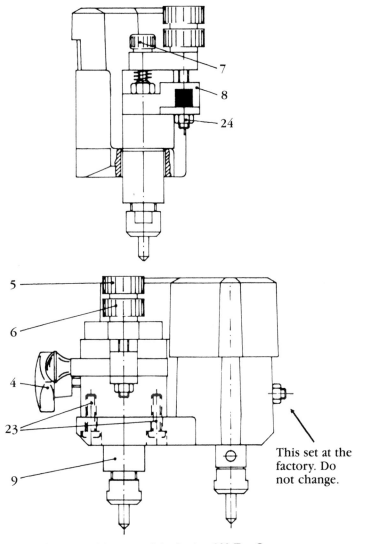

**18-19**  Adjustable parts of the Borkey 989 Top-Cut. DiMark International

### Adjusting depth of cutter guide for high-security automotive keys

Referring to FIG. 18-19, upper knob (5) controls the maximum depth of cuts. Lower knob (6) is a locking nut to hold depth settings accurately. Rear knob (7) allows the guide to operate spring-loaded or fixed.

Put two key blanks in the vises. Four-track keys or the sides of flat safe keys work well. Two-track keys do not work well.

Loosen front knobs (5) and (6) counterclockwise two to three turns. Lower the cutter/guide assembly and turn knob (7) until it is just above the surface of the left key when the cutter touches the surface of the right key. Lock the cutter/guide assembly in position, using the thumb knob at the left (4).

With the cutter touching the surface of the right key, turn knob (7) until the guide is off the surface of the left key only enough to allow a piece of normal bond paper to pass between the guide and key. Holding lower knob (6) from turning, adjust upper knob (5) clockwise until it just stops. Lock it in place using lower knob (6).

Caution: At this time the cutter is lower than the guide, and will be able to cut into parts of the machine. Do not turn the machine on at this point.

Remove the keys. Release the cutter/guide assembly so it is free, and position the cutter and guide over the face of the key vises. Put a business card on top of the right vise. Lower the cutter/guide assembly so the cutter lightly touches the business card, and lock the assembly into position using the left knob. When the depth is correctly set, the business card will catch a little as it is removed.

Before turning the machine on, move the carriage in all directions to make sure the cutter will not touch the face of the jaw on both sides of the key. Make sure the cutter is firmly tightened so that it won't slip down during cutting. Otherwise, damage might result.

## Cutting two-track keys

Secure the sample key in the left vise and the blank in the right vise, tip-stopping the keys. Move the straight side of the key disc clamps over the keys to keep them from tipping. Tighten down using the round top knobs. It is not necessary to move these again while inserting or removing keys.

Use the knurled knob at the right of the carriage to position the carriage. The spring-loaded feature is controlled by turning this knob. To cut deeper on the left side of the key, turn the knob away from you. To cut deeper on the right side of the key, turn the knob toward you.

Position cutter/guide assembly so the guide and cutter rest next to the first cut, closest to the bow of the key. Turn the machine on. Cuts will be made from bow to tip. Cutting should be accomplished in stages, especially if there are deep cuts on the key. As the deeper cuts are made into the key, the guide might bind against the high cuts. In that case, adjust the knurled knob to allow for passing those high cuts, then readjust to allow for cutting the deeper cuts individually. This might seem awkward at first, but with practice this motion will become automatic.

*Chapter* **19**

# Automotive
# lock servicing

*T*his chapter provides the information a lock-
smith needs to enter the lucrative field of automotive lock servicing. The
information can also be useful for the person who just wants to know
how to service the locks on his or her own vehicle. Some of the topics
covered here include understanding the differences among automotive
locks, removing locks from vehicles, and opening locked automobile
doors, trunks, and glove compartments.

## BASICS

Automotive lock servicing is a specialty field of locksmithing, and can be
very profitable. To be proficient in this field, you must stay informed of
the constant changes that are made to automotive locks and locking systems. Most of the changes are made to increase vehicle security and
reduce thefts.

Locksmiths are frequently called on to service automobiles that have
lock-related problems. Such problems include: foreign objects (such as a
broken piece of key) that become lodged in the lock cylinder, keys that
turn very hard in the cylinder, and keys that won't operate the lock.

A broken piece of key can be removed from an automobile lock in
the same way it can be removed from other locks. A key that is stuck in a
vehicle door lock can usually be removed by first turning the key to the
upright position (the position the key is in when it's first inserted into the
lock), then clamping a vise grip to the bow and pulling straight out.

When a key turns hard in a cylinder, it's usually the result of a bent
rod between the walls of the car door. A bent rod can be caused by the
improper use of automobile opening tools. To solve the problem, remove
the inside trim panel from the vehicle, then locate and straighten out the
bent rod.

When the key won't operate an automobile door lock, the problem could be within the lock cylinder. Solve the problem by repairing or replacing the cylinder. More often, however, a disconnected rod with the walls of the door is the cause of a key not operating a lock. If this is the case, you will need to remove the inside trim panel from the vehicle, then locate the disconnected rod and reconnect it. This often requires a new retainer clip. Figure 19-1 illustrates a tool that can be used to help you replace retainer clips in hard to reach areas.

A-1 Security Manufacturing Corp.

**19-1**   A retainer clip tool can be helpful for removing and replacing hard-to-reach retainer clips.

The procedure for removing an inside trim panel is different for different vehicles. In general, however, you need to do the following:

1. Wind the door window all the way up.
2. Remove the door handle—sometimes it's held on by a screw, other times by a retainer clip. If the handle is held on by a clip, use a door handle clip tool to quickly remove it (FIG. 19-2).

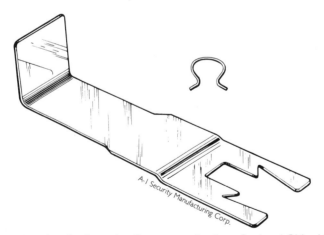

**19-2**   A door handle clip tool easily removes clips from Ford and GM vehicles.

3. Remove arm rest from the door. It is usually secured by screws and clips.
4. Remove all the screws located around the edge of the trim panel and carefully pull the panel from the door (which is now being held to the door only by retainer clips.)

When replacing the trim panel, you may need to replace some of the retainer clips that hold the trip panel to the door. Those clips often break when the trim panel is being removed.

## AUTOMOTIVE LOCK DIFFERENCES

From the outside, most automotive locks appear similar to one another, but the internal constructions often differ. The internal construction of an automotive lock depends on the following: the purpose of the lock (ignition, door, trunk, etc.), and the manufacturer, model, and year of the vehicle for which the lock was made.

In most cases, two keys are used to operate the locks on a vehicle. The primary key is usually the one that operates the vehicle's ignition and doors, while the secondary key usually operates a vehicle's trunk and glove compartment. This two-key system allows the vehicle's owner to have someone (such as a parking lot attendant) drive the car without being able to open sensitive areas such as the trunk and glove compartment.

## AMERICAN MOTORS CORPORATION

All locks for American Motors Corporation (AMC) vehicles up to the model year 1967 are standard five-disc tumbler locks. They are easy to pick. In 1968 AMC began using side bar disc tumbler locks as ignition locks on all its new vehicles, but the company continued using standard 5-disc tumbler locks for most other areas, such as doors and trunk. The 1975 to 1980 Gremlin and Pacer models use side bar disc tumbler locks for their rear compartments.

The ignition and door of an AMC vehicle is usually keyed alike. The original key code can be found on the ignition lock after the lock has been removed from the vehicle. The trunk and glove compartment locks are also keyed alike. The glove compartment lock has four disc tumblers, which are identical to the last four tumblers used in the vehicle's trunk lock. A side bar disc tumbler lock used on an AMC vehicle's rear compartment has the code number stamped on the tailpiece.

It's usually easy to impression a key to one of the door locks. The blank should be twisted very lightly, however, because the dust shutter can break. When replacing ignition locks for AMC models manufactured after 1986, use General Motors locks and keyways.

### Fitting a key

Usually key impressioning is the fastest way to fit a key for an AMC vehicle. Another method is to remove the door lock and lock pawl and read the bitting numbers stamped on the back of the tumblers. You could also remove the ignition lock by disassembling the steering wheel column, then cut a key by the code on the cylinder.

To fit a secondary key, cut a key for the glove compartment and cut one depth at a time in space No. 1 (the space closest to the bow) until you've cut that space to the depth that opens the trunk. After cutting to each depth, insert the key into the trunk lock to see if it operates the lock. You can use depth and space charts to learn the proper spacing and depth increments for the lock.

### Removing an AMC ignition lock

Ignition locks for AMC vehicles through the 1978 model year can be pulled out of the steering wheel column without first disassembling the column. Locksmithing supply houses offer tools for that purpose; most are modified versions of a dent puller. Beginning in 1979, the ignition locks for new models were bolted in; these shouldn't be removed without first disassembling the steering wheel column.

Because of the potential danger that can result in improperly reassembling a steering wheel column, you shouldn't attempt to disassemble one until you've seen the procedure done several times on the vehicle model you're planning to service. And to be on the safe side, you should have an experienced locksmith help you disassemble and reassemble a few steering wheel columns before you try doing it alone.

### AUDI

Most Audi vehicles made after 1971 use either eight- or ten-disc tumbler type locks. Usually all the locks in such a vehicle are keyed alike; in some cases, a primary key is used to operate the doors and ignition.

To fit a primary key for an Audi, use the code number found on the door lock after the lock has been removed from the vehicle. For a secondary key, use the code number on the glove compartment lock.

### BMW

Most locks used on BMWs have ten-disc tumblers, and are difficult to pick. All the locks on a BMW are keyed alike, but sometimes a secondary key is used. Many BMWs are operated with dimple keys. Some late models have a high-security deadlocking system that should be worked on only by a BMW dealer or an authorized service center. A BMW deadlocking system allows a person to lock the car only by using a key; that makes it virtually impossible to get locked out of the car.

To fit a key on a BMW, cut the key by code from the number on the door lock, or disassemble the door lock and use the tumblers to visually fit a key. The door handle is usually held on by a phillips head screw under the weatherstripping or by a screw and bolt.

### CHRYSLER

In most cases, Chrysler vehicles use pin tumbler locks on the ignition, doors, and trunks. The glove compartment locks have either three- or four-disc tumblers. The tilt/telescoping steering wheel columns found on some vehicles use a side bar disc tumbler lock.

The doors and ignition locks of a Chrysler are keyed alike and are operated by the primary key. The code number for these locks can be found on the ignition lock. Occasionally, the code number can be found on a door lock.

The secondary key operates the trunk and glove compartment locks. When the glove compartment lock has three-disc tumblers, they correspond to the three middle tumblers of the vehicle's trunk lock. When the glove compartment lock has four-disc tumblers, they correspond to the last four tumblers of the trunk lock. Many times you can find the code number for a secondary key on a trunk lock.

Beginning in 1989, some Chrysler vehicles (Plymouth Acclaim and Dodge Spirit) began using disc tumbler locks that are operated with double-sided convenience keys. These locks are designed to resist picking.

### Fitting a key

To fit a primary key for a Chrysler vehicle that uses pin tumbler locks, you can remove and disassemble a door lock and measure the bottom pins with a caliper. The measurements for Chrysler bottom pins are as follows: No. 1—.148, No. 2—.168, No. 3—.188, No. 4—.208, No. 5—.228, No. 6—.248. If a pin is worn, these measurements might not hold out. In this case, replace the pin with one that is the proper size. The new pin will be a little longer than the one it is replacing.

To fit a secondary key, fit the key to the glove compartment lock. If the key has four cuts, find the first cut of the trunk lock by cutting the key's first space one depth at a time until the key operates the trunk. If the glove compartment lock has three cuts, you can impression the first and last cuts.

When impressioning a key for a Chrysler pin tumbler lock, remember that the locks shouldn't have a No. 5 or 6 pin in the first lower pin chamber, and shouldn't have a No. 6 pin in the second lower pin chamber. When a No. 5 pin is found in the first or second lower pin chamber or a No. 6 pin is found in the second lower pin chamber, the lock was probably improperly rekeyed.

The depth and spaces for Chrysler locks that use key blank P19A/P1770U (1969–1989 ignition and doors) are as follows: Spacing No. 1—.146, No. 2—.286, No. 3—.426, No. 4—.566, No. 5—.706 (Spacing from center of one cut to center of next cut is .140). Depth No. 1—.246, No. 2—.226, No. 3—.206, No. 4—.186, No. 5—.166, No. 6—.146. (Depth drop is .020 between depths).

The depth and spacing for Chrysler locks that use key blank S19/S1770CH (1969–1989 trunk) are as follows: Spacing No. 1—.146, No. 2—.286, No. 3—.426, No. 4—.566, No. 5—.706 (Spacing from center of one cut to center of next cut is .140). Depth No. 1—.246, No. 2—.226, No. 3—.206, No. 4—.186, No. 5—.166, No. 6—.146 (Depth drop from cut depth to the next is .020).

The spacing and depth for Chrysler locks that use key blank P1789 (1989 and later Acclaim/Spirit) are as follows: Spacing No. 1—.757, No. 2—.665, No. 3—.573, No. 4—.481, No. 5—.389, No. 6—.297, No. 7—

.205 (Spacing from one space to the next is .108). Depth No. 1—.340, No. 2—.315, No. 3—.290, No. 4—.265 (Depth drop from one depth to the next is .025).

## DATSUN

Most locks used on Datsuns have 6-disc tumblers and are easy to pick. All the locks on a Datsun are keyed alike. They are operated by a double-sided convenience key. Both sides of the key are cut, but either side can operate a lock because Datsun locks have only one set of tumblers.

A key can be fitted for Datsun locks by cutting a key by the code that's written on a piece of paper glued to the glove compartment lid. The code can also be found on a door lock. A door lock can also be used to impression a key.

## HONDA

Most locks used on Hondas have six-disc tumblers. These are operated with a double-sided convenience key, and are easy to pick. All the six-disc tumbler locks on a Honda are keyed alike. A lot of pre-1976 Hondas use locks that have eight-disc tumblers.

Impressioning is usually the fastest way to fit a key for a post-1976 Honda. Another way is to use the code number found on the door lock.

A key used for a Honda manufactured in 1989 or later is 4 millimeters (.175 inches) longer from the shoulder to the bow than is a key for an older model Honda. For pre-1989 Honda Accords, Preludes (from 1982), and Civics, use key blank HD83. For Honda Accords and Preludes made in or after 1989, use key blank HD90. Key blank HD91 can be used for Honda Civics made in or after 1989. If you use the HD83 blank to cut a key for a Honda model made in or after 1989, you might have difficulty removing the key from the ignition lock.

## FORD

Up to the 1984 model year, pin tumbler locks with five sets of tumblers were used for all Ford ignitions, doors, and trunks. Glove compartment locks were either pin tumbler or disc tumbler. (Since 1981 only four-disc tumbler locks have been used on glove compartments.) In late 1984, many Ford vehicles began using only disc tumbler locks.

On pre-1982 Ford models, the ignition and doors are keyed alike and are operated by the primary key; the glove compartment and trunk are keyed alike and are operated by the secondary key. On pre-1977 models, the code for the primary key can be found on one of the door locks, usually a passenger side. On pre-1980 models, the secondary key code can be found on the glove compartment lock latch housing. Beginning in 1980, codes were no longer stamped on Ford glove compartment locks.

The primary key for a post-1980 Ford fits only the ignition lock; the secondary key operates all the other locks. Many of those locks don't have key code numbers stamped on them, but the tumblers have bitting numbers stamped on them. Those bitting numbers can be seen when a lock is disassembled.

### Fitting keys

**Pre-1984 Ford**  You can fit a primary key to a pre-1981 Ford lock by removing the door lock and measuring the bottom pins. The measurements for Ford bottom pins are as follows: No. 1—.145, No. 2—.165, No. 3—.185, No. 4—.205, No. 5—.225. To fit a primary key for a 1981 to 1984 Ford, either use the ignition lock to impression a key or remove the lock and read the bitting numbers on the tumblers.

Fit a secondary key to a pre-1981 key by cutting according to the code number on the glove compartment lock. To fit a secondary key for a 1981 to 1984 model, remove the disc tumbler lock and read the bitting numbers on the tumblers.

**Post-1984 Fords**  There are two basic locking systems for post-1984 Fords. The newer system uses only disc tumbler locks (a side bar disc tumbler lock is used for the ignition). The older system uses pin tumbler locks.

During mid-year production of the 1984 Mercury Cougar and the Ford T-Bird, Ford Motor Company began using a side bar disc tumbler ignition lock (similar to the one used by General Motors Corporation) and disc tumbler door locks. A convenience key is used to operate the locks: either side of the key blade can be used.

With the new system, one convenience key is used to operate the ignition and door locks of a vehicle. The double-sided key (Ilco blank 1184FD), sometimes called "the 10-cut key," has ten cuts on each side. The first six cuts (starting from the bow) are for the doors; the last six cuts (spaces 5 through 10) operate the ignition lock. Spaces 5 and 6 on the key correspond to tumbler depths that the ignition and doors have in common. When the key is inserted into a door, the first six cuts are aligned with the tumblers; when inserted into an ignition, the last six cuts are aligned with the tumblers.

The older locking system uses two keys to operate all the locks of a vehicle; each double-sided convenience key has only five cuts on each side. (The primary blank is Ilco 1167FD; the secondary blank is Ilco S1167FD.) In the older system, all the locks except the ignition lock are keyed alike and use the secondary key; the primary key operates only the ignition lock.

**Post-1984 Ford disk tumbler lock**  Neither the ignition nor door locks of post-1984 Fords with disc tumbler locks have key codes stamped on them. The easiest way to fit a key to those vehicles is to cut a key by code, if the code is available. The codes are stamped on key tags given to the purchaser of the vehicle. If no code is available, you can impression a key at one of the doors. If you need a key for the ignition, use the cuts of the key you made for the door to help you impression the key. Remember, the last two tumblers of the door lock have the same depth as the first two tumblers of the ignition lock. Another option is to remove and disassemble the lock.

The ignition lock must be rotated about 30 degrees to the right (to the ''on'' position) before it can be removed. This allows the retaining pin to be depressed and the pilot shaft to bypass an obstruction in the column so the lock can slide out.

You can rotate the cylinder by using a drilling jig (available from locksmith supply houses) to drill out the side bar. The jig is held in place by a setscrew; a key blank is inserted into the lock to align the jig, and the jig aligns your drill bit. The jig allows you to easily drill through the roller bearings located on either side of the lock's keyway. After drilling through the lock, you can rotate it and remove it.

Now you can rekey an uncoded ignition lock, and use that as a replacement lock. Or use the service kit supplied to Ford dealerships to replace the lock.

## GENERAL MOTORS

A General Motors (GM) vehicle ordinarily uses side bar disc tumbler locks for the ignition, door, and trunk, and standard disc tumbler locks for the glove compartment and utility compartments. Some top-of-the-line GMs have side bar disc tumbler locks on the glove compartments.

A pre-1974 GM has its ignition and door locks keyed alike, and all those locks are operated by the primary key. The code for the primary key can be found on the ignition lock. Pre-1970 models have the code for the primary lock stamped on the door locks.

A post-1973 GM uses a primary key that fits only the ignition lock; its secondary key operates the other locks. Until the late 1970s, the secondary key code number could be found on the vehicle's glove compartment lock. The glove compartment lock has four tumblers, which correspond to the last four tumblers of the trunk lock (the trunk lock has 6 tumblers).

### Fitting a key

Fit a primary key for a pre-1970 GM by removing the door lock and using the code number stamped on it. To fit a primary key for a 1970 to 1973 model, remove, disassemble, and decode the door lock. Lock decoders are available from locksmith supply houses for this purpose (FIG. 19-3). For a 1974 to 1978 model GM, save time by just pulling the ignition lock and installing a new one. Otherwise, you would need to disassemble the steering wheel column. A primary key can be fitted to a post-1978 GM model (not including vehicles with VATS or PASS-Key) by disassembling the steering wheel column, removing the lock, and using the code stamped on the lock.

To fit a key for a secondary GM lock, remove the glove compartment plug and cut the key by code. Some GM glove compartment locks are tricky to remove without using a bezel nut wrench (FIG. 19-4). The plug can be removed as follows: pick the lock open if it's locked, open the door, then pick it back to the locked position. Insert an ice pick or similar

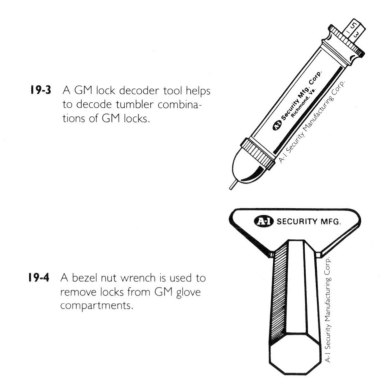

**19-3**  A GM lock decoder tool helps to decode tumbler combinations of GM locks.

**19-4**  A bezel nut wrench is used to remove locks from GM glove compartments.

instrument in the small poke hole and depress the retaining pin. You should be able to easily remove the plug. If no code is available, fit a key to the glove compartment lock and use the following GM progression method to find the remaining two cuts.

### The GM progression method

Since 1967 GM has adhered to the following three rules for making a factory original key:

- The sum total of the cut depths must equal an even number.
- There cannot be a more than two cut depth difference between any adjacent cuts.
- There can never be more than three of the same cut depths in a row.

The GM progression method is based on logically using these rules to progressively decrease the possible key cut combinations until the proper combination is determined.

The first rule refers to the fact that whenever the cut depth numbers of a GM key are added together, the sum should be an even number. A GM key might have the cut depth numbers 3-2-4-3-2 (which equals 14), for example, but not 3-2-4-3-3, because the sum of the latter is an odd number.

Suppose you obtained four cut depths from a glove compartment lock, and need to find the remaining two cuts for operating the trunk lock. If the sum of the four cut depths is an odd number, the two unknown depths must also equal an odd number, because any odd number plus any other odd number always equals an even number. If the sum of the four cut depths is an even number, the remaining two must equal an even number because any even number plus any other even number always equals an even number. Likewise, any odd number plus any even number equals an odd number.

The second of the three rules mentioned above refers to the fact that cuts directly next to each other should never differ by more than two cut depths. For example, a number 1 cut should not directly precede a number 4 depth cut.

The third of the three rules means no GM key should have three consecutive cuts of the same depth. A key bitting of 2-2-2-3-1, for example, is forbidden.

By understanding these three rules, it's easy to use the four cut depths found on a glove compartment lock to figure out the remaining two cuts for the trunk lock. First make a key with those four cuts; the first and second spaces on the key should be left uncut. There are only 25 possibilities for the first two cuts (five possible depths for two spaces equals 5 squared, or 25).

The 25 possible depths for the two remaining spaces are the following: 1,1; 1,2; 1,3; 1,4; 1,5; 2,1; 2,2; 2,3; 2,4; 2,5; 3,1; 3,2; 3,3; 3,4; 3,5; 4,1; 4,2; 4,3; 4,4; 4,5; 5,1; 5,2; 5,3; 5,4; 5,5.

Based on the second GM keying rule, six of those 25 possible depths can be ignored because they have more than two depth increments between them. They are: 1,4; 1,5; 2,5; 4,1; 5,1; 5,2. That leaves only 19 possible depth combinations in any instance where the first two depth cuts must be found for a GM.

Based on the first GM keying rule, you can immediately eliminate about half of those 19 possibilities. You would eliminate either all the odd pairs or all the even pairs—depending on whether you need a pair that equals an even number or a pair that equals an odd number. If you need a pair that equals an even number, you would have only the following 11 choices: 1,1; 2,2; 3,3; 4,4; 5,5; 3,1; 4,2; 5,3; 1,3; 2,4; 3,5. If you need a pair that equals an odd number, you would have only the following eight choices: 2,1; 1,2; 3,2; 2,3; 4,3; 5,4; 4,5.

The second GM rule would then allow you to eliminate several more of those pairs. If your four glove compartment lock cut depths are 1-1-2-3, for example, then rule two would be violated by preceding those cuts with the cuts 3,4; 5,4; or 4,5. If a 4 or 5 depth cut was next to a 1 cut on a key, the key would have adjacent cuts with more than 2 depth differences. That means only five possible cuts would be available. They are: 1,2; 2,3; 2,1; 3,2; and 4,3.

Using the same glove compartment lock cut depths as the example, you would then take the key and cut a 2 depth in the first space and a 1 depth in the second space; then try the key in the lock. If the key doesn't work, you would then progress to cutting a 3 depth in the first space and a 2 depth in the second space (both spaces would be cut a little deeper). After cutting three of the five possibilities, you would then need to use another key with the four cuts from the glove compartment on it to cut the remaining two pairs of depths on that key, beginning with the shallowest pair. One of them will operate the lock.

By using the GM progression method, you should never have to waste more than one key blank if you're searching for an odd combination for the two cuts. You should never have to waste more than two keys if you're searching for an even number for the two cuts.

There are only two reasons this method can fail: The factory made an error in keying the lock; or someone rekeyed the lock without adhering to GM's rules.

## SERVICING GENERAL MOTORS VEHICLES WITH VATS

General Motors Vehicle Anti-Theft System (VATS)—also called Personalized Automotive Security System (PASS-Key)—has been used in select GM models since 1986. The system has proven to be very successful in preventing automobile thefts.

Statistics published in December 1989 by the Highway Loss Data Institute, an organization for the insurance industry, show about a 70 percent reduction in the thefts of Corvettes, Camaros, and Firebirds since those models were equipped with VATS. Before VATS was used, those were among the most frequently stolen GM models in the United States.

VATS is an electromechanical system that consists of the following basic components: a computer module, keys that each have a resistor pellet embedded in them, an ignition cylinder, and a wire harness that connects the ignition cylinder to the computer module.

When a properly cut VATS key is inserted into the ignition cylinder, the cylinder will turn. The resistor pellet in the key will neither hinder nor aid the mechanical action of the ignition cylinder. If the key is embedded with an incorrect resistor pellet or has no resistor pellet, the computer module shuts down the vehicle's electric fuel pump, starter, and powertrain management system for about 4 minutes.

This happens because vehicles with VATS are designed to operate only when one of 15 levels of resistance is present. The 15 levels are represented in 15 different resistor pellets. When a VATS key is inserted into a VATS ignition cylinder, contacts within the cylinder touch the resistor pellet in the key and the resistor pellet's resistance value is transmitted to the VATS computer module by the wire harness connecting the cylinder to the computer module. Only if the resistance value is the right level, can the vehicle be started.

It's important to remember that the turning of the ignition cylinder is a mechanical process that is independent of the system's electronics. Any properly cut key that fits the ignition cylinder can be used to turn the cylinder to the start position. But unless the VATS control module also receives the correct resistor information, the vehicle won't start.

VATS keys look similar to other late-model GM keys, but come with a black rubber bow and contain a resistor pellet. They use standard GM depths and spacings, and fit into an A keyway. All VATS key blanks are cut in the same way other GM blanks are cut.

When cutting a key for a VATS vehicle, however, it is first necessary to determine which of 15 VATS blanks to use. That can be determined by measuring the resistor value of the pattern key (the one the customer wants duplicated) with an ohmmeter or multimeter, and comparing the reading to the VATS pellets resistor values. TABLE 19-1 shows the resistor values currently being used.

**Table 19-1    VATS Resistor Values.**

| VATS Pellet | Resistance in Ohms | B&S Part # 1986, 1987 | B&S Part # 1988, 1989 |
|---|---|---|---|
| 1 | 400 | 593581 | 594201 |
| 2 | 500 | 593582 | 594202 |
| 3 | 679 | 593583 | 594203 |
| 4 | 885 | 593584 | 594204 |
| 5 | 1,128 | 593585 | 594205 |
| 6 | 1,468 | 593586 | 594206 |
| 7 | 1,871 | 593587 | 594207 |
| 8 | 2,369 | 593588 | 594208 |
| 9 | 3,101 | 593589 | 594209 |
| 10 | 3,728 | 593590 | 594210 |
| 11 | 4,750 | 593591 | 594211 |
| 12 | 6,038 | 593592 | 594212 |
| 13 | 7,485 | 593591 | 594213 |
| 14 | 9,531 | 593594 | 594214 |
| 15 | 11,769 | 535595 | 594215 |

When VATS was used in the 1986 Corvette, each module had a predesignated resistor pellet value. That system was used through 1988. Starting with the 1988 Pontiac Trans AM GTA, General Motors began using a modified VATS. The new VATS was named the Personalized Automotive Security System. All GM vehicles with VATS manufactured after 1989 use this new system.

For locksmiths there are two major differences between the two systems. First, the keys used with the new system are 3 millimeters longer than the keys used with the old system (FIG. 19-5). The old key blanks have to be modified before they can be used to operate the new VATS ignition locks. The other big difference between the two systems is that the older system had stickers on the VATS modules showing the modules' resistor number; the new system doesn't have such stickers.

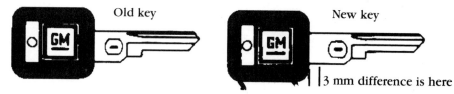

**19-5**  New VATS keys are longer than the older keys.

### Making a VATS first key

When no pattern VATS key is available for you to duplicate, you can make a VATS key as follows: First, determine the proper bitting in the same way as you would determine the bitting for non-VATS late-model GM vehicles. Then determine which VATS key blank to transfer the cuts to.

The most expensive way to determine the correct blank is to cut and try a different VATS key blank until you find one that starts the vehicle. The high costs of VATS blanks make this method impractical. A less costly method involves using an ohmmeter or multimeter, an extra VATS ignition cylinder, and 15 different VATS blanks.

Disconnect the wire harness connecting the control module to the cylinder, and connect that wire to the extra VATS cylinder. Insert one of the VATS key blanks into the extra cylinder and attempt to start the vehicle by using the correctly cut mechanical key to turn the vehicle's ignition cylinder to the start position. If the vehicle shuts down, you'll need to wait 4 minutes and repeat the procedure with another VATS key blank in the extra ignition cylinder until the vehicle starts. After the vehicle starts, transfer the cuts from the correctly cut mechanical key to the VATS key blank that allowed you to start the vehicle. Then disconnect the wire harness from your extra ignition cylinder and reconnect it to the vehicle's ignition cylinder.

### Using a VATS decoder

A VATS decoder can be very helpful for servicing vehicles with VATS. Several companies manufacture such a device. A popular model is the All-Lock A-7000, manufactured by the All-Lock Company (FIG. 19-6).

The A-7000 can be used to perform four functions: identify the correct VATS key blank from the customer's original; decode the correct VATS blank from the vehicle; diagnose steering column connection problems; and diagnose VATS computer problems.

With the A-7000, the resistor value of a VATS key can be determined simply by inserting the key into a slot in the decoder (FIG. 19-7). This feature can be useful for quickly finding the right VATS key blank to use to duplicate a VATS key.

The decoder can also help determine which VATS blank to use when no VATS key is available. First you need to cut a correct mechanical key that will turn the vehicle's ignition cylinder to the start position. Then

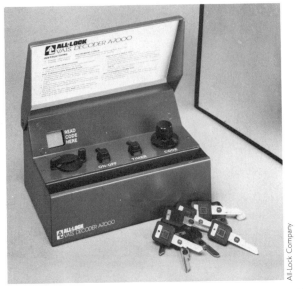

**19-6**    The All-Lock A-7000 is used to service vehicles with VATS.

**19-7**    The resistor value of a VATS key can be determined by inserting the key into the All-Lock A-7000.

All-Lock Company

**19-8** A light comes on whenever the All-Lock 7000's timer switch button is pushed.

connect the decoder's two tester connectors to the mating VATS connectors at the base of the steering column under the dash. After turning the decoder's key code switch to 1, try to start the engine with your properly cut mechanical key. If the engine doesn't start, press the 4-minute timer on the decoder (FIG. 19-8). After 4 minutes, the timer light will go out. That lets you know the vehicle should be ready to try another key code number. Then turn the decoder's key code switch to the next number and try the mechanical key again. Follow this procedure until the vehicle starts. When the vehicle starts, transfer the cuts from your mechanically correct key to the VATS blank that corresponds to the number of the key code shown on the VATS decoder.

Many locksmiths don't appreciate VATS. They don't like having to stock 15 different expensive key blanks, or having to take so much time making a first key for a vehicle. But it's likely that VATS will be used in General Motors' vehicles for many years. Anyone who wants to service automotive locks should be prepared to handle vehicles with VATS. This means obtaining the proper tools and staying informed of changes in VATS.

## OPENING LOCKED VEHICLES

Not long ago virtually any locked vehicle could be opened by inserting a flat metal tool about 3 inches long and 1 inch wide into the door between the window and the weatherstripping, then moving it up and down and from side to side until it caught something that unlocked the door.

To reduce thefts, automobile manufacturers are now making vehicles increasing difficult to open. The flat metal tool isn't nearly as effective an automobile opening tool as it used to be. In many cases, using it can damage a vehicle's locking system in a way that prevents the key from being able to operate the locks. In response to those changes, locksmiths have developed new tools and opening techniques.

To understand how to open vehicle properly, you need to understand how vehicle locks work. When you're standing outside of a vehicle, you only see the face of the lock cylinder and the lock button on the door. For vehicle opening purposes, however, there are four important parts of a door's locking system. These are: the lock cylinder, the cylinder cam, lock rods or connecting rods, and the lock button.

The cylinder cam (also called a *pawl*) is attached to the back of the cylinder. A lock rod is connected to the pawl. The lock rod (or a connecting rod) is attached to the door's lock button. When the proper key is turned in the lock cylinder, the pawl is moved up or down—depending on the direction the key is being turned—the lock rod moves with the pawl, and the lock button moves with the lock rod.

Usually when any one of the parts is moved to it's unlocked position, all the other parts are also moved to their unlocked positions. To open a locked vehicle, you need to use a tool to move one of those parts to it's unlocked position.

In most cases a locked vehicle can be opened by doing one of the following: lifting up or pushing down on the lock pawl, pulling the lock rod up, pushing the lock rod forward, lifting the lock rod button, or picking the lock. The lock pawl can usually be manipulated with an L-shaped tool. You can make such a tool be bending a 39-inch length of $^3/_{16}$-inch-diameter steel to the dimensions shown in FIG. 19-9.

Some door locks, such as those used on vehicles made by Ford Motor Company, have rigid or fixed pawls that cannot be lifted or pushed down. You can usually open a door with that type of lock by moving the lock rod.

The lock rod is connected to the pawl, and is within the walls of a door. Some lock rods are vertical and some are horizontal. In general, if a button on the door pops up when the door is being unlocked with a key, the door has a vertical lock rod. If the door is opened from the inside by sliding a button to the left or right (horizontally), the door probably has a horizontal lock rod. Doors with vertical lock rods can usually be opened by using a tool to hook the lock rod and pull it up. Doors with horizontal lock rods can usually be opened by using a tool to hook the rod and pull it forward.

The lock rod button is on top of the lock rod that pops up or slides when the lock is opened. In older model automobiles, the vertical lock rod button often had a wide head that could be hooked and lifted with a wire such as a coat hanger. Most late-model automobiles that have vertical lock rods use lock buttons that are either tapered or that sit flush with the door when the door is locked. These are difficult to hook from the top with a piece of wire; you will have better success with the tool shown in FIG. 19-9.

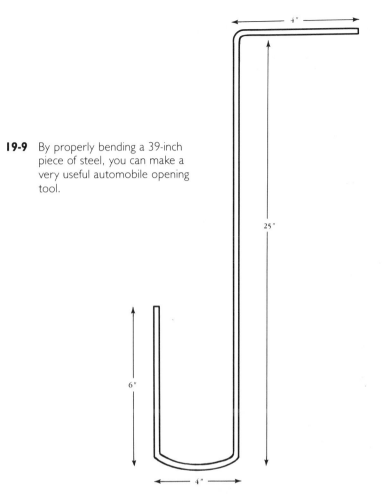

**19-9** By properly bending a 39-inch piece of steel, you can make a very useful automobile opening tool.

To use this tool to lift up a lock button, use the L end of the tool as a handle and do the following: Place a wedge between the window and weatherstripping near the center of the door. This will provide a gap into which you can insert your tool. Lower the U end of the tool below the rubber weatherstripping, and position the tip of the tool directly beneath the lock button. When the tip of the tool is touching the bottom of the lock button, lift the tool upward to unlock the door.

Wedges can be purchased from any locksmith supply house, but they are also easy to make. Use a smooth piece of plastic or wood about 4 or 5 inches long that's tapered to about $1/2$ inches thick can be used as a wedge. (Metal shouldn't be used because it can scrape paint off of a vehicle.) The narrow side of the wedge must be able to be pushed partially into the door between the window and the rubber weatherstripping; that provides the gap for the automobile opening tool. In some cases—such as when searching for a lock rod within the walls of a door—you might want to insert a thin flexible light into the gap before you insert the automobile opening tool. This can help you avoid haphazardly probing inside the door with your tool, which not only looks amateurish, but can also cause you to break a pawl, bend a rod in the door, or otherwise damage the vehicle.

Several companies publish and update manuals that provide lock opening information for hundreds of different automobile models. By using those manuals, you can quickly find out which opening methods are most efficient for the vehicle you're working on. Most publishers of those manuals, however, also sell a lot of automobile opening tools along with the manuals. In most cases knowledge is much more important for opening vehicles than are a wide assortment of tools.

A typical multi-tool automobile opening kit will contain several L tools of different sizes, several tools of different sizes for lifting a lock button rod, several tools of different sizes for pulling a lock rod, etc. But the tool shown in FIG. 19-9 can be used to quickly open virtually any vehicle you are likely to encounter.

Many publishers of automobile opening manuals will sell the manuals and tools separately. Some sell their products only to locksmiths; others also sell to other security professionals such as law enforcement officers. All-Lock Company, High Tech Tools, Pro-Lok, and Slide Lock Tool Company are publishers of some very popular automobile opening manuals.

Slide Lock Tool Company's compact manual (FIG. 19-10) shows how

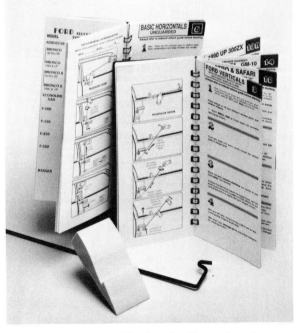

**19-10**   The basic Slide Lock Automobile Opening Kit consists of a manual, a wedge, and a Z-tool. Slide Lock Company

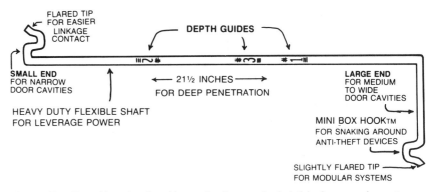

FLARED TIP
FOR EASIER
LINKAGE
CONTACT

DEPTH GUIDES

SMALL END
FOR NARROW
DOOR CAVITIES

←— 21½ INCHES —→
FOR DEEP PENETRATION

LARGE END
FOR MEDIUM
TO WIDE
DOOR CAVITIES

HEAVY DUTY FLEXIBLE SHAFT
FOR LEVERAGE POWER

MINI BOX HOOKтм
FOR SNAKING AROUND
ANTI-THEFT DEVICES

SLIGHTLY FLARED TIP ———
FOR MODULAR SYSTEMS

**19-11**  The Z-tool has depth guide marks that can be helpful when opening automobiles. Slide Lock Tool Company

to open over 620 vehicles with the company's Z-tool (FIG. 19-11). The tempered spring steel tool has a long end for wide paneled doors such as are found on large luxury cars, and a shorter end to be used on the thin-walled doors of compact and subcompact cars. The tool is very versatile because it can be bent into different configurations and then bent back into its original shape. Slide Lock Tool Company's manual has many drawings that illustrate step-by-step directions for using the tool to open many vehicles.

The tool has three numbers stamped along its shaft to aid with linkage depth locations. It is designed to hook both vertical and horizontal lock rods quickly. Because it is so thin and can be bent, the tool can also bypass most anti-theft shield guards some automobile manufacturers use to protect lock rods. Figure 19-12 shows how the Z-tool is used.

**19-12**  The Z-tool can be used to quickly open vehicles that have either horizontal or vertical lock rods.

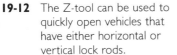

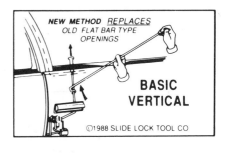

NEW METHOD *REPLACES*
OLD FLAT BAR TYPE
OPENINGS

**BASIC
VERTICAL**

©1988 SLIDE LOCK TOOL CO

OPENS ANY HORIZONTAL
LOCK SYSTEM.
ANTI THEFT GUARDED
OR NOT

**BASIC
HORIZONTAL**

©1988 SLIDE LOCK TOOL CO

Slide Lock Tool Company

# Chapter 20

# Home and office security

$S$ome people avoid taking security measures because they believe nothing can stop a determined burglar from gaining entry, or because they believe they don't own anything worth stealing. However, in many cases burglars don't break into a building because they know certain items are in it, but rather because the place is easy to enter. Unless a building has been targeted by a highly professional criminal, simple and inexpensive measures can be taken to greatly reduce the risk of an unlawful break-in.

Nearly everyone owns something a burglar will consider worth stealing. Protection of property, however, is not the only reason for taking security measures. Danger to persons inside the building is an even greater consideration. If anyone is in a building when a burglary occurs, they could be attacked or killed. And many times burglars cause a lot of damage to property.

## TYPES OF CRIMINALS

There are three basic types of criminals: the amateur, the semi-professional, and the professional. The amateur criminal is one who has little technical knowledge of locks or other security devices, and breaks into places only when there is little risk of getting caught. These types usually have a regular job, and don't steal for a living. Occasionally amateur criminals sell stolen property, but more often keep it for their own use.

The semi-professional criminal is a jack leg locksmith who steals regularly. Sometimes drug addicts become semi-professional criminals to support their addiction. Such a criminal searches for homes and offices that are easy to break into. Much of the semi-professional's income is derived from criminal activity.

Professional criminals know a great deal about locks and other security devices. They earn most or all of their income from criminal activity. The professional criminal has a good supply of tools to use for criminal purposes, and knows how to use them quickly and quietly. The professional criminal doesn't wait for opportunities to steal; they actively search for them. Their targets are properties of substantial value.

Although it's difficult to stop a professional criminal from breaking into a home or office, it's fairly easy to stop the other two types of criminals. The vast majority of burglaries are committed by amateur and semi-professional criminals. These are the types that would, for example, run into a house to grab a video cassette recorder or a compact disc player. Sometimes they will break into a place simply to vandalize it.

Simple and inexpensive security measures can be taken to dissuade the amateur and semi-professional criminal from breaking into a home or office. There are also security measures to dissuade professional criminals. These can be costly, and are usually unnecessary for homes and offices that have little valuable property in them.

## GENERAL RULES OF BUILDING SECURITY

It is neither necessary nor practical to try to completely secure every possible means of unlawful entry to a building. The goal of building security is to make attempts at unlawful entry difficult, time-consuming, and attention-getting.

Most criminals don't have a lot of knowledge about locking devices, and are apt to pass by places that are difficult for them to get into. Even criminals who know a lot about sophisticated security devices usually evaluate the risk of getting caught.

When considering the security needs of a building, also consider the needs of any adjoining building, such as a garage. Burglars don't always walk through the front door of a building. If an adjoining building looks easy to enter, often a burglar will enter it to get into a home or office.

## HOW BURGLARS ENTER BUILDINGS

Burglary statistics show that in over 98 percent of the burglaries, homes and offices were entered in one of the following ways: by going through an unlocked door or window, by kicking a door open, or by breaking or bypassing a simple lock.

Contrary to popular opinion, lock picking, key impressioning, drilling locks, and other locksmithing techniques are rarely used by burglars. Locksmithing techniques allow a person to gain entry in ways that result in as little damage as possible to a person's property, including the lock. Burglars don't care about causing damage; they only care about getting into a place as quickly as possible. That's why burglary techniques differ from locksmithing techniques.

## SPECIFIC SECURITY MEASURES

Which security measures are appropriate for a building depends on the building's location, physical characteristics, and purpose. An old barn in a remote location, for example, requires different security measures than does a home located in an urban area. Likewise, apartments within a large apartment building have different security concerns than do offices within a large office building.

One of the least expensive and most effective security measures anyone can take is to keep all doors and windows of a building locked. Many people leave their homes without locking the doors. This makes the burglar's job much easier. An unlocked home or office is too enticing for most burglars to pass by. Even if a person plans to be out of a building for a short time, the door should always be locked. Often people plan to be gone for a short period of time, but circumstances delay them from returning quickly. A few minutes is all a burglar needs to run into a building and snatch and destroy a few things.

### Doors and frames

Primary security concerns for most buildings are the doors and frames. Burglars know that most homes and offices can be entered by kicking a door in. They prefer panel doors and hollow-core doors. A typical panel door has one to eight solid wood or plywood panels; the door is stylish, but panels can be easily kicked out. Hollow-core doors, as the name implies, are hollow inside and provide very little security against kick-in attempts.

Solid-core and steel doors provide the best security against kick-in attempts. A typical solid core door consists of either a particle board core or a core made of solid blocks of laminated wood. Steel doors often have hollow cores, but the heavy-gauge steel coverings are difficult to kick through. Another secure type of door is one that has a steel covering and a solid wood interior. Such a door allows for many styling possibilities, while sacrificing little security.

To be effective, a door frame must be at least as strong as the door. When a burglar attempts to kick a door in, he doesn't care whether the door itself gives or the frame gives. The best frames are made of steel, but a wood frame can be effective if it has been properly installed.

Both the door and frame of a building can be strengthened by installing a high-security strike, a door reinforcer, and stronger hinges. (Chapter 12 explains how to properly install such devices.)

Every exterior door of a home or office should also have a wide-angle door viewer installed. The door viewer allows a person to see who's at the door without opening it. If a stranger is at the door, a person can request identification before opening it.

A door chain gives a person a false sense of security, and shouldn't be used as a substitute for a door viewer. If the person outside of a door doesn't plan to break in, a door chain isn't necessary. If he or she does plan to break in, a door chain offers little protection. Most door chains are very thin and are fastened to a door with small screws. One good kick or shoulder shove can pull a chain's holding screw out of a wall. Also, most door chains can be bypassed.

## Door locks

A door with only a key-in-knob lock on it provides very little security. As explained in Chapter 9, most can be bypassed by loiding or jimmying. But even the best key-in-knob locks can be unlocked using a hammer to knock off the knob and then using a screwdriver to withdraw the bolt from the strike. Nevertheless, many homes and offices still have exterior doors that are secured only by key-in-knob locks. Many burglars can't help noticing such doors.

A door's security can be greatly enhanced by using either a "jimmy proof" deadlock or a deadbolt. Even the least expensive models can cause a criminal to look for a less secure building to break into. If the risk of burglary is high, a high-security lock might be appropriate.

For doors that have glass in them, double cylinder locks might be helpful. Such locks prevent a burglar from sticking his hand through an opening in a door and turning the lock open. Double cylinder locks must be operated with a key to enter or exit. Such a lock is unnecessary for solid doors. During a fire, a door with a double cylinder lock can be difficult to open quickly.

Any deadbolt used should have the following features: a bolt with a hardened insert, and at least a 1-inch throw (a smaller throw is easy to jimmy); a free-spinning cylinder guard ring, to resist attempts to wrench the cylinder out of the lock; at least five sets of pin tumblers; and a pick-resistant tumbler combination. An effective tumbler combination is one in which the bottom tumbler lengths are sufficiently staggered. A lock that uses bottom pins that are all the same size or that have little size variations can opened with a pick nearly as quickly as it can be opened with a key.

You can determine a tumbler combination by examining the key for the lock. Look for staggered cut depths. If you don't like the tumbler combination of a lock, you can rekey it. (See Chapter 6.)

## Windows

Every window on the first two floors of a building should have locks. Burglars do occasionally break windows to gain entry, but they prefer not to. Breaking glass is noisy and can draw a lot of attention to a burglar. A glass cutter could be used to minimize the noise, but burglars rarely carry them. Also, glass cutters aren't as easy to properly use as many people assume.

Window glass is available in various types and thicknesses. Single-strength glass is often used; it is about .100 inch thick. The best glass to use in most cases is $7/32$-inch heavy sheet. For basement windows, it is best to use plate glass, which is thicker than heavy sheet.

Although glass is the most commonly used glazing, it isn't the most secure. A properly installed $1/2$-inch-thick sheet of acrylic plastic is practically shatter proof, and much more secure than glass.

A window's frame is also a security consideration. A jalousie window (a window with a series of glass slats fitted into metal channels) offers very little security; the glass slats can easily be removed from their channels when the window is closed. Any window frame that can be easily disassembled from outside the window shouldn't be used on the first two floors of a building.

Windows can be protected by installing bars or grates over them. Many companies manufacture highly decorative bars and grates that are easy to install. For safety reasons, such as in case of a fire, these devices should be quickly removable from inside the building.

## Lighting

One of the most cost-effective ways to protect a building from burglary is to keep the exterior well lighted at night. There should be lighting around all the building's exterior doors (front, side, back, etc.), and any adjoining building (such as a garage). Good lighting greatly increases a burglar's chances of being caught and will scare away all but the most determined of burglars.

For the lighting to be most effective, large trees and high shrubbery shouldn't be close to the building. This gives a burglar places to work without being seen.

## MAKING A SECURITY SURVEY

To make a useful security survey of a building, start by walking around the building and observing lighting, shrubbery, windows, and doors. Notice whether or not the windows and doors are strong, and whether or not they have adequate locks and hinges.

Then climb on top of the roof and look for openings that might require protection, such as skylights.

After surveying the building from the outside, go inside and carefully inspect the doors, windows, and all openings again. As you're conducting the survey, keep written notes about every item you believe could increase the building's security.

Chapter **21**

# The business of locksmithing

$A$s was explained in Chapter 1, there are many money-making opportunities for a person who has locksmithing skills. Most people begin their locksmithing career by working for a locksmithing shop. This allows a person to gain a great deal of experience, while being paid for it. Other people prefer to begin their own locksmithing business without first working for another locksmithing shop. Running your own business can give you many learning experiences, and allow you to earn a lot of money.

This chapter explains how you can start a successful locksmithing business, and how you can get a good job working for a locksmithing shop. Many related issues are also discussed, including the following: the pros and cons of joining a locksmith association, what it means to be certified, what it means to be licensed, the pros and cons of certifications and licenses, and the ethical issues of locksmithing.

## LOCKSMITHING ASSOCIATIONS

Belonging to a locksmithing association can help you improve your credibility among other locksmiths and in your community. Most of the associations offer trade journals, technical bulletins, educational classes, discounts on books and supplies, locksmith bonds, insurance, etc. They provide membership certificates you can display in your shop, and logos you can use in your advertisements. Some associations also offer certification tests.

Many successful locksmiths, however, don't belong to a locksmithing association. Some don't join because they believe the membership dues exceed the value of the services such groups provide. Other locksmiths refuse to join because they disagree with the policies or legislative activities of such organizations. The criteria for becoming a member differs among them, as do the policies, legislative activities, and membership dues. You'll need to decide for yourself whether or not joining one will be beneficial for you. Appendix F lists the addresses of many locksmithing organizations. You can contact them to obtain membership information.

Whether or not you decide to join an association, consider subscribing to a general circulation locksmith trade journal. They are only sold to locksmiths and other security professionals, but you don't need to belong to a trade association to receive them. Such publications provide a lot of current and useful technical information about a wide variety of locking devices and locksmithing tools. The addresses of two popular publications:

Locksmith Ledger International
850 Busse Highway
Park Ridge, Illinois 60068

The National Locksmith
1533 Burgundy Parkway
Streamwood, Illinois 60107

## CERTIFICATIONS

A locksmith certification signifies that a person has demonstrated a level of knowledge or proficiency that meets a school or association's criteria for being certified. Some locksmith schools and associations offer only one level of certification; others (such as Associated Locksmiths of America, Inc.) offer several levels. The significance of a certification depends on the integrity of the organization issuing it. The most respected certifications are those issued by well-respected organizations.

Locksmiths are divided on the issue of the need for certifications. Many feel they shouldn't have to meet the approval of a specific school or organization. They prefer to allow the free market determine the competency of a locksmith. An incompetent locksmith, they argue, won't be able to successfully compete against a competent one.

Others feel that locksmith certifications can benefit the locksmith industry by raising the competency level of all locksmiths, and by improving the public perception of locksmiths. Locksmiths have similar disagreements with one another about the issue of locksmith licensing.

## LICENSING

In most places, locksmiths are required only to abide by laws that apply to all businesses—zoning, taxes, building codes, etc. Some places require a person to obtain a locksmith license before offering locksmithing services. Contact your city, county, and state licensing bureaus to find out if you need to have a locksmith license.

A license differs from a certification in that a license is issued by a municipality and a certification is issued by a school or an association. Most cities and states don't offer or require locksmith licenses. Some places require only the owner, not the employees, of a locksmithing shop to be licensed.

The criteria for obtaining a locksmithing license varies from place to place. Some require an applicant to take a competency test, be finger-printed, provide photographs, and submit to a background check. Other places simply require applicants to register their name and address and pay a few dollars annually.

Some locksmiths and organizations are actively working to enact additional locksmith licensing laws throughout the United States. New bills are being proposed constantly, and many are being enacted into law. If you're a locksmith or plan to become one, you should take an interest in the current bills related to locksmiths in your area. You can write to or call your state and federal legislatures and ask to be kept informed of pro-posed bills related to locksmith licensing.

Locksmiths are divided on three major issues concerning locksmith licensing: whether or not licensing is necessary, what criteria should be required to obtain a license, and whether or not a national locksmith asso-ciation should have an integral part in qualifying all locksmiths (including nonmembers) for licensing.

The same differences of opinion occur here as with certification. Some locksmiths feel that licensing helps improve the image of the lock-smithing profession, and helps to safeguard the public from incompetent and unscrupulous locksmiths. Others believe licensing is unnecessary because current laws pertaining to all businesses adequately control lock-smiths, and that the free market effectively weeds out incompetent lock-smiths.

Proposed licensing criteria such as the licensing fees, liability insur-ance, and competency testing are also of concern to many locksmiths. The primary arguments about such matters center around whether or not those things are necessary and whether or not the cost of them will be too burdensome for locksmiths.

Some proposed locksmith licensing bills include provisions to allow certification tests offered by national locksmith trade associations to sat-isfy the locksmith competency criteria. In other words, according to some bills, a person who gets certified by a national locksmith association is to be considered competent to do locksmithing work. People in favor of such a provision believe it is in the best interest of all locksmiths for lock-smith trade associations to play an integral part in the licensing process.

Those who disagree with such a provision feel that bills authorizing an association to test applicants for a license would not benefit locksmiths who don't want to join one.

## PLANNING YOUR JOB SEARCH

Locksmithing jobs are plentiful; you should have little difficulty finding the one you want. Your job search will be most productive if you approach it in a professional manner. In a sense, when you're seeking a job you're a salesperson—you're selling yourself. You need to convince a prospective employer to hire you, whether or not the employer is currently seeking a new employee.

There are five points you should consider when planning a job search: decide where you want to work, locate potential employers, convince potential employers to meet with you, meet with potential employers, and make an employment agreement.

### Deciding where you want to work

Before contacting prospective employers, decide what type of shop you want to work in, where you want to work, and what kind of employer you want to work for.

Which cities are you willing to work in? Do you prefer to work in a large, mid-size, or small shop? Do you prefer to work for a highly experienced shop owner or for an owner who knows little about locksmithing?

You can answer those questions by considering your reasons for seeking employment. If you're primarily seeking an immediate short-term source of income, for example, your answers might be different than they would be if you were more concerned about job security or gaining useful locksmithing experience.

In a small shop you might be given a lot of responsibility, and be able to get a great deal of experience quickly. A large shop might make you feel like the lowest head on a totem pole, and you are more likely to get a lot of pressure and little respect. But a large shop might be better able to offer you more job security and a lot of opportunities for advancement.

A highly experienced locksmith can teach you a lot, but you will probably have to work a long time to learn much. Experienced locksmiths place a lot of value on their knowledge, and rarely share much of it with new employees. Often experienced locksmiths worry that if a new employee learns a lot very quickly, the employee might quit and start a competing company.

Some shop owners know very little about locksmithing. Working for them can be difficult, because they often use outdated tools, supplies, and locksmithing methods. Sometimes such an owner is a retired or wealthy person who does locksmithing as a hobby. You won't gain much useful experience working for such a person; instead, you'll probably pick up a lot of bad habits.

### Locating prospective employers

You can find prospective employers by looking through the yellow pages of telephone directories under locksmiths, and by looking through the help wanted ads in newspapers and locksmithing trade journals. You might also want to place "job wanted" ads in some locksmithing trade journals.

When creating an ad, briefly state in which cities you would like to work, your most significant qualifications (bondable, good driving record, have own tools, etc.), your name, mailing address, and telephone number. Blind advertisements—those that don't include a person's name, address, or telephone number—usually get few responses.

## Prompting prospective employers to meet with you

It's usually more advantageous for you to set up a meeting with a prospective employer than to just walk into the shop unscheduled. You might come in when the owner is very busy, which could give the person a negative impression of you. Also, it might be difficult to interest the shop owner in hiring you without any prior knowledge about you.

One way to prompt a prospective employer to meet with you is send a resume and a cover letter. It isn't always necessary to have a resume to get a locksmithing job, but it can be a highly effective selling tool. A resume allows you to project a good image of yourself to a prospective employer, and it sets the stage for your interview.

The resume is an informative document, designed to help an employer know you better. But when writing it, you need to view your resume as a document designed to promote you. Don't include everything there is to know about yourself, only information that will help an employer decide to hire you. Don't include, for example, information about being fired from a previous job. It's best to wait until you're face to face with a person before you try to explain past problems.

The resume should be neatly typed in black ink, on 20-pound bond, $8^{1}/_{2}$-×-11-inch white paper. If you can't type, write your resume by hand and have someone else type it. A resume should be kept to one page; few prospective employers like wading through long resumes.

Figure 21-1 is a sample resume. Use it as a guideline but don't be afraid to organize yours differently. There is no perfect structure for a resume; organize it in the way that allows the employer to quickly see reasons for hiring you. If your education is your strongest selling point, for example, include that information first. If your work history is your strongest selling point, include your work history first.

Include a short cover letter with your resume (FIG. 21-2). The cover letter should be directed to the owner of the shop, and should prompt him or her to review your resume. Your local library should have books on writing resumes and cover letters, which will contain more samples.

## Meeting with prospective employers

After reviewing your resume, a prospective employer will probably either call you on the telephone or write a letter. If you're not contacted within two weeks after you mailed your resume, call the owner to ask if your letter was received. If the owner has received it, request a meeting. The owner might tell you there are no job openings at the shop. In that case,

John A. Smith
123 Any Street
Any Town, Any State    01234
Ph:   012/345-6789

Job Objective:    Position as a locksmith in a large shop

CAPABILITIES:

Install, re-key, master key, and service all types of locks
Impression keys
Install and service emergency exit door hardware
Open locked automobiles quickly
Service foreign and domestic automobiles
Install a wide variety of electric and electronic security devices
Properly use most types of key duplicating and coding machines
Operate cash registers
Operate Apple and IBM computers
Accurately keep business records

Work History
1989-Present    XYZ Locksmith Shop—Locksmith
1986-89         UVW Locksmith Shop—Locksmith
1984-86         RST Locksmith Shop—Apprentice
1982-84         OPQ Department Store—Manager

Education
ABC College—Any Town, Any State—1980 majored in Business
Administration.
DEF Trade School—1981—Course in Electrical Wiring
GHI Trade School—1982—Course in Carpentry

Hobbies
Lock & Key collecting, woodworking, repairing automobiles, reading.

**21-1**   A well-designed resume can help prompt a prospective employer to meet with
you.

ask if you could still arrange a meeting so the owner can get to know you,
in the event of any future openings.

Before going to a meeting, be sure you have a positive attitude and
are fully prepared. A positive attitude requires seeing yourself as a valuable commodity. Project this image to the person you want to work for.
Do not tell an employer you "really need a job," even if the statement is
true. Few people will hire you simply because you need a job. If you seem
too needy, people will think you are incompetent.

Don't go to a meeting with the feeling that the only reason you're
there is so a prospective employer can look you over to decide if you are
hirable. You both have something to offer that the other needs. You have
your time, personality, knowledge, and skills to offer—all of which can
help the prospective employer make more money. The employer, in turn,
has knowledge, experience, and money to offer you. The purpose of the
meeting is to allow both of you to get to know each other better and to

123 Any Street
Any Town, Any State   01234
Ph:   012/345-6789
Today's Date

Ms. Mary Jones, owner
LMN Locksmith Shop
321 Another Street
Another Town, Another State   43210

Dear Ms. Jones:

I am experienced in most phases of locksmithing, and have my own service vehicle and tools. I am also bonded and have a perfect driving record. I've enclosed my resume for you to review. I'd like to meet with you to tell you more about myself and to learn more about your company.

Please call me at your earliest convenience and let me know the best time to come by your shop.

Sincerely yours,

*John A. Smith*

John A. Smith

**21-2**   A brief cover letter should be sent along with a resume.

decide if you want to work together. It might also lead to an employment agreement.

Honestly assess your strengths and weaknesses, so you'll be ready to convince the prospective employer to hire you. Consider which locksmithing tasks you're able to perform well. Also consider how well you can perform other tasks a prospective employer might want you to do, such as selling products, working with a cash register, etc. Then write a list of all your strengths, and a list of the last four places you've worked. Include dates, salaries, and the names of supervisors.

At this point, prepare explanations for any negative questions the prospective employer might ask you during the interview. If there is a gap of more than 3 months between your jobs, for example, decide how to explain that gap. If you were fired from one of your jobs or if you quit one, figure out the best way to explain difficult situations to a prospective employer and how you will avoid them in the future. Your explanations should put you in the best light possible.

Never blame other people, such as former supervisors, for problems in past jobs; this will make you seem like a crybaby or a back stabber. If there is no good explanation for why you were fired, it might be in your best interest to simply not mention that job, unless the employer is likely to find out about it anyway. Don't lie about problems you've had because a prospective employer, if he learns the truth, will probably not hire you and might tell other locksmiths about you. The locksmithing industry is fairly small and word gets around quickly.

Before the meeting learn all you can about the prospective employer and his or her shop. Stop by the shop to find out what products the company sells and what special services it offers. Call the local Better Business Bureau and Chamber of Commerce to find out how long the business has been established and how many consumer complaints have been filed against the company. Contact local, state, and national locksmithing trade associations to find out if the owner belongs to any of them.

Before the meeting also read the most current issue of a locksmithing trade journal. Study the issue to find out about major news related to the trade. Be prepared to confidently speak about such matters.

When you go to the interview wear clean clothes and be well groomed. Arrive a few minutes early. When you shake hands with the prospective employer, use a firm (but not tight) grip, look into his or her eyes, and smile. The person will probably give you a tour of the shop. That will help you assess whether or not it's a place you want to work. Allow the prospective employer to guide the meeting. If he or she offers you a cup of coffee, decline it unless the prospective employer also has one. Don't ask if you may smoke; keep your cigarettes in your pocket.

In the meeting, sit in a relaxed position and look directly at the person. Constantly looking away will make you appear insecure or dishonest. Listen intently while the employer is talking; don't interrupt. Smile a few times during the meeting. Be sure you understand a question before attempting to answer it. Whenever you have the opportunity, emphasize your strengths and what you have to offer, but don't be boastful when doing so.

If you're asked a question you feel uncomfortable answering, keep your body in a relaxed position, continue looking into the person's eyes, and answer it in the way you had planned to answer it. Don't immediately begin talking about another topic after answering the question; this will seem as if you're trying to hide something. Instead, pause after answering the question and smile at the person; he or she will then either ask you a follow up question or move on to another topic. Don't give an audible sigh of relief when you move to a new topic.

If asked about your salary requirements, don't give a figure. Instead say something like: ''I don't have a salary in mind, but I'm sure we could agree on one if we decide to work together.'' Salary is usually discussed at a second meeting. But if the prospective employer insists on discussing salary during the first meeting, let him or her state a figure first. Ask what the employer believes is a fair salary. When discussing salary always speak about ''fairness to both of us.''

The prospective employer's first offer will almost always be less than what the firm is willing to pay. You might be told that the employer doesn't know much about how well you work, so you will start at a certain salary and then will be reviewed after a short time. That might sound good, but the salary you start out with has a lot to do with how much you'll earn from the company later. You want to begin working for the highest salary possible.

If the prospective employer's salary offer is too low, you can reply that you would like to devote your full attention to the job and work hard for the company, but don't know if you could afford to do that with the salary offered. Explain that you simply want a fair salary. Pushing for the highest salary possible (within reason) will not only allow you to earn more money, but will also give the prospective employer confidence in you. You will be proving to the person that you're a good salesperson.

If the employer refuses to agree to an acceptable salary, tell him or her you need to think about the offer. Then smile, stand up, shake hands, and leave. If the person demands an immediate answer, say you want to discuss the matter with someone else, such as your spouse.

Don't get pressured into accepting an unreasonable employment agreement. It's in your best interest to give yourself time to think the matter over, and to meet with other prospective employers. If no one is willing to make a reasonable employment agreement with you, consider starting your own business.

## STARTING YOUR OWN BUSINESS

As explained in Chapter 1, there are two basic types of locksmithing shops: store front and mobile. Usually it's less expensive to begin a mobile shop than it is to establish a store front shop. For either type to be successful, you need to have sufficient working capital, obtain the tools and supplies you need, and make good marketing decisions.

Before starting a business you should seek professional assistance. Business mistakes are easy to make and can be very costly to you. First meet with an attorney, who can help you decide which organizational structure (sole proprietorship, corporation, etc.) is best for your business. An attorney can also inform you about zoning, licensing, and other legal matters.

After you've drawn up a business plan—which should include estimates about how much money you need to begin your business, and how much money you anticipate earning—meet with an accountant to review it. He or she can objectively assess your plan and offer suggestions for improving it. The accountant can also show you how to keep accurate business records.

You'll also need to meet with an insurance agent to figure out how much and which types of insurance you need. You'll probably want fire and theft insurance for your inventory and equipment. You might also want product and liability insurance.

### Tools

Chapter 10 provides a list of the tools and supplies you'll need to perform most locksmithing services. If you plan to offer limited services, you won't need to obtain all of them. Before purchasing tools, carefully decide which ones you need, and where you can get them at the best prices. Look through several catalogs to compare prices and quality.

## Telephones

Whether you have a mobile shop or a store front shop, you'll need at least one telephone. People who are locked out of their homes or cars don't walk to a locksmithing shop; they use the telephone. If they get a busy signal, they usually call another shop. That's why the telephone you use for your business should usually be free to accept incoming calls. If you're using a single telephone for both business and personal purposes, you might want to get "call waiting" service.

You could miss important calls if no one is available to answer the telephone 24 hours a day. Some people use an answering machine. However, an answering service is a better alternative. A person in an emergency situation probably won't leave a message on a machine; they will just call another locksmith.

## Advertising

Most locksmithing shops don't spend a lot of money on advertising. You might want to test various forms of advertising (newspapers, magazines, radio, etc.) to find out which ones work best for your shop. The yellow pages of a local telephone directory and word of mouth are usually the most cost-effective forms of advertising for locksmiths. Contact the telephone company to find out the deadline for placing an advertisement in the next directory.

Look through the "Locksmiths" heading in the telephone directory to see the advertisements of other shops. That will give you an idea of how you should organize your ad and how large it should be. Unless you live in a large city, you probably won't need to have a full-page advertisement.

## Pricing

A common mistake made by people who begin a locksmithing business is to charge too little for their products and services. They believe low prices will provide a competitive edge over other locksmithing businesses. But what usually happens is a new locksmithing business fails because it doesn't make enough profit.

Don't be afraid to charge fair prices for your products and services. Compete with other locksmiths on the basis of superior service and better customer relations, not on the basis of prices.

## Service vehicle

Whether you have a store front shop or a mobile shop, you'll need a service vehicle (FIG. 21-3). Most locksmiths use a van for that purpose, but a truck or a car can also be used. You'll need to properly organize your service vehicle so you can work efficiently in it (FIG. 21-4). Several companies manufacture interior units to help you better use the space in a service vehicle.

If you don't want to have your shop's name, telephone number, and address on the service vehicle, get a magnetic sign with that information.

21-3   Most locksmithing shops use a van for a service vehicle.

21-4   A well-organized service vehicle is easy to work in.

## Store front location

If you decide to operate from a store front, location will be very important. A building that many people regularly pass by will provide you with the opportunity to attract a lot of walk-in customers. People often walk into locksmithing shops on impulse, but they seldom spend much time trying to locate a particular shop.

Keep the shop clean and organize it in such a way that people will want to come in browse. Good lighting, comfortable air temperature, and attractive displays (FIG. 21-5) prompt people to come into a shop.

**21-5** Attractive displays prompt people to come into a store front to browse.

## ETHICS OF LOCKSMITHING

To develop and maintain good relationships with other locksmiths and local law enforcement agents, you should strictly adhere to the following rules of ethics:

- Always safeguard your locksmithing tools.
- Never duplicate a master key or a key stamped "Do not duplicate" for anyone without first obtaining written permission from a person who has legal control of the key.
- Never unlock a vehicle or building without first assuring yourself that the person hiring you to unlock it is authorized to do so.

- Keep a written record of the model, year, and license plate number of every vehicle you are hired to unlock. Also keep a written record of the name and driver's license number of every person who hired you to unlock a vehicle.

- Keep a written record of the address of every building you have been hired to unlock. Also keep a written record of the name and driver's license (or number from other positive identification) of every person who hired you to unlock a building.

- Don't perform locksmithing services that you're not qualified to perform.

- Perform each locksmithing job to the best of your ability, regardless of the price you're charging the customer.

*Appendix* **A**

# Lockset
# function charts

| Schlage Number | A. N. S. I. No. | Grade | |
|---|---|---|---|

## Non-Keyed Locks *ANSI A156.2 Series 4000*

| A10S | F75 | 2 |
| D10S■▲ | | 1 |
| F10N | | 2 |

**Passage Latch:** Both knobs always unlocked.

| D12D | F89 | 1 |

**Exit Lock:** Unlocked by knob inside only. Outside knob always fixed.

| A20S | | |

**Closet Latch:** Outside knob and inside thumbturn are always unlocked.

| A25D | | |
| D25D■ | | |

**Exit Lock:** Blank plate outside. Inside knob always unlocked. (Specify door thickness, 1⅜" or 1¾".)

| A30D | F77 | 2 |
| D30D | | 1 |
| F30N | | 2 |

**Patio Lock:** Push-button locking. Turning inside knob releases button. Closing door on A & D Series also releases button.

| A40S | F76 | 2 |
| D40S■▲ | | 1 |
| F40N | | 2 |

**Bath/Bedroom Privacy Lock:** Push-button locking. Can be opened from outside with small screwdriver or flat narrow tool. Turning inside knob releases push-button. Closing door on A, C and D Series also releases button, preventing lock-out.

| A43D | F79 | 2 |

**Communicating Lock:** Turn button in outer knob locks and unlocks knob and inside thumbturn.

| A44S | | |
| D44S | | |

**Hospital Privacy Lock:** Push-button locking. Unlocked from outside by turning emergency turn-button. Rotating inside knob or closing door releases inside button.

| A170 | | |
| D170■ | | |
| F170N | | |

**Single Dummy Trim:** Single dummy trim for one side of door. Used for door pull or as matching inactive trim.

| Schlage Number | A. N. S. I. No. | Grade | |
|---|---|---|---|

## Keyed Locks

| F51N | F81 | 2 |

**Entrance Lock:** Unlocked by key from outside when outer knob is locked by turn-button in inside knob. Inside knob always unlocked.

| D50PD■ | F82 | 1 |
| (ATH & OLY designs only) | | |

**Entrance/Office Lock:** Push button locking. Pushing button locks outside lever until unlocked with key or by turning inside lever.

## Keyed Locks

| A53PD | F81 | 2 |
| D53PD▲ | F82 | 1 |

**Entrance Lock:** Turn/Push button locking. Pushing and turning button locks outside knob requiring use of key until button is manually unlocked. Push button locking: Pushing button locks outside knob until unlocked by key or by turning inside knob.

| A55PD | F92 | 2 |
| D55PD | | 1 |

**Service Station Lock:** Unlocked by key from outside when outer knob is locked by universal button in inside knob. Closing door releases button. Outside knob may be fixed by rotating universal button.

| D60PD■ | F88 | 1 |

**Vestibule Lock:** Unlocked by key from outside when outside knob is locked by key inside knob. Inside knob is always unlocked.

| D66PD■† | F91 | 1 |

**Store Lock:** Key in either knob locks or unlocks both knobs.

| A70PD | F84 | 2 |
| D70PD■▲ | | 1 |

**Classroom Lock:** Outside knob locked and unlocked by key. Inside knob always unlocked.

| D72PD† | F80 | 1 |

**Communicating Lock:** Key in either knob locks or unlocks each knob independently.

| A73PD | F90 | 2 |
| D73PD■ | | 1 |

**Dormitory Lock:** Locked or unlocked by key from outside. Push-button locking from inside. Turning inside knob or closing door releases button.

| D76PD | F85 | 1 |

**Classroom Hold-Back Lock:** Outside knob locked or unlocked by key. Inside knob always unlocked. Latch may be locked in retracted position by key.

| A79PD | | |

**Communicating Lock:** Locked or unlocked by key from outside. Blank plate inside.

| A80PD | F86 | 2 |
| D80PD■▲ | | 1 |
| F80N | | 2 |

**Storeroom Lock:** Outside knob fixed. Entrance by key only. Inside knob always unlocked.

| D82PD† | F87 | 1 |

**Institution Lock:** Both knobs fixed. Entrance by key in either knob.

| A85PD | F93 | 2 |
| D85PD | | 1 |

**Hotel-Motel Lock:** Outside knob fixed. Entrance by key only. Push-button in inside knob activates visual occupancy indicator, allowing only emergency masterkey to operate. Rotation of inside spanner-button provides lockout feature by keeping indicator thrown.

■ Available functions for Athens and Olympiad designs.
▲ Available functions for "C" Series non-ferrous locks.

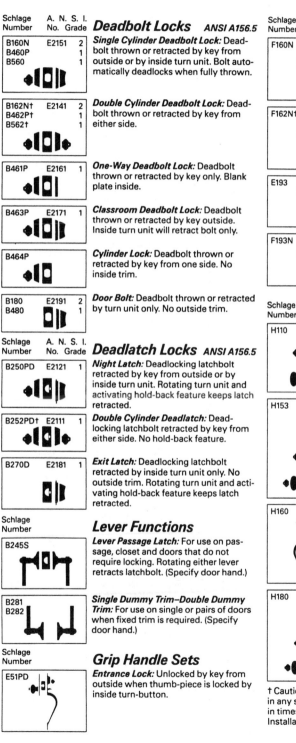

| Schlage Number | A.N.S.I. No. | Grade |
|---|---|---|

## Deadbolt Locks  ANSI A156.5

**B160N** / **B460P** / **B560** — E2151 (2, 1, 1)
**Single Cylinder Deadbolt Lock:** Deadbolt thrown or retracted by key from outside or by inside turn unit. Bolt automatically deadlocks when fully thrown.

**B162N†** / **B462P†** / **B562†** — E2141 (2, 1, 1)
**Double Cylinder Deadbolt Lock:** Deadbolt thrown or retracted by key from either side.

**B461P** — E2161 (1)
**One-Way Deadbolt Lock:** Deadbolt thrown or retracted by key only. Blank plate inside.

**B463P** — E2171 (1)
**Classroom Deadbolt Lock:** Deadbolt thrown or retracted by key outside. Inside turn unit will retract bolt only.

**B464P**
**Cylinder Lock:** Deadbolt thrown or retracted by key from one side. No inside trim.

**B180** / **B480** — E2191 (2, 1)
**Door Bolt:** Deadbolt thrown or retracted by turn unit only. No outside trim.

## Deadlatch Locks  ANSI A156.5

**B250PD** — E2121
**Night Latch:** Deadlocking latchbolt retracted by key from outside or by inside turn unit. Rotating turn unit and activating hold-back feature keeps latch retracted.

**B252PD†** — E2111
**Double Cylinder Deadlatch:** Deadlocking latchbolt retracted by key from either side. No hold-back feature.

**B270D** — E2181
**Exit Latch:** Deadlocking latchbolt retracted by inside turn unit only. No outside trim. Rotating turn unit and activating hold-back feature keeps latch retracted.

## Lever Functions

**B245S**
**Lever Passage Latch:** For use on passage, closet and doors that do not require locking. Rotating either lever retracts latchbolt. (Specify door hand.)

**B281** / **B282**
**Single Dummy Trim–Double Dummy Trim:** For use on single or pairs of doors when fixed trim is required. (Specify door hand.)

## Grip Handle Sets

**E51PD**
**Entrance Lock:** Unlocked by key from outside when thumb-piece is locked by inside turn-button.

| Schlage Number | | |
|---|---|---|

## Grip Handle Sets

**F160N**
**Entrance Lock:** Deadbolt thrown or retracted by key from outside or by inside turn unit Latch retracted by thumbpiece from outside or by inside knob.

**F162N†**
**Double Cylinder Entrance Lock:** Deadbolt thrown or retracted by key from either side. Latch retracted by thumbpiece from outside or by inside knob.

**E193**
**Outside and Inside Dummy Trim:** For use as door pull or as dummy trim on an inactive pair of doors. Fixed thumbpiece and inside knob. Thru bolted dummy cylinder.

**F193N**
**Outside and Inside Dummy Trim:** For use as door pull or as dummy trim on inactive leaf of pair of doors. Fixed thumbpiece and inside knob. Dummy cylinder with inside plate.

## Interconnected Locks  ANSI A156.12

| Schlage Number | A.N.S.I. No. | Grade |
|---|---|---|

**H110** — F95 (4)
**Entrance—Single Locking:** Deadbolt thrown or retracted by key in upper lock from outside or by inside turn unit. Latchbolt retracted by knob from either side. Turning inside knob retracts deadbolt and latchbolt simultaneously for immediate exit.

**H153** — F97 (4)
**Entrance—Double Locking:** Deadbolt thrown or retracted by key in upper lock from outside or by inside turn unit. Deadlatch retracted by key in outer knob when locked by pushing turn-button in inner knob. Outer knob may be fixed in locked position by rotating turn-button. Inside knob retracts deadbolt and deadlatch simultaneously for immediate exit.

**H160** — F102 (4)
**Entrance, Single Locking Decorative Grip Handle Trim Outside:** Deadbolt thrown or retracted by key in upper lock from outside or by inside turn unit. Latchbolt retracted by grip outside or knob from inside. Turning inside knob retracts deadbolt and latchbolt simultaneously for immediate exit.

**H180**
**Storeroom Lock:** Bolt may be operated by key from outside or by turn unit from inside. Bolt automatically deadlocks when fully thrown. Lock may be opened by key from outside. Inside knob will retract both latch and deadbolt. Latch automatically deadlocks when door is closed, inside knob always free for immediate exit. Outer knob always fixed.

† Caution: Double cylinder locks on residences and any door in any structure which is used for egress are a safety hazard in times of emergency and their use is not recommended. Installation should be in accordance with existing codes only.

## Interconnected Locks

ANSI A156.12

| Schlage Number | A. N. S. I. No. | Grade |
|---|---|---|
| H185 | F100 | 4 |

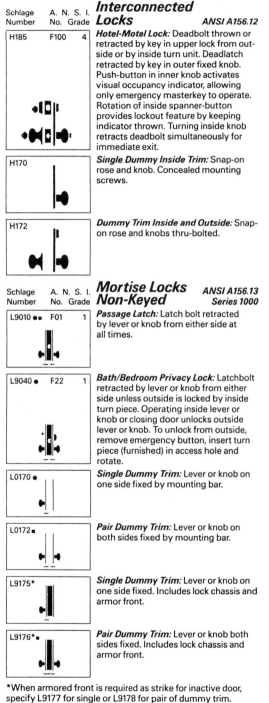

**Hotel-Motel Lock:** Deadbolt thrown or retracted by key in upper lock from outside or by inside turn unit. Deadlatch retracted by key in outer fixed knob. Push-button in inner knob activates visual occupancy indicator, allowing only emergency masterkey to operate. Rotation of inside spanner-button provides lockout feature by keeping indicator thrown. Turning inside knob retracts deadbolt simultaneously for immediate exit.

| H170 | | |

**Single Dummy Inside Trim:** Snap-on rose and knob. Concealed mounting screws.

| H172 | | |

**Dummy Trim Inside and Outside:** Snap-on rose and knobs thru-bolted.

## Mortise Locks Non-Keyed

ANSI A156.13
Series 1000

| Schlage Number | A. N. S. I. No. | Grade |
|---|---|---|
| L9010 | F01 | 1 |

**Passage Latch:** Latch bolt retracted by lever or knob from either side at all times.

| L9040 | F22 | 1 |

**Bath/Bedroom Privacy Lock:** Latchbolt retracted by lever or knob from either side unless outside is locked by inside turn piece. Operating inside lever or knob or closing door unlocks outside lever or knob. To unlock from outside, remove emergency button, insert turn piece (furnished) in access hole and rotate.

| L0170 | | |

**Single Dummy Trim:** Lever or knob on one side fixed by mounting bar.

| L0172 | | |

**Pair Dummy Trim:** Lever or knob on both sides fixed by mounting bar.

| L9175* | | |

**Single Dummy Trim:** Lever or knob on one side fixed. Includes lock chassis and armor front.

| L9176* | | |

**Pair Dummy Trim:** Lever or knob both sides fixed. Includes lock chassis and armor front.

*When armored front is required as strike for inactive door, specify L9177 for single or L9178 for pair of dummy trim. Specify door hand.

## Keyed Locks

| Schlage Number | A. N. S. I. No. | Grade |
|---|---|---|
| L9050 | F04 | 1 |

**Office and Inner Entry Lock:** Latchbolt retracted by lever or knob from either side unless outside is made inoperative by key outside or by rotating inside turn piece. When outside is locked, latchbolt is retracted by key outside or by lever or knob inside. Outside lever or knob remains locked until thumbturn is returned to vertical or by counter clockwise rotation of key. Auxiliary latch deadlocks latchbolt when door is closed.

| L9060 | F09 | 1 |

**Apartment Entrance Lock:** Latchbolt retracted by lever or knob from either side unless outside is locked by key from inside. When locked, latchbolt retracted by key outside or lever or knob inside. Auxiliary latch deadlocks when door is closed.

| L9070 | F05 | 1 |

**Classroom Lock:** Latchbolt retracted by lever or knob from either side unless outside is locked by key. Unlocked from outside by key. Inside lever or knob always free for immediate exit. Auxiliary latch deadlocks latchbolt when door is closed.

| L9080 | F07 | 1 |

**Storeroom Lock:** Latchbolt retracted by key outside or by lever or knob inside. Outside lever or knob always inoperative. Auxiliary latch deadlocks latchbolt when door is closed.

| L9080EL | | |

**Storeroom Lock:** Electrically locked. Outside lever or knob continuously locked by 24V AC or DC. Latchbolt retracted by key outside or by lever or knob inside. Switch or power failure allows outside lever or knob to retract latchbolt. Auxiliary latch deadlocks latchbolt when door is closed. Inside lever or knob always free for immediate exit.

| L9080EU | | |

**Storeroom Lock:** Electrically unlocked. Outside lever or knob unlocked by 24V AC or DC. Latchbolt retracted by key outside or lever or knob inside. Auxiliary latch deadlocks latchbolt when door is closed. Inside lever or knob always free for immediate exit.

| L9082 | | |

**Institution Lock:** Latchbolt retracted by key from either side. Lever or knob on both sides always inoperative. Auxiliary latch deadlocks latchbolt when door is closed.

| L9453 | F20 | 1 |

**Entrance Lock:** Latchbolt retracted by lever or knob from either side unless outside is locked by 20° rotation of thumbturn. Deadbolt thrown or retracted by 90° rotation of thumbturn. When locked, key outside or lever or knob inside retracts deadbolt and latchbolt simultaneously. Outside lever or knob remains locked until thumbturn is restored to vertical position. Throwing deadbolt automatically locks outside lever or knob. Auxiliary latch deadlocks latchbolt when door is closed.

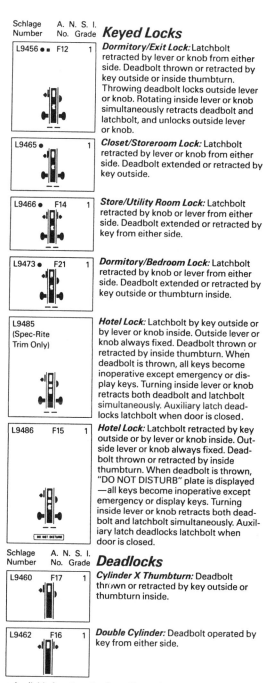

| Schlage Number | A. N. S. I. No. | Grade |
|---|---|---|
| L9456 ●■ | F12 | 1 |

## Keyed Locks

**Dormitory/Exit Lock:** Latchbolt retracted by lever or knob from either side. Deadbolt thrown or retracted by key outside or inside thumbturn. Throwing deadbolt locks outside lever or knob. Rotating inside lever or knob simultaneously retracts deadbolt and latchbolt, and unlocks outside lever or knob.

**L9465 ●  1**

**Closet/Storeroom Lock:** Latchbolt retracted by lever or knob from either side. Deadbolt extended or retracted by key outside.

**L9466 ●  F14  1**

**Store/Utility Room Lock:** Latchbolt retracted by knob or lever from either side. Deadbolt extended or retracted by key from either side.

**L9473 ●  F21  1**

**Dormitory/Bedroom Lock:** Latchbolt retracted by knob or lever from either side. Deadbolt extended or retracted by key outside or thumbturn inside.

**L9485** (Spec-Rite Trim Only)

**Hotel Lock:** Latchbolt by key outside or by lever or knob inside. Outside lever or knob always fixed. Deadbolt thrown or retracted by inside thumbturn. When deadbolt is thrown, all keys become inoperative except emergency or display keys. Turning inside lever or knob retracts both deadbolt and latchbolt simultaneously. Auxiliary latch deadlocks latchbolt when door is closed.

**L9486  F15  1**

**Hotel Lock:** Latchbolt retracted by key outside or by lever or knob inside. Outside lever or knob always fixed. Deadbolt thrown or retracted by inside thumbturn. When deadbolt is thrown, "DO NOT DISTURB" plate is displayed—all keys become inoperative except emergency or display keys. Turning inside lever or knob retracts both deadbolt and latchbolt simultaneously. Auxiliary latch deadlocks latchbolt when door is closed.

| Schlage Number | A. N. S. I. No. | Grade |
|---|---|---|

## Deadlocks

**L9460  F17  1**

**Cylinder X Thumbturn:** Deadbolt thrown or retracted by key outside or thumbturn inside.

**L9462  F16  1**

**Double Cylinder:** Deadbolt operated by key from either side.

● Available functions for Spec-Rite series.
■ Available functions for grip handle sets.

---

| Schlage Number | A. N. S. I. No. | Grade |
|---|---|---|
| L9463 | | 1 |

## Keyed Locks

**Classroom Lock:** Deadbolt thrown or retracted by key from outside. Inside cylinder turn retracts deadbolt but cannot project it.

**L9464  F18  1**

**Cylinder Lock:** Deadbolt thrown or retracted by key from one side. No trim on opposite side.

**Schlage Number**

**TH885  TL8485**

## Intellis® Locks

**Hotel Lock:** Latchbolt unlocked by key card outside or by lever or knob inside. Outside lever or knob always rotates but does not retract latchbolt until activated by key card. Deadbolt thrown or retracted by inside thumbturn. When thrown, all key cards become inoperative except key cards encoded with the deadbolt override feature. Display light indicates that deadbolt is thrown when key card is passed through lock. Turning inside lever or knob retracts both deadbolt and latchbolt simultaneously. Auxiliary latch deadlocks latchbolt when door is closed.

**TH880  TL8080**

**Storeroom Lock:** Outside lever or knob always locked. Key card unlocks lever or knob for one entry. Inside lever or knob always free for immediate exit. Auxiliary latch deadlocks latchbolt when door is closed.

**General Purpose/Classroom Lock*:** Latchbolt retracted by lever or knob from either side unless outside is locked by key card. Unlocked from outside by key card. Inside lever or knob always free for immediate exit. Auxiliary latch deadlocks latchbolt when door is closed.

**All "T" Series Locks**

**Passage Latch*:** Latchbolt is retracted by lever or knob from either side at all times until reprogrammed with a key card.

**Maximum Security Lock*:** Outside knob or lever always locked. Two key cards necessary for entry. Inside lever or knob always free for immediate exit. Auxiliary latch deadlocks latchbolt when door is closed.

**All "T" Series Locks When Suffix 5 Is Specified**

**"Smart" Lock*:** Unlocks/relocks twice in a 24 hour period. Programmable for a 7 day cycle. Eliminates need for manually unlocking and relocking at specific times. Inside lever or knob always free for immediate exit. Auxiliary latch deadlocks latchbolt when door is closed.

*These functions are field programmable. Consult the Intellis "User Reference Manual," or contact your Intellis representative.

*Appendix* **B**

# Comparative key blank list

| AMERICAN | ESP |
|---|---|
| 1J | AM1 |
| 2J | AM3 |

| ARCO | ESP |
|---|---|
| 30 | R30 |

| ARROW | ESP |
|---|---|
| K2 | AR1 |
| K2C | AR5 |
| K7 | AR4 |

| BEST | ESP |
|---|---|
| 1A1A1 | BE2 |
| 1A1A1 | BE2-A |
| 1A1B1 | BE2-B |
| 1A1C1 | BE2-C |
| 1A1D1 | BE2-D |
| 1A1DD1 | BE2-DD |
| 1A1E1 | BE2-E |
| 1A1F1 | BE2-F |
| 1A1G1 | BE2-G |
| 1A1H1 | BE2-H |
| 1A1J1 | BE2-J |
| 1A1K1 | BE2-K |
| 1A1L1 | BE2-L |
| 1A1M1 | BE2-M |
| 1A1N1 | BE2-N |
| 1A1Q1 | BE2-Q |

| BRIGGS & STRATTON | ESP |
|---|---|
| 32318 | B10 |
| 32319 | B11 |
| 32671 | Y152 |
| 32725 | Y152 |
| 32838 | H27 |
| 32839 | H26 |
| 32842 | H27 |
| 32849 | H26 |
| 32900 | B24 |
| 32903 | Y152 |
| 32915 | Y152 |
| 42107 | B4 |
| 75101 | B1 |
| 75102 | B2 |
| 320032 | H26 |
| 320045 | B5 |
| 320085 | H26 |
| 320158 | H50 |
| 320181 | H27 |
| 320259 | Y149 |
| 320296 | B40 |
| 320297 | B41 |
| 320321 | H50 |

| BRIGGS & STRATTON | ESP |
|---|---|
| 320368 | B42 |
| 320369 | B43 |
| 320380 | Y152 |
| 320381 | Y149 |
| 320404 | B44 |
| 320405 | B45 |
| 320433 | Y149 |
| 320434 | Y152 |
| 320509 | RA4 |
| 320510 | RA3 |
| 320588 | B48 |
| 320589 | B49 |
| 320652 | B50 |
| 320653 | B51 |
| 321355 | B46 |
| 321356 | B47 |
| GAS | GAS |

| CHICAGO | ESP |
|---|---|
| K4 | CG1 |
| K4L | CG2 |
| K4R | CG3 |
| K4W | CG6 |
| K5 | CG16 |

| COLE NATIONAL | ESP |
|---|---|
| AM1 | AM1 |
| AM4 | AM3 |
| AR1 | AR1 |
| AR4 | AR4 |
| AR5 | AR5 |
| B1 | B1 |
| B2 | B2 |
| B4 | B4 |
| B5 | B5 |
| B10 | B10 |
| B11 | B11 |
| B24 | B24 |
| B40 | B40 |
| B41 | B41 |
| B42 | B42 |
| B43D | B43 |
| B44E | B44 |
| B45H | B45 |
| B46J | B46 |
| B47K | B47 |
| B48A | B48 |
| B49B | B49 |
| B50C | B50 |
| B51D | B51 |
| BE2 | BE2 |
| BN1 | BN1 |

| COLE NATIONAL | ESP |
|---|---|
| BN7 | BN7 |
| BO1 | HL2 |
| CG1 | CG1 |
| CG2 | CG2 |
| CG4 | CG3 |
| CG6 | CG6 |
| CG15 | CG16 |
| CO1 | CO1 |
| CO2 | CO2 |
| CO3 | CO3 |
| CO4 | CO4 |
| CO5 | CO5 |
| CO6 | CO6 |
| CO7 | CO7 |
| CO10 | CO10 |
| CO26 | CO26 |
| CO34 | CO34 |
| CO35 | CO35 |
| CO36 | CO36 |
| CO44 | CO44 |
| CO45 | CO45 |
| CO62 | CO62 |
| CO65 | CO87 |
| CO66 | CO66 |
| CO67 | CO97 |
| CO68 | CO68 |
| CO89 | CO89 |
| CO90 | CO90 |
| CO91 | CO91 |
| CO92 | CO92 |
| CO94 | CO94 |
| CO95 | CO95 |
| CO105 | CO106 |
| CO105 | HL1 |
| DE6 | DE6 |
| DE8 | DE8 |
| EA1 | EA1 |
| EA27 | EA27 |
| EL3 | EL3 |
| EL10 | EL10 |
| EL11 | EL11 |
| EN1 (U2) | UN3 |
| EN2 (FO272) | UN18 |
| EN11 (FO248) | UN16 |
| EN13 (FO249) | YS2 |
| EN15 | MG1 |
| ER1 | ER1 |
| FR2 (FO373) | FT36 |
| FR2 | PP10 |
| FR2 | RN11 |
| GE1 (FO417) | V26K |
| GE11 (FO466) | V28M |
| GE12 (FO467) | V29R |

| COLE NATIONAL | ESP |
|---|---|
| GE21 | YS1 |
| GE24 (FO460) | MB33 |
| GE25 (FO461) | MB34 |
| GE26 | PA2 |
| GE30 (FO468) | V30F |
| GE30 (FO468) | V30R |
| GE31 (FO394) | OP19 |
| GE34 | PA6 |
| GE57 (FO383) | OP11 |
| H7 | FC2 |
| H26 | H26 |
| H27 | H27 |
| H32 | H50 |
| H33 | H51 |
| HO3 | HO3 |
| IN1 | IN1 |
| IN3 | IN3 |
| IN8 | IN8 |
| IN8 | ES8 |
| IN10 | IN28 |
| IN21 | IN18 |
| IT3 | FT37 |
| IT4 | FT38 |
| IT6 | WS2 |
| JA1 | GAS |
| JA10 | MZ12 |
| JA11 (FO469) | DA22 |
| JA11 (FO469) | SR1 |
| JA62 (FO444) | TR23 |
| JA63 (FO454) | TR20 |
| JA64 | TR25 |
| JA65 | MZ9 |
| JA72 | MZ10 |
| JA73 | DA21 |
| JA77 | HD69 |
| JA79 | FC5 |
| K2 | K2 |
| 54KL | L37 |
| KW1 | KW1 |
| L1 | IN33 |
| L1 | L1 |
| L4 | L4 |
| L5 | L5 |
| M1 | M1 |
| M2 | M2 |
| M3 | M4 |
| M10 | M10 |
| M11 | M11 |
| M12 | M12 |
| NA5 | NA6 |
| NA12 | NA12 |
| NA13 | NA24 |
| NH1 | NH1 |

| COLE NATIONAL | ESP |
|---|---|
| RA3 | RA3 |
| RA4 | RA4 |
| RO1 | RO1 |
| RO4 | RO4 |
| RU1 | RU1 |
| RU2 | RU2 |
| RU4 | RU4 |
| RU16 | RU16 |
| RU18 | RU18½ |
| RU19 | RU19 |
| RU20 | RU45 |
| RU21 | RU46 |
| S1 | S1 |
| S3 | S3 |
| S4 | S4 |
| S6 | S6 |
| S16 | S16 |
| S22 | S22 |
| S31 | S31 |
| S32 | S32 |
| S43 | S68 |
| S44 | S69 |
| SC1 | SC1 |
| SC4 | SC4 |
| SC6 | SC6 |
| SC7 | SC7 |
| SC8 | SC8 |
| SC9 | SC9 |
| SC10 | SC10 |
| SC22 | SC22 |
| SE1 | SE1 |
| SK1 | SK1 |
| SW5 (FO418) | VL3 |
| SW12 | VL5 |
| SW13 | VL4 |
| T4 | T7 |
| WE1 | WE1 |
| WK1 | WK2 |
| WR2 | WR3 |
| WR5 | WR4 |
| XL7 | XL7 |
| Y1 | Y1 |
| Y2 | Y2 |
| Y3 | Y3 |
| Y4 | Y4 |
| Y11 | Y11 |
| Y12 | Y12 |
| Y14 | Y14 |
| Y52 | Y52 |
| Y54 | Y54 |
| Y103 | Y103 |
| Y104 | Y104 |
| Y107 | DC2 |
| Y145 | Y220 |

| COLE NATIONAL | ESP |
|---|---|
| Y149 | Y149 |
| Y152 | Y152 |

| CORBIN | ESP |
|---|---|
| X1-27-5 | CO3 |
| X1-27-6 | CO4 |
| X1-57A1-6 | CO35 |
| X1-57A2-6 | CO45 |
| X1-57B1-6 | CO44 |
| X1-57B2-6 | CO34 |
| X1-57D1-6 | CO57 |
| X1-57D2-6 | CO58 |
| Z1-59A1-6 | CO91 |
| Z1-59A2-6 | CO92 |
| Z1-59B1-6 | CO94 |
| Z1-59B2-6 | CO95 |
| Z1-59C1-6 | CO98 |
| Z1-59C2-6 | CO99 |
| Z1-59D1-6 | CO51 |
| Z1-59D2-6 | CO52 |
| Z1-059AB-5 | CO90 |
| Z1-059AB-6 | CO89 |
| Z1-60-5 | CO87 |
| Z1-60-6 | CO66 |
| X1-67-6 | CO36 |
| Z1-67-5 | CO7 |
| X1-77-6 | CO62 |
| Z1-80-5 | CO97 |
| X1-97-5 | CO5 |
| X1-97-6 | CO6 |
| X1-99-5 | CO1 |
| X1-99-6 | CO2 |
| 5865 JVR | CO68 |
| 8658 JVR | CO26 |
| 8618C-CR | CO10 |
| 8618C-R14 | CO106 |

| CURTIS | ESP |
|---|---|
| AA2 | AA2 |
| AM1 | AM1 |
| AM3 | AM3 |
| AR1 | AR1 |
| AR4 | AR4 |
| AR5 | AR5 |
| B1 | B1 |
| B2 | B2 |
| B4 | B4 |
| B5 | B5 |
| B10 | B10 |
| B11 | B11 |
| B24 | B24 |
| B40 | B40 |
| B41 | B41 |

| CURTIS | ESP |
|---|---|
| B42 | B42 |
| B43 | B43 |
| B44 | B44 |
| B45 | B45 |
| B46 | B46 |
| B47 | B47 |
| B48 | B48 |
| B49 | B49 |
| B50 | B50 |
| B51 | B51 |
| BE2 | BE2 |
| BM3 | BM3 |
| BN1 | BN1 |
| BN7 | BN7 |
| BO1 | HL2 |
| CG1 | CG1 |
| CG2 | CG2 |
| CG4 | CG3 |
| CG6 | CG6 |
| CG16 | CG16 |
| CO1 | CO1 |
| CO2 | CO2 |
| CO3 | CO3 |
| CO4 | CO4 |
| CO5 | CO5 |
| CO6 | CO6 |
| CO7 | CO7 |
| CO10 | CO10 |
| CO26 | CO26 |
| CO35 | CO35 |
| CO36 | CO36 |
| CO44 | CO44 |
| CO45 | CO45 |
| CO62 | CO62 |
| CO68 | CO68 |
| CO87 | CO87 |
| CO89 | CO89 |
| CO91 | CO91 |
| CO92 | CO92 |
| CO94 | CO94 |
| CO95 | CO95 |
| CO97 | CO97 |
| CO98 | CO98 |
| CO102 | CO99 |
| CO106 | CO106 |
| CO106 | HL1 |
| DA21 | DA21 |
| DA22 | DA22 |
| DA23 | DA23 |
| DA24 | DA24 |
| DC2 | DC2 |
| DE6 | DE6 |
| DE8 | DE8 |
| EA1 | EA1 |

| CURTIS | ESP |
|---|---|
| EA27 | EA27 |
| EL3 | EL3 |
| EL3 | EL10 |
| EL11 | EL11 |
| ER1 | ER1 |
| FC2 | FC2 |
| FC5 | FC5 |
| FC6 | FC6 |
| FC7 | FC7 |
| FT36 | FT36 |
| FT37 | FT37 |
| FT38 | FT38 |
| H26 | H26 |
| H27 | H27 |
| H50 | H50 |
| H51 | H51 |
| H54 | H54 |
| HD69 | HD69 |
| HD70 | HD70 |
| HD70 | HD71 |
| HD71 | HD70 |
| HD71 | HD71 |
| HO3 | HO3 |
| IN1 | IN1 |
| IN3 | IN3 |
| IN8 | IN8 |
| IN8 | ES8 |
| IN18 | IN18 |
| IN28 | IN28 |
| IN29 | IN29 |
| IN33 | IN33 |
| K2 | K2 |
| KW1 | KW1 |
| L1 | L1 |
| L4 | L4 |
| L5 | L5 |
| M1 | M1 |
| M2 | M2 |
| M4 | M4 |
| M10 | M10 |
| M11 | M11 |
| M12 | M12 |
| MB33 | MB33 |
| MB34 | MB34 |
| MG1 | MG1 |
| MZ9 | MZ9 |
| MZ10 | MZ10 |
| MZ12 | MZ12 |
| NA6 | NA6 |
| NA12 | NA12 |
| NA24 | NA24 |
| NH1 | NH1 |
| OP11 | OP11 |
| OP19 | OP19 |

| CURTIS | ESP | CURTIS | ESP | DOMINION | ESP | DOMINION | ESP |
|---|---|---|---|---|---|---|---|
| PA2 | PA2 | UN18 | UN18 | GA1 | GAS | WT7 | PA6 |
| PA5 | PA5 | V26 | V26K | MA1 | M10 | 8 | Y4 |
| PA6 | PA6 | V28 | V28M | MS12 | DC2 | 9 | Y1 |
| PA7 | PA7 | V29 | V29R | OM1 | JE1 | 9A | Y2 |
| PP10 | PP10 | V30 | V30F | XL1 | XL7 | 9B | Y220 |
| PP10 | RN11 | V30 | V30R | O1 | CO5 | C9 | Y3 |
| RA3 | RA3 | V32 | V32FB | O1AB | CO6 | CE9 | FC7 |
| RA4 | RA4 | VL3 | VL3 | O1EA | CO4 | DM9 | WS2 |
| RN11 | PP10 | VL4 | VL4 | O1EB | CO3 | K9 | Y1E |
| RN11 | RN11 | VL5 | VL5 | O1EG | CO36 | MZ9 | MZ9 |
| RO1 | RO1 | WE1 | WE1 | O1EN | CO7 | NE9 | FT36 |
| RO4 | RO4 | WE3 | WE3 | O1GM | CO90 | O9 | S1 |
| RU1 | RU1 | WK2 | WK2 | O1GH | CO97 | SB9 | SB11 |
| RU2 | RU2 | WR3 | WR3 | O1MA | CO44 | 10 | S3 |
| RU4 | RU4 | WR4 | WR4 | O1MB | CO45 | 10N | S68 |
| RU16 | RU16 | WS2 | MZ12 | RO1ED | CO34 | A10N | S69 |
| RU18 | RU18 | WS2 | WS2 | RO1EE | CO35 | HF10 | OP11 |
| RU18½ | RU18½ | XL7 | XL7 | RO1EG | CO62 | MZ10 | MZ10 |
| RU19 | RU19 | Y1 | Y1 | RO1EN | SK1 | U10 | S4 |
| RU20 | RU20 | Y1E | Y1E | UO1A1 | CO91 | 11 | RU1 |
| RU21 | RU21 | Y2 | Y2 | UO1A2 | CO92 | 11D1 | RU45 |
| RU45 | RU45 | Y3 | Y3 | UO1B1 | CO94 | 11GH | RU19 |
| RU46 | RU46 | Y4 | Y4 | UO1B2 | CO95 | 11P | RU4 |
| S1 | S1 | Y11 | Y11 | UO1C1 | CO98 | A11D1 | RU46 |
| S3 | S3 | Y12 | Y12 | UO1C2 | CO99 | A11P | RU16 |
| S4 | S4 | Y14 | Y14 | UO1D1 | CO51 | 12 | RU2 |
| S6 | S6 | Y52 | Y52 | UO1D2 | CO52 | MZ12 | MZ12 |
| S16 | S16 | Y54 | Y54 | UO1GM | CO89 | NE12 | VL3 |
| S22 | S22 | Y103 | Y103 | AC2 | R30 | TA13 | TR23 |
| S22½ | S22½ | Y104 | Y104 | MA2 | M11 | 14 | EA1 |
| S32 | S31 | Y149 | Y149 | UN2 | MG1 | U14T | EA27 |
| S31 | S32 | Y152 | Y152 | O3M | CO106 | TA17 | TR20 |
| S68 | S68 | Y220 | Y220 | O3M | HL1 | DT18 | DA22 |
| S69 | S69 | YS1 | YS1 | MA4 | M12 | DT18 | SR1 |
| SB11 | SB11 | YS2 | YS2 | MZ4 | FC5 | DT19 | DA21 |
| SC1 | SC1 | | | O4 | L1 | HF19 | V26K |
| SC4 | SC4 | **DEXTER** | **ESP** | O4A | L4 | TA20 | TR25 |
| SC6 | SC6 | 2 | DE6 | O4AL | L5 | YM20 | YS1 |
| SC7 | SC7 | 2L | DE8 | O4KL | L37 | DT21 | DA23 |
| SC8 | SC8 | | | MZ5 | FC6 | DT22 | DA24 |
| SC9 | SC9 | **DOMINION** | **ESP** | SF5 | FT37 | NE21 | AA2 |
| SC10 | SC10 | OO | CO1 | BG6 | BN1 | 22 | SE1 |
| SC22 | SC22 | OOAB | CO2 | KL6 | FC2 | 22EZ | NH1 |
| SE1 | SE1 | OOBH | HL2 | SF6 | FT38 | 22W | AR1 |
| SK1 | SK1 | OOG | CO10 | WT6 | PA5 | D22 | AR5 |
| SR1 | SR1 | OOV | CO26 | 7E | Y52 | HL22W | AR4 |
| T7 | T7 | OOZ | CO87 | BG7 | BN7 | NE22 | PP10 |
| TR20 | TR20 | SOOV | CO68 | O7KMA | S6 | NE22 | RN11 |
| TR23 | TR23 | UOOZ | CO66 | O7KMB | S16 | HF24 | MB33 |
| TR25 | TR25 | RUKM | RU18 | O7LA | S22½ | NE24 | VL5 |
| TR26 (DISC) | TR25 | RUKM | RU18½ | O7RMA | S31 | HF25 | MB34 |
| TR29 | TR29 | CA1 | V30F | O7RMB | S32 | NE25 | VL4 |
| UN3 | UN3 | CA1 | V30R | U7E | Y54 | YM26 | YS2 |
| UN16 | UN16 | EG1 | EL10 | UO7LA | S22 | HF28 | BM3 |

| DOMINION | ESP | DOMINION | ESP | HURD | ESP | ILCO | ESP |
|---|---|---|---|---|---|---|---|
| TA28 | TR29 | P98C | B50 | 9128 | H50 | 62DT | DA22 |
| HF30 | PA2 | P98E | B44 | 9129 | H51 | 62DT | SR1 |
| HF31 | PA7 | P98J | B46 | 9338 | H50 | 62DU | DA21 |
| HF32 | V29R | S98B | B49 | 9422 | H51 | 62FS | UN18 |
| HF33 | V28M | S98D | B51 | 9423 | H27 | AB62C | AB1 |
| HF36 | V32FB | S98H | B45 | 9424 | H50 | VO62 | VL3 |
| 41C | AM1 | S98K | B47 | 9427 | H51 | VW67 | V26K |
| 41E | CG16 | U98B | B41 | 9428 | H50 | HD70 | HD70 |
| 41G | CG1 | U98D | B43 | 9431 | H51 | HD70 | HD71 |
| 41GA | CG2 | U98LA | B11 | 9432 | H50 | HD71 | HD70 |
| 41GR | CG3 | 114A | BE2 | 9433 | H27 | HD71 | HD71 |
| 41N | CG6 | 119 | ER1 | 9524 | H27 | VW71 | V28M |
| 41X | AM3 | U122 | Y11 | 9525 | H26 | VW71A | V29R |
| HD44 | HD69 | U122A | Y12 | 9544 | H27 | 73VB | V30F |
| 54F | IN1 | U122AR | Y14 | 9545 | H26 | 73VB | V30R |
| 54UN | IN29 | U122B | Y103 | 9546 | H27 | VO73 | VL5 |
| H54KD | DE6 | U122BR | Y104 | 9547 | H26 | VO73S | VL4 |
| H54WA | WR3 | 123 | WE1 | 9549 | H26 | 997E | Y52 |
| HD54 | HD70 | 127DP | H27 | | | 0997E | Y54 |
| HD54 | HD71 | 127ES | H26 | **ILCO** | **ESP** | 997X | Y6 |
| HL54KD | DE8 | 141GE | T7 | BMW1 | BM3 | 998 | Y4 |
| L54B | IN8 | 145 | SC1 | DC1 | DC2 | 999 | Y1 |
| L54B | ES8 | 145E | SC8 | MG1 | MG1 | 999A | Y2 |
| L54WA | WR4 | 145F | SC7 | YS1 | YS1 | 999B | Y220 |
| X54F | IN18 | A145 | SC4 | AA2 | AA2 | 999N | Y1E |
| X54FN | IN28 | A145E | SC9 | FC2 | FC2 | C999 | Y3 |
| X54K | IN3 | A145F | SC10 | YS2 | YS2 | 1000 | CO1 |
| X54MT | IN33 | 167FD | H51 | MZ4 | FC5 | 1000AB | CO2 |
| HD54 | HD70 | S167FD | H50 | MZ5 | FC6 | 1000G | CO10 |
| HD55 | HD71 | 170 | HO3 | PA5 | PA5 | 1000V | CO26 |
| 62DP | UN3 | H175 | WK2 | PO5 | PA2 | S1000V | CO68 |
| 62DR | UN18 | 176 | KW1 | PA6 | PA6 | 1001 | CO5 |
| U62VP | OP19 | 770U | Y152 | PO6 | PA7 | 1001AB | CO6 |
| L64A | NA12 | S770U | Y149 | FC7 | FC7 | 1001ABM | CO90 |
| L64BH | NA24 | 707A | SC6 | MZ9 | MZ9 | 1001EA | CO4 |
| R64D | NA6 | 707W | SC22 | MZ10 | MZ10 | 1001EB | CO3 |
| 69 | RO1 | 970AM | RA4 | OM10 | JE1 | 1001EG | CO36 |
| M69C | RO4 | S970AM | RA3 | OP11 | OP11 | 1001EH | CO87 |
| 79B | K2 | | | MZ12 | MZ12 | 1001EN | CO7 |
| UN90 | UN16 | **EAGLE** | **ESP** | MZ12 | WS2 | 1001GH | CO97 |
| 92 | M1 | 11309 | EA1 | MB15 | MB34 | 1001MA | CO44 |
| 92B | M2 | 11929B | EA27 | UN16 | UN16 | 1001MB | CO45 |
| 92V | M4 | | | MB18 | MB33 | A1001ABM | CO89 |
| 96L | EL3 | **EARLE** | **ESP** | OP19 | OP19 | A1001AH | CO91 |
| E96LN | EL11 | 7000K | ER1 | TR26 | TR25 | A1001BH | CO94 |
| 98M | B1 | | | FT37 | FT37 | A1001C1 | CO98 |
| 98X | B4 | **HARLOC** | **ESP** | FT38 | FT38 | A1001C2 | CO99 |
| H98A | B40 | 700 | EA27 | F44 | FT36 | A1001D1 | CO51 |
| H98C | B42 | | | HO44 | HD69 | A1001D2 | CO52 |
| H98DB | B5 | **HOLLYMADE** | **ESP** | RE61F | PP10 | A1001DH | CO95 |
| H98LA | B10 | 1015CA | HO3 | RE61F | RN11 | A1001EH | CO66 |
| H98M | B2 | | | T61C | TR20 | A1001FH | CO92 |
| H98X | B24 | **HURD** | **ESP** | T61F | TR23 | R1001ED | CO34 |
| P98A | B48 | 9124 | H27 | 62DP | UN3 | R1001EE | CO35 |

| ILCO | ESP | ILCO | ESP | ILCO | ESP | KEIL | ESP |
|------|-----|------|-----|------|-----|------|-----|
| R1001EG | CO62 | X1054K | IN3 | 1145A | SC4 | 64CA | CO62 |
| R1001EN | SK1 | R1064D | NA6 | 1145E | SC8 | 64D | CO7 |
| 1003M | HL1 | 1069 | RO1 | 1145F | SC7 | 64EH | CO87 |
| 1003M | CO106 | 1069FL | RO4 | A1145E | SC9 | 64G | CO36 |
| R1003M | HL2 | 1069LA | NA12 | A1145F | SC10 | 64GH | CO97 |
| 1004 | L1 | 1069LC | NA24 | 1167FD | H51 | 64MA | CO44 |
| 1004A | L4 | 1079B | K2 | S1167FD | H50 | 64N | CO90 |
| 1004AL | L5 | 1092 | M1 | 1170B | HO3 | A64AH | CO91 |
| 1004KL | L37 | 1092B | M2 | 1175N | WK2 | A64BH | CO94 |
| 1007LA | S22½ | 1092D | M12 | 1176 | KW1 | A64C1 | CO98 |
| N1007KMA | S6 | 1092H | M11 | 1177N | NH1 | A64C2 | CO99 |
| N1007KMB | S16 | 1092N | M10 | 1179 | AR1 | A64D1 | CO51 |
| N1007RMA | S31 | 1092V | M4 | 1179A | AR4 | A64D2 | CO52 |
| N1007RMB | S32 | 1096L | EL3 | 1179C | AR5 | A64DH | CO95 |
| 01007LA | S22 | C1096CN | EL10 | 1180 | XL7 | A64EH | CO66 |
| 1009 | S1 | C1096LN | EL11 | 1184FD | H54 | A64FH | CO92 |
| 1010 | S3 | 1098DB | B5 | 1307A | SC6 | A64N | CO89 |
| 1010N | S68 | 1098GX | GAS | 1307W | SC22 | 65 | CO5 |
| L1010N | S69 | 1098M | B1 | S1770U | Y149 | 65A | CO6 |
| 01010 | S4 | 1098X | B4 | P1770U | Y152 | 65E | CO35 |
| 1011 | RU1 | H1098A | B40 | 1970AM | RA4 | 66A | CO10 |
| 1011D1 | RU45 | H1098C | B42 | S1970AM | RA3 | 66N | CO26 |
| 1011D41 | RU19 | H1098LA | B10 | | | 66NS | CO68 |
| 1011P | RU4 | H1098M | B2 | **KEIL** | **ESP** | 83 | RU2 |
| 1011PZ | RU18 | H1098X | B24 | 2A | Y220 | 84 | RU1 |
| A1011D1 | RU46 | O1098B | B41 | 2B | Y1 | 84A | SE1 |
| A1011D41 | RU20 | O1098D | B43 | 2C | Y2 | 84AA | ER1 |
| A1011M | RU21 | O1098LA | B11 | 2KK | K2 | A87N | RU21 |
| A1011P | RU16 | P1098A | B48 | 6P | Y11 | 88 | RU4 |
| A1011PZ | RU18½ | P1098C | B50 | 6S | Y12 | 88A | RU16 |
| 1012 | RU2 | P1098CV | B63-C | 6SS | Y14 | 88AZ | RU18½ |
| 1014 | EA1 | P1098E | B44 | 6TB | BN1 | 88D1 | RU45 |
| 1014C | HR1 | P1098J | B46 | 6V | Y103 | 88D41 | RU19 |
| X1014F | EA27 | S1098B | B49 | 6VV | Y104 | 88Z | RU18 |
| 1022 | SE1 | S1098D | B51 | 7A | Y4 | A88D1 | RU46 |
| 1041C | AM1 | S1098H | B45 | 8GL | Y149 | A88D41 | RU20 |
| 1041G | CG1 | S1098K | B47 | 8HM | Y152 | A96LA | S22½ |
| 1041GA | CG2 | P1098WC | B64-C | 11 | Y6 | B96LA | S22 |
| 1041GR | CG3 | 1114A | BE2 | 12B | YS2 | 99LN | S16 |
| 1041N | CG6 | 1119 | ER1 | 12E | Y54 | N99RN | S32 |
| 1041T | CG16 | K1122D | BN1 | 17N | NH1 | 100LN | S6 |
| 1046 | AM3 | O1122 | Y11 | 24F | Y3 | N100RN | S31 |
| 1054F | IN1 | O1122A | Y12 | 59D | H27 | 102 | S3 |
| 1054FN | IN28 | O1122AR | Y14 | 59E | H26 | 102N | S68 |
| 1054MT | IN33 | O1122B | Y103 | 60D | CO1 | 102NA | S69 |
| 1054UN | IN29 | O1122BR | Y104 | 60E | CO2 | A102 | S4 |
| 1054WB | WR3 | O1122R | Y13 | 62CG | H50 | 103 | S1 |
| A1054WB | WR4 | 1123 | WE1 | 62HG | H51 | 122 | L1 |
| D1054K | DE6 | 1123S | WE3 | 63M | CO106 | 122A | L4 |
| D1054KA | DE8 | 1127DP | H27 | 63M | HL1 | 122C | L5 |
| K1054B | BN7 | 1127ES | H26 | 63MR | HL2 | 125B | AR1 |
| L1054B | IN8 | 1131R | R30 | 64 | CO3 | 125BA | AR4 |
| L1054B | ES8 | 1141GE | T7 | 64A | CO4 | 125BC | AR5 |
| X1054F | IN18 | 1145 | SC1 | 64C | SK1 | 153GB | B41 |

| KEIL | ESP |
|---|---|
| 153GD | B43 |
| 153GH | B45 |
| 153GK | B47 |
| 153H | RO1 |
| 153HA | B40 |
| 153HE | B44 |
| 153HJ | B46 |
| 153PG | B11 |
| 153PH | B10 |
| L153GB | B49 |
| L153HA | B48 |
| L153HC | B42 |
| N153GD | B51 |
| N153HC | B50 |
| 154P | BE2 |
| H154NL | RA4 |
| H154R | B24 |
| H154SS | B2 |
| R154NL | RA3 |
| R154S | B5 |
| R154SS | B1 |
| R154X | B4 |
| 155BN | EL10 |
| 155F | WE1 |
| 155FS | WE3 |
| 155GE | T7 |
| 155S | M1 |
| 155W | EL3 |
| 155WN | EL11 |
| 155X | M2 |
| D155K | DE6 |
| D155KA | DE8 |
| R155D | NA6 |
| 158N | M10 |
| 158V | M4 |
| 159AA | IN1 |
| 159H | SC1 |
| 159J | SC4 |
| 159K | IN3 |
| 159WB | WR3 |
| A159WB | WR4 |
| E159 | SC8 |
| E159A | SC9 |
| F159 | SC7 |
| F159A | SC10 |
| X159AA | IN18 |
| D161VW | V26K |
| G161VW | V28M |
| H161VW | V29R |
| 168B | EA1 |
| X168 | EA27 |
| 169FL | RO4 |
| 169LA | NA12 |
| 170 | HO3 |

| KEIL | ESP |
|---|---|
| 175N | WK2 |
| 176 | KW1 |
| 180AJ | AM3 |
| 180BC | CG16 |
| 180E | AM1 |
| 180FS | CG1 |
| 180GA | CG2 |
| 180GR | CG3 |
| 180S | CG6 |
| 181N | IN8 |
| 181N | ES8 |
| 181NB | BN7 |
| 202C | SC6 |
| 202W | SC22 |

| KWIKSET | ESP |
|---|---|
| 1268 | KW1 |

| LOCKWOOD | ESP |
|---|---|
| B308 | L1 |
| B310 | L4 |
| B346 | L5 |

| LORI CORP. | ESP |
|---|---|
| LOR27 | EA27 |

| MASTER | ESP |
|---|---|
| 1K | M1 |
| 7K | M2 |
| 15K | M10 |
| 17K | M11 |
| 81KR | M4 |
| 150K | M12 |

| NATIONAL (EZ) | ESP |
|---|---|
| 9407 | NH1 |

| NATIONAL (ROCKFORD) | ESP |
|---|---|
| 68-619-1 | RO4 |
| 68-635-1 | RO1 |
| 68-676-11 | NA24 |
| 411-31 | NA6 |
| 676-1 | NA12 |

| RUSSWIN | ESP |
|---|---|
| 5D1R | RU45 |
| 6D1R | RU46 |
| 5D41 | RU19 |
| 6D41 | RU20 |
| 752R | RU2 |
| 852R | RU1 |
| 960GGM | RU18½ |

| RUSSWIN | ESP |
|---|---|
| 961 | RU16 |
| 980BGGM | RU18 |
| 981B | RU4 |
| 59812 | RU21 |

| SAFE | ESP |
|---|---|
| 7525 | RU2 |

| SARGENT | ESP |
|---|---|
| 265K | S3 |
| 265R | S1 |
| 265U | S4 |
| 270LN | S16 |
| 270RN | S32 |
| 275LA | S22½ |
| 275S | S68 |
| 6270LN | S6 |
| 6270RN | S31 |
| 6275LA | S22 |
| 6275S | S69 |

| SCHLAGE | ESP |
|---|---|
| 100C | SC1 |
| 100E | SC8 |
| 100F | SC7 |
| 101C | SC4 |
| 101E | SC9 |
| 101F | SC10 |
| 180 | SC6 |
| 200 | SC22 |
| 920A | SC6 |
| 923C | SC1 |
| 923E | SC8 |
| 924C | SC4 |
| 924E | SC9 |
| 924F | SC7 |
| 924F | SC10 |
| 927W | SC22 |

| SEGAL | ESP |
|---|---|
| K9 | SE1 |

| SKILLMAN | ESP |
|---|---|
| SK100 | SK1 |

| STAR | ESP |
|---|---|
| AD1 | PA6 |
| 5AR2 | AR1 |
| 6AR2 | AR4 |
| 5AU1 | CO106 |
| 5AU1 | HL1 |
| 5AU2 | NA12 |
| 6BE1 | BE2 |
| BN1 | BN1 |

| STAR | ESP |
|---|---|
| 5BO1 | HL2 |
| HBR1 | B2 |
| OBR1 | B1 |
| OBR1DB | B5 |
| HBR2 | B10 |
| OBR2 | B11 |
| HBR3 | B24 |
| OBR3 | B4 |
| HBR5 | B40 |
| OBR5 | B41 |
| HBR5M | B42 |
| OBR7 | B43 |
| HBR9E | B44 |
| OBR9H | B45 |
| HBR10J | B46 |
| OBR10K | B47 |
| HBR11 | RA4 |
| OBR11 | RA3 |
| HBR12A | B48 |
| OBR12B | B49 |
| HBR14C | B50 |
| OBR14D | B51 |
| CG1 | CG1 |
| CG2 | CG3 |
| CG6 | CG2 |
| 5CG7 | CG16 |
| 5CO1 | CO7 |
| 6CO1 | CO36 |
| 5CO2 | CO3 |
| 6CO2 | CO4 |
| 5CO3 | CO1 |
| 5CO4 | CO5 |
| 5CO5 | SK1 |
| 5CO6 | CO62 |
| CO7 | CO10 |
| LCO7 | CO68 |
| 5CO11 | CO26 |
| 6CO11 | CO87 |
| 5CO12 | CO66 |
| 6CO12 | CO90 |
| 5CO13 | CO89 |
| 6CO16 | CO97 |
| 6CO16 | CO51 |
| 6CO16 | CO52 |
| 6CO16 | CO98 |
| CP1 | CO99 |
| CP2 | FC2 |
| CP3 | YS1 |
| 5DA2 | YS2 |
| DA3 | MZ12 |
| DA4 | DA21 |
| DA4 | DA22 |
| | SR1 |
| LDC1 | DC2 |

| STAR | ESP |
|---|---|
| 5DE3 | DE6 |
| 6DE3 | DE8 |
| 5EA1 | EA27 |
| 5EA2 | EA1 |
| 5EL1 | EL3 |
| 5EL3 | EL11 |
| 5EL4 | EL10 |
| 5ER1 | ER1 |
| HFD4 | H27 |
| OFD4 | H26 |
| HFD10 | H51 |
| OFD10 | H50 |
| HFD12 | H54 |
| 5FT1 | FT36 |
| HN2 | HD70 |
| HN2 | HD71 |
| HN3 | HD70 |
| HN3 | HD71 |
| 5HO1 | HO3 |
| 5IL1 | IN1 |
| 7IL2 | L37 |
| 5IL4 | IN3 |
| IL5 | IN8 |
| IL5 | ES8 |
| 5IL7 | IN18 |
| 51L9 | IN28 |
| IL10UN | IN29 |
| 5IL11 | IN33 |
| JU1 | AM1 |
| 5JU2 | AM3 |
| 5KE1 | K2 |
| 5KW1 | KW1 |
| 5LO1 | L1 |
| 6LO1 | L4 |
| 7LO1 | L5 |
| MA1 | M1 |
| 4MA2 | M2 |
| 5MA3 | M4 |
| 5MA5 | M10 |
| 5MA6 | M11 |
| 5MA7 | M12 |
| MZ1 | MZ9 |
| MZ2 | MZ10 |
| 5NA1 | NH1 |
| HPL68 | Y152 |
| OPL68 | Y149 |
| RO1 | RO1 |
| RO3 | RO4 |
| 5RO4 | NA6 |
| 5RU1 | RU1 |
| 5RU2 | RU4 |
| 6RU2 | RU16 |
| 5RU5 | RU18 |
| 6RU5 | RU18½ |

| STAR | ESP |
|---|---|
| 5RU6 | RU2 |
| 5RU7 | RU45 |
| 6RU7 | RU46 |
| 5RU8 | RU19 |
| 6RU8 | RU20 |
| 6RU9 | RU21 |
| 5SA1 | S4 |
| 5SA2 | S1 |
| 5SA3 | S16 |
| 6SA3 | S6 |
| 5SA5 | S3 |
| 5SA6 | S32 |
| 6SA6 | S31 |
| 5SA7 | S68 |
| 6SA7 | S69 |
| 5SE1 | SE1 |
| 5SH1 | SC1 |
| 6SH1 | SC4 |
| SH2 | SC6 |
| 5SH4 | SC8 |
| 6SH4 | SC9 |
| 5SH5 | SC7 |
| 6SH5 | SC10 |
| SH6 | SC22 |
| 5TA4 | T7 |
| TO1 | TR20 |
| UN3 | UN18 |
| VW2 | V26K |
| VW3 | V28M |
| VW4 | V30F |
| VW4 | V30R |
| VW5 | V29R |
| 5WE1 | WE1 |
| 5WK1 | WK2 |
| 5WR2 | WR3 |
| 6WR2 | WR4 |
| 4YA1 | Y220 |
| 5YA1 | Y1 |
| 5YA1E | Y1E |
| 6YA1 | Y2 |
| 5YA2 | Y3 |
| 5YA3 | Y4 |
| 5YA6 | Y52 |
| 6YA6 | Y54 |
| YJ1 | Y12 |
| YJ2 | Y14 |
| YJ3 | Y11 |
| YJ4 | Y103 |

| TAYLOR | ESP |
|---|---|
| X1 | MZ12 |
| X4 | FC5 |
| X5 | FC6 |
| X6 | DA22 |

| TAYLOR | ESP |
|---|---|
| X6 | SR1 |
| 7E | Y52 |
| 7NX | Y11 |
| 7X | Y6 |
| O7B | Y103 |
| O7BR | Y104 |
| O7E | Y54 |
| O7NX | Y12 |
| RO7NX | Y14 |
| R7NX | Y13 |
| X9 | V30F |
| X9 | V30R |
| 12GM | Y4 |
| X12 | PP10 |
| X12 | RN11 |
| 14 | Y1 |
| 14A | Y2 |
| 14YM | Y3 |
| E14 | Y1E |
| X14S | Y220 |
| XL16 | XL7 |
| P19 | Y152 |
| S19 | Y149 |
| 20 | CO1 |
| 20AB | CO2 |
| 20G | CO10 |
| 20V | CO26 |
| 20VS | CO68 |
| X20 | FC2 |
| 21A | CO4 |
| 21EB | CO3 |
| 21EG | CO36 |
| 21EN | CO7 |
| B21EF | CO35 |
| B21EG | CO44 |
| B21EH | CO45 |
| B21EJ | CO34 |
| R21EG | CO62 |
| R21EN | SK1 |
| X21 | YS1 |
| 22B | HL2 |
| 22GM | CO90 |
| 22Z2 | CO87 |
| A22A1 | CO91 |
| A22A2 | CO92 |
| A22B1 | CO94 |
| A22B2 | CO95 |
| A22C1 | CO98 |
| A22C2 | CO99 |
| A22D1 | CO51 |
| A22D2 | CO52 |
| A22GM | CO89 |
| A22Z2 | CO66 |

| TAYLOR | ESP |
|---|---|
| K22 | CO97 |
| R22B | CO106 |
| R22B | HL1 |
| X22 | YS2 |
| 23 | CO5 |
| 23AB | CO6 |
| X26 | MZ9 |
| P27 | H51 |
| S27 | H50 |
| X27 | MZ10 |
| X29 | VL5 |
| X30 | VL4 |
| 31R | R30 |
| X32 | PA7 |
| 35 | L1 |
| 35A | L4 |
| 35AL | L5 |
| 40F | JE1 |
| 41C | AM1 |
| 41G | CG1 |
| 41GA | CG2 |
| 41GR | CG3 |
| 41N | CG6 |
| 41R | CG16 |
| 41RB | IN29 |
| J41 | AM3 |
| X44 | HD69 |
| 43LA | S22½ |
| O43LA | S22 |
| 48 | S1 |
| 48KM | S16 |
| 48KMR | S32 |
| O48KM | S6 |
| O48KMR | S31 |
| 50 | S3 |
| O50 | S4 |
| 51S | S68 |
| 51SA | S69 |
| X51 | AA2 |
| X52 | SB11 |
| X53 | V32FB |
| 54DR | DE6 |
| 54F | IN1 |
| 54FN | IN28 |
| 54KL | L37 |
| 54KS | KW1 |
| 54WA | WR4 |
| 54WB | WR3 |
| A54DR | DE8 |
| L54B | IN8 |
| L54B | ES8 |
| L54N | BN7 |
| L54P | BN1 |
| X54 | DC2 |

| TAYLOR | ESP | TAYLOR | ESP | TAYLOR | ESP |
|--------|-----|--------|-----|--------|-----|
| X54F | IN18 | 96L | EL3 | X165 | HD84 |
| X54K | IN3 | 96LN | EL11 | 170 | HO3 |
| 55 | RU1 | 98M | B1 | 174BA | NA12 |
| 55P | RU4 | 98X | B4 | 174BN | NA24 |
| 56 | RU2 | H98C | B42 | R174D | NA6 |
| 57M | RU18 | H98DB | B5 | X174 | TR40 |
| 57MA | RU18½ | H98LA | B10 | 175W | WK2 |
| 57PA | RU16 | H98M | B2 | X176 | MIT1 |
| 57R | RU19 | H98X | B24 | X178 | MZ16 |
| 57-1D | RU45 | O98B | B41 | X181 | HD90 |
| A57-1D | RU46 | O98D | B43 | X182 | HD91 |
| 61 | EA1 | O98H | B45 | X183 | HD92 |
| X61F | EA27 | O98K | B47 | X185 | SUZ15 |
| X61FR | HR1 | O98LA | B11 | X186 | SUZ17 |
| 62 | RO1 | X98A | B40 | X192 | B72 |
| 62DL | UN3 | X98E | B44 | 307A | SC6 |
| 62H | RO4 | X98J | B46 | 307W | SC22 |
| F68XR | UN16 | 99A | RA4 | | |
| S71B | UN18 | 99B | RA3 | **WEISER** | **ESP** |
| X71 | HD70 | 102 | SE1 | 1556 | WR3 |
| X71 | HD71 | N102 | NH1 | 1559 | WR4 |
| F74T | FT36 | 111GE | T7 | | |
| O76R | OP11 | 114A | BE2 | **WESLOCK** | **ESP** |
| V78JK | V26K | X114 | DA24 | 4425 | WK2 |
| 79HK | K2 | X115 | DA23 | | |
| F79-1 | FT37 | X116 | RN24 | **YALE** | **ESP** |
| F79-3 | FT38 | 119 | ER1 | 8 | Y1 |
| M79S | MB33 | X121 | DC3 | E8 | Y1E |
| M79T | MB34 | X123 | DA25 | 9½ | Y220 |
| O79JB | OP19 | X123 | SR5 | 11 | Y2 |
| V79D | VL3 | 127DP | H27 | 12½ | Y52 |
| B80NR | BM3 | 127ES | H26 | 12½ | Y54 |
| T80R | TR20 | X128 | HD79 | 111GMK | Y4 |
| T80V | TR23 | X129 | HD80 | 385 | Y3 |
| A81M | PA2 | X130 | HD81 | 9114 | Y11 |
| A81R | PA5 | X131 | MZ13 | 9278 | Y12 |
| A81S | PA6 | 133 | WE1 | 9279 | Y14 |
| F81E | WS2 | 133S | WE3 | 9290 | Y103 |
| M81G | MG1 | 135 | AR1 | 9882 | Y104 |
| T81B | TR25 | 135A | AR4 | | |
| V81V | V29R | 135C | AR5 | | |
| V81W | V28M | X137 | TR33 | | |
| X86 | FC7 | X143 | B53 | | |
| P91A | B48 | 145 | SC1 | | |
| P91C | B50 | 145A | SC4 | | |
| S91B | B49 | 145E | SC9 | | |
| S91D | B51 | 145ES | SC8 | | |
| 92 | M1 | 145F | SC10 | | |
| 92B | M2 | 145FS | SC7 | | |
| 92G | M12 | X145 | B55 | | |
| 92N | M10 | X146 | B56 | | |
| 92T | M11 | X151 | TR39 | | |
| U92A | M4 | X159 | TR37 | | |
| 96CN | EL10 | X160 | HY2 | | |

*Appendix* **C**

# Depth and space charts

Manufacturer: Abus
Key Series: Padlocks
Key Blank No.: 244L & 2441R
No. of Steps: 5, 1−5
Drop: .025
Cuts Start At: .155
Spacing: .120
Depth No. 1: .260
Depth No. 2: .235
Depth No. 3: .210
Depth No. 4: .185
Depth No. 5: .160


Manufacturer: Abus
Key Series: Padlocks
Key Blank No.: 42 (Orig.)
No. of Steps: 5, 1−5
Drop: .025
Cuts Start At: .145
Spacing: .140
Depth No. 1: .260
Depth No. 2: .235
Depth No. 3: .210
Depth No. 4: .185
Depth No. 5: .160

Manufacturer: Abus
Key Series: Padlocks
Key Blank No.: 5525 (Orig.)
No. of Steps: 3, 1−3
Drop: .030
Cuts Start At: .100
Spacing: .125
Depth No. 1: .180
Depth No. 2: .150
Depth No. 3: .120

Manufacturer: American
Key Series: Padlocks
Key Blank No.: 1046
No. of Steps: 8, 1−8
Drop: Var.
Cuts Start At: .156
Spacing: .125
Depth No. 1: .2840
Depth No. 2: .2684
Depth No. 3: .2523
Depth No. 4: .2372
Depth No. 5: .2215
Depth No. 6: .2059
Depth No. 7: .1903
Depth No. 8: .1747

Manufacturer: American
Key Series: P4–P6 Padlocks
Key Blank No.: 1046
No. of Steps: 5, 1–5
Drop: Var.
Cuts Start At: .155
Spacing: .130
Depth No. 1: .260
Depth No. 2: .235
Depth No. 3: .210
Depth No. 4: .185
Depth No. 5: .165


Manufacturer: Amerock
Key Series: All
Key Blank No.: 1179A
No. of Steps: 10, 0–9
Drop: .014
Cuts Start At: .265
Spacing: .155
Depth No. 0: .315
Depth No. 1: .301
Depth No. 2: .287
Depth No. 3: .273
Depth No. 4: .259
Depth No. 5: .245
Depth No. 6: .231
Depth No. 7: .217
Depth No. 8: .203
Depth No. 9: .189


Manufacturer: Arrow
Key Series: Standard
Key Blank No.: 1179
No. of Steps: 9, 1–9
Drop: .018
Cuts Start At: .230
Spacing: .160
Depth No. 1: .320
Depth No. 2: .302
Depth No. 3: .284
Depth No. 4: .266
Depth No. 5: .248
Depth No. 6: .230
Depth No. 7: .212
Depth No. 8: .194
Depth No. 9: .176

Manufacturer: Arrow
Key Series: New
Key Blank No.: 1179
No. of Steps: 10, 0–9
Drop: .014
Cuts Start At: .265
Spacing: .155
Depth No. 0: .335
Depth No. 1: .321
Depth No. 2: .307
Depth No. 3: .293
Depth No. 4: .279
Depth No. 5: .265
Depth No. 6: .251
Depth No. 7: .237
Depth No. 8: .223
Depth No. 9: .209

Manufacturer: Arrow
Key Series: Old
Key Blank No.: 1179
No. of Steps: 7, 0–6
Drop: .020
Cuts Start At: .265
Spacing: .155
Depth No. 0: .335
Depth No. 1: .315
Depth No. 2: .295
Depth No. 3: .275
Depth No. 4: .255
Depth No. 5: .235
Depth No. 6: .215

Manufacturer: Craftsman
Key Series: Key-in knob
Key Blank No.: 1096CN
No. of Steps: 10, 0–9
Drop: .018
Cuts Start At: .160
Spacing: .156
Depth No. 0: .323
Depth No. 1: .305
Depth No. 2: .287
Depth No. 3: .269
Depth No. 4: .251
Depth No. 5: .233
Depth No. 6: .215
Depth No. 7: .197
Depth No. 8: .179
Depth No. 9: .161

Manufacturer: Dexter
Key Series: After 1969
Key Blank No.: D1054K
No. of Steps: 10, 0−9
Drop: .015
Cuts Start At: .216
Spacing: .155
Depth No. 0: .320
Depth No. 1: .305
Depth No. 2: .290
Depth No. 3: .275
Depth No. 4: .260
Depth No. 5: .245
Depth No. 6: .230
Depth No. 7: .215
Depth No. 8: .200
Depth No. 9: .185

Manufacturer: Dexter
Key Series: Close Pin
Key Blank No.: 1054KD
No. of Steps: 8, 0−7
Drop: .020
Cuts Start At: .180
Spacing: .125
Depth No. 0: .325
Depth No. 1: .305
Depth No. 2: .285
Depth No. 3: .265
Depth No. 4: .245
Depth No. 5: .225
Depth No. 6: .205
Depth No. 7: .185

Manufacturer: Falcon
Key Series: A,E,M,R,S,X
Key Blank No.: 1054WB
No. of Steps: 10, 0−9
Drop: .018
Cuts Start At: .237 (from tip to bow)
Spacing: .156
Depth No. 0: .315
Depth No. 1: .297
Depth No. 2: .279
Depth No. 3: .261
Depth No. 4: .243
Depth No. 5: .225
Depth No. 6: .207
Depth No. 7: .189

Depth No. 8: .171
Depth No. 9: .153

Manufacturer: Harloc
Key Series: Large Pin
Key Blank No.: X1014F
No. of Steps: 10, 1−0
Drop: .018
Cuts Start At: .216
Spacing: .155
Depth No. 1: .320
Depth No. 2: .302
Depth No. 3: .284
Depth No. 4: .266
Depth No. 5: .248
Depth No. 6: .230
Depth No. 7: .212
Depth No. 8: .194
Depth No. 9: .176
Depth No. 0: .158

Manufacturer: Ilco
Key Series: FN, XK, MT
Key Blank No.: 1054FN
No. of Steps: 10, 0−9
Drop: .018
Cuts Start At: .277
Spacing: .156
Depth No. 0: .320
Depth No. 1: .302
Depth No. 2: .284
Depth No. 3: .266
Depth No. 4: .248
Depth No. 5: .230
Depth No. 6: .212
Depth No. 7: .194
Depth No. 8: .176
Depth No. 9: .158

Manufacturer: Ilco
Key Series: XR
Key Blank No.: 1154G
No. of Steps: 7, 1−7
Drop: .018
Cuts Start At: .162
Spacing: .140
Depth No. 1: .270
Depth No. 2: .252
Depth No. 3: .234
Depth No. 4: .216

Depth No. 5: .198
Depth No. 6: .180
Depth No. 7: .162

Manufacturer: Ilco
Key Series: #308 Padlocks
Key Blank No.: K1054AX
No. of Steps: 7, 1–7
Drop: .018
Cuts Start At: .165
Spacing: .140
Depth No. 1: .270
Depth No. 2: .252
Depth No. 3: .234
Depth No. 4: .216
Depth No. 5: .198
Depth No. 6: .180
Depth No. 7: .162

Manufacturer: Ilco
Key Series: AH & F Small Pin
Key Blank No.: X1054A & JK
No. of Steps: 7, 1–7
Drop: .018
Cuts Start At: .162
Spacing: .125
Depth No. 1: .270
Depth No. 2: .252
Depth No. 3: .234
Depth No. 4: .216
Depth No. 5: .198
Depth No. 6: .180
Depth No. 7: .162

Manufacturer: Ilco
Key Series: Large Pin
Key Blank No.: X1054
No. of Steps: 10, 0–9
Drop: .018
Cuts Start At: .277
Spacing: .156
Depth No. 0: .320
Depth No. 1: .302
Depth No. 2: .284
Depth No. 3: .266
Depth No. 4: .248
Depth No. 5: .230
Depth No. 6: .212
Depth No. 7: .194

Depth No. 8: .176
Depth No. 9: .158

Manufacturer: Juncunc
Key Series: Pin Tumbler
Key Blank No.: 1046
No. of Steps: 8, 1–8
Drop: .015
Cuts Start At: .156
Spacing: .125
Depth No. 1: .285
Depth No. 2: .270
Depth No. 3: .255
Depth No. 4: .240
Depth No. 5: .225
Depth No. 6: .210
Depth No. 7: .195
Depth No. 8: .180

Manufacturer: Kwikset
Key Series: Old
Key Blank No.: 1176
No. of Steps: 4
Drop: .031
Cuts Start At: .247
Spacing: .150
Depth No. 1: .328
Depth No. 3: .297
Depth No. 5: .266
Depth No. 7: .235

Manufacturer: Kwikset
Key Series: New
Key Blank No.: 1176
No. of Steps: 7, 1–7
Drop: .023
Cuts Start At: .247
Spacing: .150
Depth No. 1: .328
Depth No. 2: .305
Depth No. 3: .282
Depth No. 4: .259
Depth No. 5: .236
Depth No. 6: .213
Depth No. 7: .190

Manufacturer: Lustre Line
Key Series: Key-in-knob
Key Blank No.: 1176
No. of Steps: 7, 1–7

Drop: .020
Cuts Start At: .170
Spacing: .190
Depth No. 1: .310
Depth No. 2: .290
Depth No. 3: .270
Depth No. 4: .250
Depth No. 5: .230
Depth No. 6: .210
Depth No. 7: .190

Manufacturer: Master
Key Series: 1K, 77K, 15, 17, 81
Key Blank No.: 1092, 1092V, etc.
No. of Steps: 8, 0−7
Drop: .015
Cuts Start At: .185
Spacing: .125
Depth No. 0: .275
Depth No. 1: .260
Depth No. 2: .245
Depth No. 3: .230
Depth No. 4: .215
Depth No. 5: .200
Depth No. 6: .185
Depth No. 7: .170

Manufacturer: Master
Key Series: 7K
Key Blank No.: 1092B
No. of Steps: 7, 0−6
Drop: .0155
Cuts Start At: .132
Spacing: .125
Depth No. 0: .212
Depth No. 1: .1965
Depth No. 2: .181
Depth No. 3: .1655
Depth No. 4: .150
Depth No. 5: .1345
Depth No. 6: .119

Manufacturer: Master
Key Series: #150 & #160 Large
Key Blank No.: 1092N & 1092NR
No. of Steps: 9, 0−8
Drop: .015
Cuts Start At: .150
Spacing: .129
Depth No. 0: .270

Depth No. 1: .255
Depth No. 2: .240
Depth No. 3: .225
Depth No. 4: .210
Depth No. 5: .195
Depth No. 6: .180
Depth No. 7: .165
Depth No. 8: .150

Manufacturer: Master
Key Series: #19 Padlocks
Key Blank No.: F76 (Orig.)
No. of Steps: 8, 0−7
Drop: .025
Cuts Start At: .213
Spacing: .156
Depth No. 0: .370
Depth No. 1: .345
Depth No. 2: .320
Depth No. 3: .295
Depth No. 4: .270
Depth No. 5: .245
Depth No. 6: .220
Depth No. 7: .195

Manufacturer: Schlage
Key Series: Pin Tumblers
Key Blank No.: 1145C, etc.
No. of Steps: 10, 0−9
Drop: .015
Cuts Start At: .231
Spacing: .156
Depth No. 0: .335
Depth No. 1: .320
Depth No. 2: .305
Depth No. 3: .290
Depth No. 4: .275
Depth No. 5: .260
Depth No. 6: .245
Depth No. 7: .230
Depth No. 8: .215
Depth No. 9: .200

Manufacturer: Weiser
Key Series: New
Key Blank No.: 1054WB
No. of Steps: 10, 0−9
Drop: .018
Cuts Start At: .237 (from top shoulder)

Spacing: .155
Depth No. 0: .315
Depth No. 1: .297
Depth No. 2: .279
Depth No. 3: .261
Depth No. 4: .243
Depth No. 5: .225
Depth No. 6: .207
Depth No. 7: .189
Depth No. 8: .171
Depth No. 9: .153

Manufacturer: Weiser
Key Series: Old
Key Blank No.: 1054WB
No. of Steps: 10, 0−9
Drop: .018
Cuts Start At: .237 (from top shoulder)
Spacing: .156
Depth No. 0: .320
Depth No. 1: .302
Depth No. 2: .284
Depth No. 3: .266
Depth No. 4: .248
Depth No. 5: .230
Depth No. 6: .212
Depth No. 7: .194
Depth No. 8: .176
Depth No. 9: .158

Manufacturer: Yale
Key Series: IN & AL Series
Key Blank No.: 01122, etc.
No. of Steps: 7, 1−7
Drop: .018
Cuts Start At: .175
Spacing: .135
Depth No. 1: .260
Depth No. 2: .242

Depth No. 3: .224
Depth No. 4: .206
Depth No. 5: .188
Depth No. 6: .170
Depth No. 7: .152

Manufacturer: Yale
Key Series: Standard #8
Key Blank No.: 999 (#8 Orig.)
No. of Steps: 10, 0−9
Drop: .019
Cuts Start At: .200
Spacing: .165
Depth No. 0: .320
Depth No. 1: .301
Depth No. 2: .282
Depth No. 3: .263
Depth No. 4: .244
Depth No. 5: .225
Depth No. 6: .206
Depth No. 7: .187
Depth No. 8: .168
Depth No. 9: .149

Manufacturer: Yale
Key Series: Large Sectional
Key Blank No.: 998, etc.
No. of Steps: 8, 0−7
Drop: .025
Cuts Start At: .200
Spacing: .165
Depth No. 0: .320
Depth No. 1: .295
Depth No. 2: .270
Depth No. 3: .245
Depth No. 4: .220
Depth No. 5: .195
Depth No. 6: .170
Depth No. 7: .145

# *Appendix* **D**

# **ANSI/BHMA finish numbers**

| New No. | Description | Material | Old No. |
|---|---|---|---|
| 600 | primed for painting | steel | USP |
| 601 | bright japanned | steel | US1B |
| 603 | zinc plated | steel | US2G |
| 604 | zinc plated and dichromate sealed | brass | — |
| 605 | bright brass | brass | US3 |
| 606 | satin brass | brass | US4 |
| 607 | oxidized satin brass, oil rubbed | brass | — |
| 608 | oxidized satin brass, relieved | brass | — |
| 609 | satin brass, blackened, satin relieved | brass | US5 |
| 610 | satin brass, blackened, bright relieved | brass | US7 |
| 611 | bright bronze | bronze | US9 |
| 612 | satin bronze | bronze | US10 |
| 613 | dark oxidized satin bronze, oil rubbed | bronze | US10B |
| 614 | oxidized satin bronze, relieved | bronze | — |
| 615 | oxidized satin bronze, relieved, waxed | bronze | — |
| 616 | satin bronze, blackened, satin relieved | bronze | US11 |
| 617 | darkened oxidized satin bronze, bright relieved | bronze | US13 |

| | | | |
|---|---|---|---|
| 618 | bright nickel plated | brass/ bronze | US14 |
| 619 | satin nickel plated | brass/ bronze | US15 |
| 620 | satin nickel plated, blackened, satin relieved | brass/ bronze | US15A |
| 621 | nickel plated, blackened, relieved | brass/ bronze | US17A |
| 622 | flat black coated | brass/ bronze | US19 |
| 623 | light oxidized statuary bronze | bronze | US20 |
| 624 | dark oxidized statuary bronze | bronze | US20A |
| 625 | bright chromium plated | brass/ bronze | US26 |
| 626 | satin chromium plated | brass/ bronze | US26D |
| 627 | satin aluminum | aluminum | US27 |
| 628 | satin aluminum, clear anodized | aluminum | US28 |
| 629 | bright stainless steel | stainless steel | US32 |
| 630 | satin stainless steel | stainless steel | US32D |
| 631 | flat black coated | steel | US19 |
| 632 | bright brass plated | steel | US3 |
| 633 | satin brass plated | steel | US4 |
| 634 | oxidized satin brass, oil rubbed | steel | — |
| 635 | oxidized satin brass, relieved | steel | — |
| 636 | satin brass plated, blackened, bright relieved | steel | US7 |
| 637 | bright bronze plated | steel | US9 |
| 638 | satin brass plated, blackened, satin relieved | steel | US5 |
| 639 | satin bronze plated | steel | US10 |
| 640 | oxidized satin bronze plated, oil rubbed | steel | US10B |
| 641 | oxidized satin bronze plated, relieved | steel | — |
| 642 | oxidized satin bronze plated, relieved, waxed | steel | — |
| 643 | satin bronze plated, blackened, satin relieved | steel | US11 |
| 644 | dark oxidized satin bronze plated, bright relieved | steel | US13 |
| 645 | bright nickel plated | steel | US14 |
| 646 | satin nickel plated | steel | US15 |
| 647 | satin nickel plated, blackened, satin relieved | steel | US15A |

| 648 | nickel plated, blackened, relieved | steel | US17A |
| 649 | light oxidized bright bronze plated | steel | US20 |
| 650 | dark oxidized statuary bronze plated, clear coated | steel | US20A |
| 651 | bright chrome plated | steel | US26 |
| 652 | satin chromium plated | steel | US26D |
| 653 | bright stainless steel | stainless steel | — |
| 654 | satin stainless steel | stainless steel | — |
| 655 | light oxidized satin bronze, bright relieved, clear coated | bronze | US13 |
| 656 | light oxidized satin bronze plated, bright relieved | steel | US13 |
| 657 | dark oxidized copper plated, satin relieved | steel | — |
| 658 | dark oxidized copper plated, bright relieved | steel | — |
| 659 | light oxidized copper plated, satin relieved | steel | — |
| 660 | light oxidized copper plate, bright relieved | steel | — |
| 661 | oxidized satin copper, relieved | steel | — |
| 662 | satin brass plated, browned, satin relieved | steel | — |
| 663 | zinc plated with clear chromate seal | steel | — |
| 664 | cadmium plated with chromate seal | steel | — |
| 665 | cadmium plated with iridescent dichromate | steel | — |
| 666 | bright brass plated | aluminum | US3 |
| 667 | satin brass plated | aluminum | US4 |
| 668 | satin bronze plated | aluminum | US10 |
| 669 | bright nickel plated | aluminum | US14 |
| 670 | satin nickel plated | aluminum | US15 |
| 671 | flat black coated | aluminum | US19 |
| 672 | bright chromium plated | aluminum | US26 |
| 673 | aluminum clear coated | aluminum | — |
| 674 | primed for painting | zinc | USP |
| 675 | dichromate sealed | zinc | — |
| 676 | flat black coated | zinc | US19 |
| 677 | bright brass plated | zinc | US3 |
| 678 | satin brass plated | zinc | US4 |
| 679 | bright bronze plated | zinc | US9 |

| | | | |
|---|---|---|---|
| 680 | satin bronze plated | zinc | US10 |
| 681 | bright chromium plated | zinc | US26 |
| 682 | satin chromium plated | zinc | US26D |
| 683 | oxidized satin brass plated, oil rubbed | zinc | — |
| 684 | black chrome, bright | brass, bronze | — |
| 685 | black chrome, satin | brass, bronze | — |
| 686 | black chrome, bright | steel | — |
| 687 | black chrome, satin | steel | — |
| 688 | satin aluminum, gold anodized | aluminum | US4 |
| 689 | aluminum plated | any | US28 |
| 690 | dark bronze painted | any | US20 |
| 691 | light bronze painted | any | US10 |
| 692 | tan painted | any | — |
| 693 | black painted | any | — |
| 694 | medium bronze painted | any | — |
| 695 | dark bronze painted | any | — |
| 696 | satin brass painted | any | US4 |
| 697 | bright brass plated | plastic | US3 |
| 698 | satin brass plated | plastic | US4 |
| 699 | satin bronze plated | plastic | US10 |
| 700 | bright chromium plated | plastic | US26 |
| 701 | satin chromium plated | plastic | US26D |
| 702 | satin chromium plated | aluminum | US26D |
| 703 | oxidized satin bronze plated, oil rubbed | aluminum | US10B |
| 704 | oxidized satin bronze plated, oil rubbed | zinc | US10B |

(Courtesy of Builders Hardware Manufacturers Assn., Inc.)

*Appendix* **E**

# Manufacturers and distributors of locksmithing equipment and supplies

A-1 SECURITY MFG. CORP.
3528 Mayland Court
Richmond, VA 23233

AAA-GLENN LOCKSMITH SUPPLY
P.O. Box 6520
Erie, PA 16512

ABUS LOCK CO.
218 W. Cummings Park
P.O. Box 2367
Woburn, MA 01888

ACCREDITED LOCK SUPPLY CO.
1161 Patterson Plank Rd.
Secaucus, NJ 07094

ADAMS RITE MFG. CO.
4040 S. Capitol Ave.
City of Industry, CA 91749

AEGIS LOCKWORKS LTD.
7020 82 Ave.
Edmonton, Alberta T6B 0E7

ALARM LOCK SYSTEMS INC.
P.O. Box 2001
Pine Brook, NJ 07058

ALL-LOCK CO.
Sub. of General
   Auto. Specialty Co., Inc.
P.O. Box 3042
North Brunswick, NJ 08902

AMERICAN LOCK CO./JUNKUNC BROS.
3400 W. Exchange Rd.
Crete, IL 60417

ARROW
Essex Industries, Inc.
555 Long Wharf Dr.
New Haven, CT 06511

BELWITH INTERNATIONAL
18071 Arenth Ave.
P.O. Box 8430
City of Industry, CA 91748

BEST LOCK CORP.
P.O. Box 50444
6165 E. 75th St.
Indianapolis, IN 46250

BRIGGS & STRATTON
  TECHNOLOGIES
P.O. Box 702
Milwaukee, WI 53201

COLE NATIONAL CORP.
29001 Cedar Rd.
Cleveland, OH 44124

CORKEY CONTROL SYSTEMS, INC.
20705 S. Western Ave., #108
Torrance, CA 90501

DETEX CORP.
302 Detex Dr.
New Braunfels, TX 78130

DEXTER LOCK
Subsidiary of Master Lock Co.
300 Webster Rd.
Auburn, AL 36830

DIMARK INTERNATIONAL, INC.
3233 Skyway Dr./Corp. H6R 3D
Santa Maria, CA 93455

DOM SECURITY LOCKS
225 Episcopal Rd.
Berlin, CT 06037

ESP CORP.
Engineered Security Products
375 Harvard St.
Leominster, MA 01453

FALCON LOCK
7345 Orangewood
Garden Grove, CA 92641

FOLGER ADAM CO.
16300 W. 103rd St.
Lemont, IL 60439

FORT LOCK CORP.
3000 N. River Road
River Grove, IL 60171

FRAMON MANUFACTURING CO., INC.
909 Washington Ave.
Alpena, MI 49707

HPC INC.
3999 N. 25th Ave.
Schiller Park, IL 60176

ILCO UNICAN CORP.
400 Jeffreys Rd.
Rocky Mount, NC 27804

ILCO UNICAN INC.
Dominion Lock Division
7301 Decarie Blvd.
Montreal, Que. H4P 2G7
Canada

ITL TOOLS LTD.
12779 80th Ave., #105
Surrey, BC
V3W 3A6 Canada

KWIKSET CORPORATION
A Black & Decker Company
516 E. Santa Ana St.
Anaheim, CA 92805

LORI CORP.
Old Turnpike Rd.
P.O. Box 490
Southington, CT 06489

LUCKY LINE PRODUCTS, INC.
7890 Dunbrook Rd.
San Diego, CA 92126

M.A.G. ENG. & MFG. CO., INC.
15261 Transistor Lane
Huntington Beach, CA 92649

MAKITA U.S.A., INC.
14930-C Northam St.
La Mirada, CA 90638

MASTER LOCK CO.
2600 N. 32nd St.
Milwaukee, WI 53210

MEDECO SECURITY LOCKS, INC.
P.O. Box 3075
Salem, VA 24153

MILWAUKEE ELECTRIC TOOL
CORPORATION
13135 W. Lisbon Rd.
Brookfield, WI 53005

MRL, INC.
7640 Fullerton Rd.
Springfield, VA 22153

PRECISION PRODUCTS, INC.
P.O. Box 4441
Lexington, KY 40544

PRESO-MATIC LOCK CO., INC.
3048 Industrial 33rd St.
Ft. Pierce, FL 34946-8694

R & D TOOL CO.
7705 R.C. Gorman Ave., NE
Albuquerque, NM 87122-2738

RAINBOW KEY CO.
1025A S. Santa Fe Ave.
Vista, CA 92083

ROFU INTERNATIONAL
CORPORATION
3725 Old Conejo Rd.
Newbury Park, CA 91320

SCHLAGE LOCK CO.
P.O. Box 3324
San Francisco, CA 94119

SECURITECH GROUP
54−45 44th St.
Maspeth, NY 11378

SECURITRON
1815 W. 205th St., #105
Torrance, CA 90501

SERRUBEC INC.
2069 Chartier Ave.
Montreal, Dorval
Que. H9P 1H3
Canada

SIMPLEX ACCESS CONTROLS CORP.
P.O. Box 4114
2941 Indiana Ave.
Winston Salem, NC 27115-4114

SKIL CORP.
4300 W. Peterson
Chicago, IL 60646

SLIDE LOCK TOOL CO.
P.O. Box 386
Louisville, TN 37777

STANLEY HARDWARE
195 Lake St.
New Britain, CT 06050

STAR KEY & LOCK MFG. CO., INC.
1274 Flushing Ave.
Brooklyn, NY 11237

TANN CANADA LTD.
Zeiss Div.
725 Main St. E.
Milton, Ont. L9T-3Z3
Canada

ULTRA HARDWARE PRODUCTS
Box 679
9246 Commerce Hwy.
Pennsauken, NJ 08110

U.S. LOCK CORP.
77 Rodeo Dr.
Edgewood, NY 11717

VON DUPRIN, INC.
2720 Tobey Dr.
P.O. Box 6023
Indianapolis, IN 46206

WESLOCK
AJP Industries Company
13344 S. Main St.
Los Angeles, CA 90061

YALE SECURITY DIVISION
P.O. Box 25288
Charlotte, NC 28229-8010

*Appendix* **F**

# Locksmithing-related organizations

AMERICAN SOCIETY FOR INDUSTRIAL SECURITY
1655 N. Ft. Meyer Dr., Suite 1200
Arlington, VA 22209

ASSOCIATED LOCKSMITHS OF AMERICA, INC.
3003 Live Oak St.
Dallas, TX 75204

DOOR AND HARDWARE INSTITUTE
7711 Old Springhouse Rd.
McLean, VA 22102-3474

INSTITUTIONAL LOCKSMITHS' ASSOCIATION
P.O. Box 108
Woodville, MA 01784-0108

LOCKSMITHS GUILD
850 Busse Hwy.
Park Ridge, IL 60068

NATIONAL LOCKSMITH SUPPLIERS ASSOCIATION
1900 Arch St.
Philadelphia, PA 19103

SAFE & VAULT TECHNICIAN'S ASSOCIATION
5083 Danville Rd.
Nicholasville, KY 40356

SAFECRACKERS INTERNATIONAL/NATIONAL ANTIQUE SAFE ASSN.
P.O. Box 110099
1142 Nucla St.
Aurora, CO 80011

VEHICLE SECURITY ASSOCIATION
5100 Forbes Blvd.
Lanham, MD 20706

# Glossary

**ac** Alternating current. Electrical current that reverses its direction of flow at regular intervals. For practical purposes, alternating current is the type of electricity that flows throughout a person's house and is accessed by the use of wall sockets.

**access code** The symbolic data, usually in the form of numbers or letters, that allows entry into an access controlled area without prompting an alarm condition.

**access control** Procedures and devices designed to control or monitor entry into a protected area. Many times access control is used to refer to electronic and electromechanical devices that control or monitor access.

**Ace lock** A term sometimes used to refer to any tubular key lock. The term is more properly used to refer to the Chicago Ace Lock, the first brand name for a tubular key lock.

**actuator** A device, usually connected to a cylinder, which, when activated, causes a lock mechanism to operate.

**adjustable mortise cylinder** Any mortise cylinder whose length can be adjusted for a better fit in doors of varying thickness.

**AFTE** Association of Firearm and Toolmark Examiners.

**AHC** Architectural Hardware Consultant, as certified by the Door and Hardware Institute.

**all-section key blank** The key section that enters all keyways of a multiplex key system.

**ALOA** Associated Locksmiths of America.

**ampere** (or amp) A unit of electrical current.

**angularly bitted key** A key that has cuts made into the blade at various degrees of rotation from the perpendicular.

**annunciator**   A device, often used in an alarm system, that flashes lights, makes noises, or otherwise attracts attention.

**ANSI**   American National Standards Institute.

**anti-passback**   A feature in some electronic access control systems designed to make it difficult for a person who has just gained entry into a controlled area to allow another person to also use the card to gain entry.

**antipick latch**   A spring latch fitted with a parallel bar that is depressed by the strike when the door is closed. When the bar is depressed it prevents the latch from responding to external pressure from lock picking tools.

**armored front**   A plate covering the bolts or set screws holding a cylinder to its lock. These bolts are normally accessible when the door is ajar.

**ASIS**   American Society for Industrial Security.

**associated change key**   A change key that is related directly to particular master key(s) through the use of constant cuts.

**associated master key**   A master key that has particular change keys related directly to its combination through the use of constant cuts.

**ASTM**   American Society for Testing and Materials.

**audit trail**   A record of each entry and exit within an access controlled area.

**auxiliary code**   A secondary or temporary access code.

**back plate**   A thin piece of metal, usually with a concave portion, used with machine screws to fasten certain types of cylinders to a door.

**backset**   The horizontal distance from the edge of a door to the center of an installed lock cylinder, keyhole, or knob hub. On locks with rabbeted fronts, it is measured from the upper step at the center of the lock face.

**ball bearing**   1. A metal ball used in the pin stack to accomplish some types of hotel or construction keying. 2. A ball, usually made of steel, used by some lock manufacturers as the bottom element in the pin stack in one or more pin chambers. 3. Any metal ball used as a tumbler's primary component.

**barrel key**   A key with a bit projecting from a hollow cylindrical shank. The hollow fits over a pin in a lock keyway and helps keep the key aligned. The key is also known as a hollow post key or pipe key.

**bell key**   A key whose cuts are in the form of wavy grooves milled into the flat sides of the key blade. The grooves usually run the entire length of the blade.

**BHMA**   Builders Hardware Manufacturers Association.

**bible**   That portion of the cylinder shell normally housing the pin chambers, especially those of key-in-knob cylinders or certain rim cylinders.

**bicentric cylinder**   A cylinder that has two independent plugs, usually with different keyways. Both plugs are operable from the same face of the cylinder. It is designed for use in extensive master key systems.

**bi-directional cylinder**   A cylinder that can be operated in a clockwise and counterclockwise direction with a single key.

**binary cut key**   A key whose combination only allows for two possibilities in each bitting position: cut/no cut.

**binary type cylinder or lock**   A cylinder or lock whose combination only allows for two bitting possibilities in each bitting position.

**bit**   1. The part of the key that serves as the blade, usually for use in a warded or lever tumbler lock. 2. To cut a key.

**bit key**   A key with a bit projecting from a solid cylindrical shank. The key is sometimes referred to as a ''skeleton key.'' A bit key is used to operate a bit key lock.

**bit key lock**   A lock operated by a bit key.

**bitting**   1. The number(s) representing the dimensions of the key cut(s). 2. The actual cut(s) or combination of a key.

**bitting depth**   The depth of a cut that is made into the blade of a key.

**bitting list**   A listing of all the key combinations used within a system. The combinations are usually arranged in order of the blind code, direct code, and/or key symbol.

**bitting position**   The location of a key cut.

**blade**   The portion of a key that contains the cuts and/or milling.

**blank**   See *key blank*.

**blank**   Uncut key.

**blind code**   A designation, unrelated to the bitting, assigned to a particular key combination for future reference when additional keys or cylinders are needed.

**block master key (BM)**   The one pin master key for all combinations listed as a block in the standard progression format.

**bottom of blade**   The portion of the blade opposite the cut edge of a single-bitted key.

**bottom pin**   A tumbler, usually cylindrical (also may be conical, ball-shaped or chisel-pointed), that makes contact with the key.

**bow**   The portion of the key that serves as a grip or handle.

**bow stop**   A type of stop located near the key bow.

**broach**   1. A tool used to cut the keyway into the cylinder plug. 2. To cut the keyway into a cylinder plug with a broach.

**building master key**   A master key that operates all or most master keyed locks in a given building.

**bypass key**   The key that operates a key override cylinder.

**cam**   A flat actuator or locking bolt attached to the rear of a cylinder perpendicular to its plug and rotated by the key.

**cam lock**   A complete locking assembly in the form of a cylinder whose cam is the actual locking bolt.

**cap**   1. A spring cover for a single-pin chamber. 2. A part that can serve as a plug retainer and/or a holder for the tailpiece. 3. To install a cap.

**capping block**   A holding fixture for certain interchangeable cores that aids in the installation of the caps.

**case**   (of a cylinder) See *shell*.

**case** (of a lock) The box that houses the lock-operating mechanism.

**case ward** A ward or obstruction integral to the case of a warded lock.

**central processing unit (CPU)** Also called a central processor; the section in a digital computer that contains the logic and internal memory units.

**chamber** Any cavity in a cylinder plug and/or shell that houses the tumblers.

**change key** A key that operates only one cylinder or one group of keyed alike cylinders in a keying system.

**changeable bit key** A key that can be recombined by exchanging and/or rearranging portions of its bit or blade.

**circuit** A complete path through which electricity flows to perform work.

**circuit breaker** A device designed to protect a circuit by automatically breaking (or opening) the circuit when current flow becomes excessive.

**CK** 1. Change key. 2. Control key.

**clutch** The part of a profile cylinder that transfers rotational motion from the inside or outside element to a common cam or actuator.

**code** A designation assigned to a particular key combination for reference when additional keys or cylinders might be needed.

**code key** A key cut to a specific code rather than duplicated from a pattern key. It may or may not conform to the lock manufacturer's specifications.

**code original key** A code key that conforms to the lock manufacturer's specifications.

**combinate** To set a combination in a lock, cylinder, or key.

**combination** The group of numbers that represent the bitting of a key and/or the tumblers of a lock or cylinder.

**combination lock** A lock that may or may not be operated with a key, but can be operated by inserting a combination of numbers, letters, or other symbols by rotating a dial or by pushing buttons.

**combination wafer** A type of disc tumbler used in certain binary type disc tumbler key-in-knob locks. Its presence requires that a cut be made in that position of the operating key(s).

**compensate drivers** To select longer or shorter top pins, depending on the length of the rest of the pin stack, in order to achieve a uniform pin stack height.

**complementary keyway** Usually a disc tumbler keyway used in master keying. It accepts keys of different sections whose blades contact different bearing surfaces of the tumblers.

**complex circuit** A combination of series and parallel circuits.

**composite keyway** A keyway that has been enlarged to accept more than one key section, often key sections of more than one manufacturer.

**concealed shell cylinder** A specially constructed (usually mortise) cylinder. Only the plug face is visible when the lock trim is in place.

**conductor**   Material, such as copper wire, used to direct current flow.

**connecting bar**   A flat bar attached to the rear of the plug in a rim lock to operate the locking bar mechanism.

**constant cut**   Any bitting(s) that are identical in corresponding positions from one key to another in a keying system. They usually serve to group these keys together within a given level of keying, and/or link them with keys of other levels.

**construction breakout key**   A key used by some manufacturers to render all construction master keys permanently inoperative.

**construction core**   An interchangeable or removable core designed for use during the construction phase of a building. The cores are normally keyed alike and, upon completion of construction, they are to be replaced by the permanent system's cores.

**construction master key (CMK)**   A key normally used by construction personnel for a temporary period during building construction. It may be rendered permanently inoperative without disassembling the cylinder.

**construction master keyed**   Of or pertaining to a cylinder that is operated temporarily by a construction master key.

**control cut**   Any bitting that operates the retaining device of an interchangeable or removable core.

**control key**   1. A key whose only purpose is to remove and/or install an interchangeable or removable core. 2. A bypass key used to operate and/or reset some combination type locks. 3. A key that allows disassembly of some removable cylinder locks.

**control lug**   The part of an interchangeable or removable core retaining device that locks the core into its housing.

**control sleeve**   The part of an interchangeable core retaining device surrounding the plug.

**controlled cross keying**   A condition in which two or more different keys of the same level of keying and under the same higher level key(s) operate one cylinder by design. For example, XAA1 can be operated by AA2 but not by AB1. This condition could severely limit the security of the cylinder and the maximum expansion of the system when more than a few of these different keys operate a cylinder, or when more than a few differently cross keyed cylinders per system are required.

**core**   A complete unit, often with a figure-8 shape, which usually consists of the plug, shell, tumblers, springs, plug retainer, and spring cover(s). It is primarily used in removable and interchangeable core cylinders and locks.

**corrugated key**   A key with pressed longitudinal corrugations in its shank to correspond to a compatibly shaped keyway.

**CPL**   Certified Professional Locksmith (as certified by Associated Locksmiths of America).

**CPP**   Certified Protection Professional (as certified by American Society for Industrial Security).

**crash bar**  See *panic bar*.

**cross keying**  The deliberate process of combinating a cylinder (usually in a master key system) to two or more different keys that would not normally be expected to operate together. See also *controlled cross keying* and *uncontrolled cross keying*.

**CSI**  Construction Specifiers Institute.

**current**  The flow of electricity. Current is measured in amperes.

**cut**  An indentation, notch, or cutout made in a key blank in order to make it operate a lock. See *bitting*.

**cut**  To make cuts into a key blade.

**cut angle**  A measurement, usually expressed in degrees, for the angle between the two sides of a key cut.

**cut edge**  A key that has been bitted or combinated.

**cut root**  The bottom of a key cut.

**cut root shape**  The shape of the bottom of a key cut. It might have a flat or radius of a specific dimension, or be a perfect V.

**cutter**  The part of a key machine that makes the cuts into the key blank.

**cylinder**  A complete operating unit that usually consists of the plug, shell, tumblers, springs, plug retainer, a cam/tailpiece or other actuating device, and all other necessary operating parts.

**cylinder blank**  A dummy cylinder that has a solid face and no operating parts.

**cylinder clip**  A spring steel device used to secure some types of cylinders.

**cylinder collar**  A plate or ring installed under the head of a cylinder to improve appearance and/or security.

**cylinder guard**  A protective cylinder mounting device.

**cylinder key**  A broad generic term including virtually all pin and disc tumbler keys.

**dc**  Direct current. Electric current that flows in one direction. For practical purposes, direct current is the type of electricity obtained from batteries.

**deadbolt**  A lockbolt having no spring action, usually rectangular and actuated by a key or turn knob.

**deadlatch**  A lock with a beveled latchbolt that can be automatically or manually locked against end pressure when projected.

**declining step key**  A key whose cuts are progressively deeper from bow to tip.

**decode**  To determine a key combination by physical measurement of a key and/or cylinder parts.

**degree of rotation**  A specification for the angle at which a cut is made into a key blade as referenced from the perpendicular.

**department master key**  A master key that operates all or most master keyed locks of a given department.

**depth key set**  A set of keys used to make a code original key on a key duplicating machine to a lock manufacturer's given set of key bitting specifications. Each key is cut with the correct spacing to one depth only in all bitting positions, with one key for each depth.

**derived series**  A series of blind codes and bittings that are directly related to those of another bitting list.

**DHI**  Door and Hardware Institute.

**dimple**  A key cut in a dimple key.

**dimple key**  A key that has cuts drilled or milled into its blade surfaces. The cuts normally do not change the blade silhouette.

**direct code**  A designation assigned to a particular key that includes the actual combination of the key.

**disc tumbler**  1. A flat tumbler that must be drawn into the cylinder plug by the proper key so that none of its extremities extends into the shell. 2. A flat, usually rectangular tumbler with a gate that must be aligned with a sidebar by the proper key.

**display key**  A special change key in a hotel master key system that will allow access to one designated guest room, even if the lock is in the shut-out mode. It might also act as a shut-out key for that room.

**dmm**  Digital multimeter. A device used to measure current, resistance, and voltage.

**double bitted key**  A key bitted on two opposite surfaces.

**double pin**  To place more than one master pin in a single pin chamber.

**driver**  See *top pin*.

**driver spring**  A spring placed on top of the pin stack to exert pressure on the pin tumblers.

**drop**  A pivoting or swinging dust cover.

**dummy cylinder**  A non-functional facsimile of a rim or mortise cylinder used for appearance only, usually to conceal a cylinder hole.

**duplicate key**  Any key reproduced from a pattern key.

**dust cover**  A device designed to prevent foreign matter from entering a mechanism through the keyway.

**dustproof cylinder**  A cylinder designed to prevent foreign matter from entering either end of the keyway.

**effective plug diameter**  The dimension obtained by adding the root depth of a key cut to the length of its corresponding bottom pin, which establishes a perfect shear line. This will not necessarily be the same as the actual plug diameter.

**ejector hole**  A hole found on the bottom of certain interchangeable cores under each pin chamber. It provides a path for the ejector pin.

**ejector pin**  A tool used to drive all the elements of a pin chamber out of certain interchangeable cores.

**electric strike**  An electrically controlled solenoid and mechanical latching device.

**electromagnetic lock**  A locking device that uses magnetism to keep it in a locked position.

**emergency key**  The key that operates a privacy function lockset.

**emergency master key**  A special master key that usually operates all guest room locks in a hotel master key system at all times, even in the shut-out mode. This key may also act as a shut-out key.

**emf**  Electromotive force (also called voltage). The force needed to cause current to flow within a circuit.

**EMK**   Emergency master key.

**ENG**   Symbol for engineer's key.

**escutcheon**   A surface-mounted trim that enhances the appearance and/or security of a lock installation.

**extractor key**   A tool that normally removes a portion of a two-piece key or blocking device from a keyway.

**face plate**   A mortise lock cover plate exposed in the edge of the door.

**factory original key**   The cut key furnished by the lock manufacturer for a lock or cylinder.

**fail safe lock**   A lock that automatically unlocks during a power failure.

**fail secure lock**   A lock that automatically locks during a power failure.

**fence**   A projection on a lock bolt that prevents movement of the bolt unless it can enter gates of properly aligned tumblers.

**finish**   A material, coloring and/or texturing specification.

**fireman's key**   A key used to override normal operation of elevators, bringing them to the ground floor.

**first generation duplicate**   A key that was duplicated using a factory original key or a code original key as a pattern.

**first key**   Any key produced without the use of a pattern key.

**five-column progression**   A process wherein key bittings are obtained by using the cut possibilities in five columns of the key bitting array.

**five-pin master key**   A master key for all combinations obtained by progressing five bitting positions.

**flexible head mortise cylinder**   An adjustable mortise cylinder that can be extended against spring pressure to a slightly longer length.

**floor master key**   A master key that operates all or most master keyed locks on a particular floor of a building.

**following tool**   See *plug follower*.

**four-column progression**   A process wherein key bittings are obtained by using the cut possibilities in four columns of the key bitting array.

**four-pin master key**   A master key for all combinations obtained by progressing four bitting positions.

**gate**   A notch cut into the edge of a tumbler to accept a fence or sidebar.

**graduated drivers**   A set of top pins of different lengths. Usage is based on the height of the rest of the pin stack, in order to achieve a uniform pin stack height.

**grand master key (GMK)**   The key that operates two or more separate groups of locks, which are each operated by a different master key.

**grand master key system**   A master key system that has exactly three levels of keying.

**grand master keyed**   Of or pertaining to a lock or cylinder that is or is to be keyed into a grand master key system.

**great grand master key (GGMK)**   The key that operates two or more separate groups of locks, which are each operated by a different grand master key.

**great grand master key system**   A master key system that has exactly four levels of keying.

**great grand master keyed**   Of or pertaining to a lock or cylinder that is or is to be keyed into a great grand master key system.

**great great grand master key (GGGMK)**   The key that operates two or more separate groups of locks, which are each operated by different great grand master keys.

**great great grand master key system**   A master key system that has five or more levels of keying.

**great great grand master keyed**   Of or pertaining to a lock or cylinder which is or is to be keyed into a great great grand master key system.

**ground**   An electrical connection to a metallic object that is either buried in the earth or is connected to a metallic object buried in the earth.

**guard key**   A key that must be used in conjunction with a renter's key to unlock a safe deposit lock. It is usually the same for every lock within an installation.

**guest key**   A key in a hotel master key system that is normally used to unlock only the one guest room for which it was intended, but will not operate the lock in the shut-out mode.

**guide**   That part of a key machine that follows the cuts of a pattern key or template during duplication.

**hardware schedule**   A listing of the door hardware used on a particular job. It includes the types of hardware, manufacturers, locations, finishes, and sizes. It should include a keying schedule specifying how each locking device is to be keyed.

**high-security cylinder**   A cylinder that offers a greater degree of resistance to any or all of the following: picking, impressioning, key duplication, drilling or other forms of forcible entry.

**high-security key**   A key for a high-security cylinder.

**hold open cylinder**   A cylinder provided with a special cam that will hold a latch bolt in the retracted position when so set by the key.

**holding fixture**   A device that holds cylinder plugs, cylinders, housings, and/or cores to facilitate the installation of tumblers, springs and/or spring covers.

**hollow driver**   A top pin hollowed out on one end to receive the spring, typically used in cylinders with extremely limited clearance in the pin chambers.

**horizontal group master key (HGM)**   The two-pin master key for all combinations listed in all blocks in a line across the page in the standard progression format.

**housekeeper's key (HKP)**   A selective master key in a hotel master key system that may operate all guest and linen rooms and other housekeeping areas.

**housing**   That part of a locking device designed to hold a core.

**impression**   1. The mark made by a tumbler on its key cut. 2. To fit a key by the impression technique.

**impression technique**   A means of fitting a key directly to a locked cylinder by manipulating a blank in the keyway and cutting the blank where the tumblers have made marks.

**incidental master key**   A key cut to an unplanned shearline created when the cylinder is combinated to the top master key and a change key.

**increment**   A usually uniform increase or decrease in the successive depths of a key cut, which must be matched by a corresponding change in the tumblers.

**indicator**   A device that provides visual evidence that a deadbolt is extended or that a lock is in the shut out mode.

**individual key**   An operating key for a lock or cylinder that is not part of a keying system.

**insulator**   Materials such as rubber and plastics that provide resistance to current flow. They are used to cover conductors and electrical devices to prevent unwanted current flow.

**interchangeable core (IC)**   A key removable core that can be used in all or most of the core manufacturer's product line. No tools other than the control key are required for removal of the core.

**interlocking pin tumbler**   A type of pin tumbler that is designed to be linked together with all other tumblers in its chamber when the cylinder plug is in the locked position.

**jumbo cylinder**   A rim or mortise cylinder $1^{1}/_{2}$ inches in diameter.

**k**   Symbol for "keys" used after a numerical designation of the quantity of the keys requested to be supplied with the cylinders: 1k, 2k, 3k, etc. It is usually found in hardware/keying schedules.

**KA**   Keyed alike. This symbol indicates cylinders that are to be operated by the same key(s)—for example: KA1, KA2, etc. KA/2, KA/3, etc. is the symbol used to indicate the quantity of locks or cylinders in keyed alike groups. These groups are usually formed from a larger quantity.

**KBA**   Key bitting array.

**KD**   Keyed different.

**key**   A properly combinated device that is, or most closely resembles, the device specifically intended by the lock manufacturer to operate the corresponding lock.

**key bitting array**   A matrix (graphic) of all possible bittings for change keys and master keys as related to the top master key.

**key bitting specifications**   The technical data required to bit a given key blank or family of key blanks to the lock manufacturer's dimensions.

**key bitting punch**   A manually operated device that stamps or punches the cuts into the key blade, rather than grinding or milling them.

**key blank**   Any material manufactured to the proper size and configuration that allows its entry into the keyway of a specific locking device. A key blank has not yet been combinated or cut.

**key changeable**   Of or pertaining to a lock or cylinder that can be recombinated without disassembly, by use of a key. The use of a tool might also be required.

**key coding machine**   A key machine designed for the production of code keys. It might or might not also serve as a duplicating machine.

**key control**   1. Any method or procedure that limits unauthorized acqui-
sition of a key and/or controls distribution of authorized keys. 2. A
systematic organization of keys and key records.

**key cut(s)**   The portion of the key blade that remains after being cut and
that aligns the tumbler(s).

**key cut profile**   The shape of a key cut, including the cut angle and the
cut root shape.

**key duplicating machine**   A key machine that is designed to make copies
from a pattern key.

**key gauge**   A device (usually flat) with a cutaway portion indexed with a
given set of depth or spacing specifications. Used to help determine
the combination of a key.

**key-in-knob cylinder**   A cylinder used in a key-in-knob lockset.

**key interchange**   An undesirable condition, usually in a master key system, whereby a key unintentionally operates a cylinder or lock.

**key machine**   Any machine designed to cut keys.

**key manipulation**   Manipulation of an incorrect key in order to operate a
lock or cylinder.

**key milling**   The grooves machined into the length of the key blade to
allow its entry into the keyway.

**key override**   A provision allowing interruption or circumvention of
normal operation of a combination lock or electrical device.

**key override cylinder**   A lock cylinder installed in a device to provide a
key override function.

**key pull position**   Any position of the cylinder plug at which the key can
be removed.

**key records**   Records that typically include some or all of the following:
bitting list, key bitting array, key system schematic, end user, number
of keys/cylinders issued, names of persons to whom keys were
issued, hardware/keying schedule.

**key retaining**   1. Of or pertaining to a lock that must be locked before its
key can be removed. 2. Of or pertaining to a cylinder or lock that
may prevent removal of a key without the use of an additional key
and/or tool.

**key section**   The exact cross-sectional configuration of a key blade as
viewed from the bow toward the tip.

**keyswitch**   A switch that is operated with a key.

**key system schematic**   A drawing with blocks utilizing keying symbols,
usually illustrating the hierarchy of all keys within a master key system. It indicates the structure and total expansion of the system.

**key trap core/cylinder**   A special core or cylinder designed to capture
any key to which it is combinated, once that key is inserted and
turned slightly.

**keyed**   1. Combinated. 2. Having provision for operation by key.

**keyed alike**   Of or pertaining to two or more locks or cylinders that have
the same combination. They may or may not be part of a keying system.

**keyed random**    Of or pertaining to a cylinder or group of cylinders selected from a limited inventory of different key changes. Duplicate bittings may occur.

**keying**    Any specification for how a cylinder or group of cylinders are combinated in order to control access.

**keying conference**    A meeting of the end user and the keying system supplier at which the keying and levels of keying, including future expansion, are determined and specified.

**keying kit**    A compartmented container that holds an assortment of tumblers, springs, and/or other parts.

**keying schedule**    A detailed specification of the keying system listing how all cylinders are to be keyed and the quantities, markings, and shipping instructions of all keys and/or other parts.

**keying symbol**    A designation used for a lock or cylinder combination in the standard key coding system: AA1, XAA1, X1X, etc.

**keyway**    1. The opening in a lock or cylinder shaped to accept a key bit or blade of a proper configuration. 2. The exact cross-sectional configuration of a keyway as viewed from the front. It is not necessarily the same as the key section.

**keyway unit**    The plug of certain binary type disc tumbler key-in-knob locks.

**KR**    1. Keyed random. 2. Key retaining.

**KWY**    Keyway.

**layout tray**    A compartmented container used to organize cylinder parts during keying or servicing.

**lazy cam/tailpiece**    A cam or tailpiece designed to remain stationary while the cylinder plug is partially rotated (or vice-versa).

**LCD**    Liquid crystal display.

**levels of keying**    The divisions of a master key system into hierarchies of access, as shown in the TABLE G-1. Note that the standard key coding system has been expanded to include key symbols for systems of more than four levels of keying.

**lever tumbler**    A flat, spring-loaded tumbler that pivots on a post. It contains a gate that must be aligned with a fence to allow movement of the bolt.

**loading tool**    A tool that aids installation of cylinder components into the cylinder shell.

**lock plate compressor**    A tool designed to depress the lock plate of an automobile steering column. The tool (also called lock plate remover) is used when disassembling automobile steering columns.

**lockout**    Any situation in which the normal operation of a lock or cylinder is prevented.

**lockout key**    A key made in two pieces. One piece is trapped in the keyway by the tumblers when inserted and blocks entry of any regular key. The second piece is used to remove the first piece.

### Table G-I  Master keying system.

#### TWO LEVEL SYSTEM

| level of keying | Key name | abb. | Key symbol |
|---|---|---|---|
| Level  II | master key | MK | AA |
| Level  I | change key | CK | 1AA, 2AA, etc. |

#### THREE LEVEL SYSTEM

| level of keying | Key name | abb. | Key symbol |
|---|---|---|---|
| Level  III | grand master key | GMK | A |
| Level  II | master key | MK | AA, AB, etc. |
| Level  I | change key | CK | AA1, AA2, etc. |

#### FOUR LEVEL SYSTEM

| level of keying | Key name | abb. | Key symbol |
|---|---|---|---|
| Level  IV | great grand master key | GGMK | GGMK |
| Level  III | grand master key | GMK | A, B, etc. |
| Level  II | master key | MK | AA, AB, etc. |
| Level  I | change key | CK | AA1, AA2, etc. |

#### FIVE LEVEL SYSTEM

| level of keying | Key name | abb. | Key symbol |
|---|---|---|---|
| Level  V | great great grand master key | GGGMK | GGGMK |
| Level  IV | great grand master key | GGMK | A, B, etc. |
| Level  III | grand master key | GMK | AA, AB, etc. |
| Level  II | master key | MK | AAA, AAB, etc. |
| Level  I | change key | CK | AAA1, AAA2, etc. |

#### SIX LEVEL SYSTEM

| level of keying | Key name | abb. | Key symbol |
|---|---|---|---|
| Level  VI | great great grand master key | GGGMK | GGGMK |
| Level  V | great grand master key | GGMK | A, B, etc. |
| Level  IV | grand master key | GMK | AA, AB, etc. |
| Level  III | master key | MK | AAA, AAB, etc. |
| Level  II | sub-master key | SMK | AAAA, AAAB, etc. |
| Level  I | change key | CK | AAAA1, AAAA2, etc. |

**mA**   Milliampere.

**MACS**   Maximum adjacent cut specification.

**maid's master key**   The master key in a hotel master key system given to the maid. It operates only cylinders of the guest rooms and linen closets in the maid's designated area.

**maison key system**   A keying system in which one or more cylinders are operated by every key (or relatively large numbers of different keys) in the system—for example, main entrances of apartment buildings operated by all individual suite keys of the building.

**manipulation key**   Any key other than a correct key that can be variably positioned and/or manipulated in a keyway to operate a lock or cylinder.

**master disc**   A special disc tumbler with multiple gates to receive a sidebar.

**master key(MK)**   1. A key that operates all the master keyed locks or cylinders in a group, with each lock or cylinder operated by its own change key. 2. To combinate a group of locks or cylinders such that each is operated by its own change key as well as by a master key for the entire group.

**master key changes**   The number of different usable change keys available under a given master key.

**master key system**   1. Any keying arrangement that has two or more levels of keying. 2. A keying arrangement that has exactly two levels of keying.

**master keyed**   Of or pertaining to a cylinder or group of cylinders that are combined so that all may be operated by their own change keys and by additional keys known as master keys.

**master keyed only**   Of or pertaining to a lock or cylinder that is combinated only to a master key.

**master lever**   A lever tumbler that can align some or all other levers in its lock so that lever gates are at the fence. It is typically used in locker locks.

**master pin**   1. Usually a cylindrical tumbler, flat on both ends, placed between the top and bottom pin to create an additional shear line. 2. A pin tumbler with multiple gates to accept a sidebar.

**master ring**   A tube-shaped sleeve located between the plug and shell of certain cylinders to create a second shear line. Normally the plug shear line is used for change key combinations and the shell shear line is used for master key combinations.

**master ring lock/cylinder**   A lock or cylinder equipped with a master ring.

**master wafer**   A ward used in certain binary type disc tumbler key-in-knob locks.

**maximum adjacent cut specification**   The maximum allowable depths to which opposing cuts can be made without breaking through the key blade. This is typically a consideration with dimple keys.

**metal oxide varister**   A voltage-dependent resistor.

**miscut**   Of or pertaining to a key that has been cut incorrectly.

**MOCS**   Maximum opposing cut specification.

**mogul cylinder**   A very large pin tumbler cylinder whose pins, springs, key, etc. are also proportionately increased in size. it is frequently used in prison locks.

**mortise**   An opening made in a door to receive a lock or other hardware.

**mortise cylinder**   A threaded cylinder typically used in mortise locks of American manufacture.

**MOV**   Metal oxide varister.

**multi-section key blank**   A key section that enters more than one, but not all, keyways in a multiplex key system.

**multiple gating**   A means of master keying by providing a tumbler with more than one gate.

**multiplex key blank**   Any key blank that is part of a multiplex key system.

**multiplex key system**   1. A series of different key sections that may be used to expand a master key system by repeating bittings on additional key sections. The keys of one section will not enter the keyway of another key section. This type of system always includes another key section which will enter more than one, or all of the keyways. 2. A keying system that uses such keyways and key sections.

**multitester**   A device designed to measure current, resistance, and voltage. The VOM and the DMM are two types of multitesters.

**mushroom pin**   A pin tumbler, usually a top pin, that resembles a mushroom. It is typically used to increase pick resistance.

**NC** (or **N/C**)   Normally closed.

**NCK**   No change key. This symbol is primarily used in hardware schedules.

**negative locking**   Locking achieved solely by spring pressure or gravity, which prevents a key cut too deeply from operating a lock cylinder.

**nickel-cadmium**   A long-life rechargeable battery or cell oftentimes used as a backup power supply.

**night latch**   An auxiliary rim lock with a spring latchbolt.

**NMK**   Not master keyed. This keying symbol is suffixed in parentheses to the regular key symbol. It indicates that the cylinder is not to be operated by the master key(s) specified in the regular key symbol— for example: AB6(NMK).

**NO** (or **N/O**)   Normally open.

**non-key-retaining (NKR)**   Of or pertaining to a lock whose key can be removed in both the locked and unlocked positions.

**non-keyed**   Having no provision for key operation. Note: This term also includes privacy function locksets operated by an emergency key.

**non-original key blank**   Any key blank other than an original.

**normally closed switch**   A switch whose contacts normally remain closed when electrical current isn't flowing through it.

**normally open switch**   A switch whose contacts normally remain open when electrical current isn't flowing through it.

**NSLA**   National Locksmith Suppliers Association.

**odometer method**   A means of progressing key bittings using a progression sequence of right to left.

**ohm**   A unit of measure of resistance to electrical current flow.

**Ohm's law**   The description of the relationship between voltage, current, and resistance in an electrical circuit. Ohm's law states that a resistance of 1 ohm passes through a current of one ampere, in response to 1 volt. Mathematically expressed, $E = IR$, with E as voltage in volts, I as current in amperes, and R as resistance in ohms.

**one-bitted**   Of or pertaining to a cylinder that is combinated to keys cut to the manufacturer's reference number one bitting.

**one-column progression**   A process wherein key bittings are obtained by using the cut possibilities in one column of the bitting array.

**one pin master key**   A master key for all combinations obtained by progressing only one bitting position.

**operating key**   Any key that will properly operate a lock or cylinder to lock or unlock the lock mechanism, and is not a control key or reset key.

**original key blank**   A key blank supplied by the lock manufacturer to fit that manufacturer's specific product.

**page master key**   The three-pin master key for all combinations listed on a page in the standard progression format.

**panic bar**   A door-mounted exit bar designed to allow fast egress from the inside of the door and resistance to entry from the outside of the door.

**paracentric**   Of or pertaining to a keyway with one or more wards on each side projecting beyond the vertical center line of the keyway to hinder picking.

**parallel circuit**   An electrical circuit that provides more than one path for current to flow.

**PASS-Key**   Personalized Automotive Security System. See *VATS*.

**pass key**   A master key or skeleton key.

**pattern key**   1. An original key kept on file to use in a key duplicating machine when additional keys are required. 2. Any key used in a key duplicating machine to create a duplicate key.

**peanut cylinder**   A mortise cylinder of 3/4-inch diameter.

**pick**   1. A tool or instrument, other than the specifically designed key, made for the purpose of manipulating tumblers in a lock or cylinder into the locked or unlocked position through the keyway without obvious damage. 2. To manipulate tumblers in a keyed lock mechanism through the keyway without obvious damage, by means other than the specifically designed key.

**pick key**   A type of manipulation key, cut or modified to operate a lock or cylinder.

**pin**   To install pin tumblers into a cylinder and/or cylinder plug.

**pin chamber**   The corresponding hole drilled into the cylinder shell and/or plug to accept the pin(s) and spring.

**pin kit**   A type of keying kit for a pin tumbler mechanism.

**pin stack**   All the tumblers in a given pin chamber.

**pin stack height**   The measurement of a pin stack, often expressed in units of the lock manufacturer's increment or as an actual dimension.

**pin tumbler**   Usually a cylindrical shaped tumbler. Three types are normally used: bottom pin, master pin and top pin.

**pin tweezers**   A tool used in handling tumblers and springs.

**pinning block**   A holding fixture that assists in the loading of tumblers into a cylinder or cylinder plug.

**pinning chart**   A numerical diagram that indicates the sizes and order of installation of the various pins into a cylinder. The sizes are usually indicated by a manufacturer's reference number, which equals the quantity of increments a tumbler represents.

**plug**   The part of a cylinder containing the keyway, with tumbler chambers usually corresponding to those in the cylinder shell.

**plug follower**   A tool used to allow removal of the cylinder plug while retaining the top pins, springs, and/or other components within the shell.

**plug holder**   A holding fixture that assists in the loading of tumblers into a cylinder plug.

**plug retainer**   The cylinder component that secures the plug in the shell.

**plug spinner**   A tool designed to quickly spin a plug clockwise or counterclockwise into an unlocked position, when the lock has been picked, into a position that doesn't allow the lock to open.

**positional master keying**   A method of master keying typical of certain binary type disc tumbler key-in-knob locks and of magnetic and dimple key cylinders. Of all possible tumbler positions within a cylinder, only a limited number contain active tumblers. The locations of these active tumblers are rotated among all possible positions to generate key changes. Higher level keys must have more cuts or magnets than lower level keys.

**positive locking**   The condition brought about when a key cut that is too high forces its tumbler into the locking position. This type of locking does not rely on gravity or spring pressure.

**post**   The part of a bit key to which the bit is attached.

**practical key changes**   The total number of usable different combinations available for a specific cylinder or lock mechanism.

**prep key**   A type of guard key for a safe deposit box lock with only one keyway. It must be turned once and withdrawn before the renter's key will unlock the unit.

**privacy key**   A key that operates an SKD cylinder.

**profile cylinder**   A cylinder with a uniform cross section that slides into place and is held by a mounting screw. It is typically used in mortise locks of non-U.S. manufacturers.

**progress**   To select possible key bittings, usually in numerical order, from the key bitting array.

**progression**   A logical sequence of selecting possible key bittings, usually in numerical order from the key bitting array.

**progression column**   A listing of the key bitting possibilities available in one bitting position as displayed in a column of the key bitting array.

**progression list**   A bitting list of change keys and master keys arranged in sequence of progression.

**progressive**   Any bitting position that is progressed rather than held constant.

**proprietary**   Of or pertaining to a keyway and key section assigned exclusively to one end user by the lock manufacturer. It may also be protected by law from duplication.

**radiused blade bottom**   The bottom of a key blade that has been radiused to conform to the curvature of the cylinder plug it is designed to enter.

**random master keying**   Any undesirable process of master keying that uses unrelated keys to create a system.

**rap**   1. To unlock a plug from its shell by striking sharp blows to the spring side of the cylinder while applying tension to the plug. 2. To unlock a padlock shackle from its case by striking sharp blows to the sides in order to disengage the locking dogs.

**read key**   A key that allows access to the sales and/or customer data on certain types of cash control equipment, such as a cash register.

**recombinate**   To change the combination of a lock, cylinder, or key.

**recore**   To rekey by installing a different core.

**register groove**   The reference point on the key blade from which some manufacturers locate the bitting depths.

**register number**   1. A reference number, typically assigned by the lock manufacturer to an entire master key system. 2. A blind code assigned by some lock manufacturer to higher level keys in a master key system.

**rekey**   To change the existing combination of a cylinder or lock.

**relay**   A type of switching device, usually electronic or electromechanical.

**removable core**   A key removable core that can only be installed in one type of cylinder housing—for example, rim cylinder or mortise cylinder or key-in-knob locks.

**removable cylinder**   A cylinder that can be removed from a locking device by a key and/or tool.

**removal key**   The part of a two-piece key that is used to remove its counterpart from a keyway.

**renter's key**   A key that must be used together with a guard key, prep key, or electronic release to unlock a safe deposit lock. It is usually different for every unit within an installation.

**repin**   To replace pin tumblers, with or without changing the existing combination.

**reset key**   1. A key used to set some types of cylinders to a new combination. Many of these cylinders require the additional use of tools and/or the new operating key to establish the new combination. 2. A key that allows the tabulations on various types of cash control equipment such as cash registers to be cleared from the records of the equipment.

**resistance**    Opposition to electrical current flow.

**resistor**    A component that resists electrical current flow in a dc circuit.

**restricted**    Of or pertaining to a keyway and corresponding key blank whose sale and/or distribution is limited by the lock manufacturer in order to reduce unauthorized key proliferation.

**retainer clip tool**    A tool designed to install or remove retainer clips from automobiles.

**reversible key**    A symmetrical key that may be inserted either up or down to operate a lock.

**reversible lock**    A lock in which the latchbolt can be turned over and adapted to doors of either hand, opening in or out.

**rim cylinder**    A cylinder typically used with surface applied locks and attached with a back plate and machine screws. It has a tailpiece to actuate the lock mechanism.

**RL**    Registered Locksmith (as certified by Associated Locksmiths of America).

**RM**    Row master key.

**root depth**    The dimension from the bottom of a cut on a key to the bottom of the blade.

**rose**    A usually circular escutcheon.

**rotary tumbler**    A circular tumbler with one or more gates. Rotation of the proper key aligns the tumbler gates at a sidebar, fence, or shackle slot.

**rotating constant**    One or more cut(s) in a key of any level that remain constant throughout all levels and are identical to the top master key cuts in their corresponding positions. The positions where the top master key cuts are held constant may be moved, always in a logical sequence.

**rotating constant method**    A method used to progress key bittings in a master key system, wherein at least one cut in each key is identical to the corresponding cut in the top master key. The identical cut is moved to different locations in a logical sequence until each possible planned position has been used.

**row master key**    The one pin master key for all combinations listed on the same line across the page in the standard progression format.

**S/A**    Sub-assembled.

**SAVTA**    Safe and Vault Technicians Association.

**scalp**    A thin piece of metal that is usually crimped or spun onto the front of a cylinder. It determines the cylinder's finish and may also serve as the plug retainer.

**second generation duplicate**    A key reproduced from a first generation duplicate.

**security collar**    A protective cylinder collar.

**segmented follower**    A plug follower that is sliced into sections, which are introduced into the cylinder shell one at a time. It is typically used with profile cylinders.

**selective key system**    A key system in which every key has the capacity of being a master key. It is normally used for applications requiring a limited number of keys and extensive cross keying.

**selective master key**  An unassociated master key that can be made to operate any specific lock in the entire system, in addition to the regular master key and/or change key for the cylinder, without creating key interchange.

**sequence of progression**  The order in which bitting positions are progressed to obtain change key combinations.

**series circuit**  An electrical circuit that provides only one path for current flow.

**series wafer**  A type of disc tumbler used in certain binary type disc tumbler key-in-knob locks. Its presence requires that no cut be made in that position on the operating key(s).

**set-up key**  A key used to calibrate some types of key machines.

**set-up plug**  A type of loading tool shaped like a plug follower. It contains pin chambers and is used with a shove knife to load springs and top pins into a cylinder shell.

**seven-column progression**  A process wherein key bittings are obtained by using the cut possibilities in seven columns of the key bitting array.

**seven-pin master key**  A master key for all combinations obtained by progressing seven bitting positions.

**shackle**  The usually curved portion of a padlock that passes though a hasp and snaps into the padlock's body.

**shackle spring**  The spring inside the body of a padlock that allows the shackle to pop out of the body when in the unlocked position.

**shank**  The part of a bit key between the shoulder and the bow.

**shear line**  A location in a cylinder at which specific tumbler surfaces must be aligned, removing obstruction(s) that prevent the plug from moving.

**shell**  The part of the cylinder that surrounds the plug and that usually contains tumbler chambers corresponding to those in the plug.

**shim**  1. A thin piece of material used to unlock the cylinder plug from the shell by separating the pin tumblers at the shear line, one at a time. 2. To unlock a cylinder plug from its shell using a shim.

**shoulder**  Any key stop other than a tip stop.

**shouldered pin**  A bottom pin whose diameter is larger at the flat end to limit its penetration into a counter-bored chamber.

**shove knife**  A tool used with a set-up plug that pushes the springs and pin tumblers into the cylinder shell.

**shut-out key**  Usually used in hotel keying systems, a key that will make the lock inoperative to all other keys in the system, except the emergency master key, display key, and some types of shut-out keys.

**shut-out mode**  The state of a hotel function lockset that prevents operation by all keys except the emergency master key, display key, and some types of shut-out keys.

**sidebar**  A primary or secondary locking device in a cylinder. When locked, it extends along the plug beyond its circumference. It must enter gates in the tumblers in order to clear the shell and allow the plug to rotate.

**simplex key section**   A single independent key section that cannot be used in a multiplex key section.

**single-key section**   An individual key section that can be used in a multiplex key system.

**single-step progression**   A progression using a one increment difference between bittings of a given position.

**six-column progression**   A process wherein key bittings are obtained by using the cut possibilities in six columns of the key bitting array.

**six pin master key**   A master key for all combinations obtained by progressing six bitting positions.

**SKD**   Symbol for "single keyed," normally followed by a numerical designation in the standard key coding system—SKD1, SKD2, etc. It indicates that a cylinder or lock is not master keyed but is part of the keying system.

**skeleton key**   A warded lock key cut especially thin to bypass the wards in several warded locks so the locks can be opened. The term is sometimes mistakenly used when referring to any bit key.

**spacing**   The dimensions from the stop to the center of the first cut and/or to the centers of successive cuts.

**special application cylinder**   Any cylinder other than a mortise, rim, key-in-knob, or profile cylinder.

**split pin master keying**   A method of master keying a pin tumbler cylinder by installing master pins into one or more pin chambers.

**spool pin**   Usually a top pin that resembles a spool, typically used to increase pick resistance.

**spring cover**   A device for sealing one or more pin chambers.

**standard key coding system**   An industry standard and uniform method of designating all keys and/or cylinders in a master key system. The designation automatically indicates the exact function and keying level of each key and/or cylinder in the system, usually without further explanation.

**standard progression format**   A systematic method of listing and relating all change key combinations in a master key system. The listing is divided into segments known as blocks, horizontal groups, vertical groups, rows, and pages, for levels of control.

**step pin**   A spool or mushroom pin that has had a portion of its end machined to a smaller diameter than the opposite end. It is typically used as a top pin to improve pick resistance by some manufacturers of high-security cylinders.

**stepped tumbler**   A special (usually disc) tumbler used in master keying. It has multiple bearing surfaces for blades of different key sections.

**stop (of a key)**   The part of a key from which all cuts are indexed and which determines how far the key enters the keyway.

**strike**   The part of a locking arrangement that receives the bolt, latch, or fastener, when the lock is in the locked position. The strike (sometimes called a keeper) is usually recessed in a door frame.

**sub-master key (SMK)**   The master key level immediately below the master key in a system of six or more levels of keying.

**switch**    A device used for opening and closing a circuit.

**tailpiece**    An actuator attached to the rear of the cylinder, parallel to the plug, typically used on rim, key-in-knob, or special application cylinders.

**theoretical key changes**    The total possible number of different combinations available for a specific cylinder or lock mechanism.

**three-column progression**    A process wherein key bittings are obtained by using the cut possibilities in three columns of the key bitting array.

**three-pin master key**    A master key for all combinations obtained by progressing three bitting positions.

**thumb turn cylinder**    A cylinder with a turn knob rather than a keyway and tumbler mechanism.

**tip**    The portion of the key that enters the keyway first.

**tip stop**    A type of stop located at or near the tip of the key.

**tolerance**    The deviation allowed from a given dimension.

**top master key (TMK)**    The highest level master key in a master key system.

**top of blade**    The bitted edge of a single bitted key.

**top pin**    A cylindrical tumbler, usually flat on both ends, that is installed directly under the spring in the pin stack.

**torque wrench**    A device used to apply pressure on a cylinder while its tumblers are being manipulated by a pick. It is also used to turn the plug to the unlocked position after the lock has been picked.

**total position progression**    A process used to obtain key bittings in a master key system wherein bittings of change keys differ from those of the top master key in all bitting positions.

**transformer**    A device that transfers electrical energy from one circuit to another without direct connection between them.

**try-out key**    A manipulation key that is usually part of a set, used for a specific series, keyway, and/or brand of lock.

**tubular key**    A key with a tubular blade. The key cuts are made into the end of the blade, around the circumference.

**tubular key lock**    A type of lock with tumblers arranged in a circle, often used on vending machines and coin-operated washing machines.

**tumbler**    A movable obstruction of varying size and configuration in a lock or cylinder that makes direct contact with the key or another tumbler, and prevents an incorrect key or torquing device from activating the lock or other mechanism.

**tumbler spring**    Any spring that acts directly on a tumbler.

**two-column progression**    A process wherein key bittings are obtained by using the cut possibilities in two columns of the key bitting array.

**two-pin master key**    A master key for all combinations obtained by progressing two bitting positions.

**two-step progression**    A progression using a two increment difference between bittings of a given position.

**UL**    Underwriters Laboratories.

**unassociated change key**    A change key that is not related directly to a particular master key through the use of certain constant cuts.

**unassociated master key** A master key that does not have change keys related to its combination through the use of constant cuts.

**uncombinated** 1. Of or pertaining to a cylinder that is supplied without keys, tumblers and springs. 2. Of or pertaining to a lock, cylinder, or key in which the combination has not been set.

**uncontrolled cross keying** A condition in which two or more different keys under different higher level keys operate one cylinder by design—for example: XAA1 operated by AB, AB1. Note: This condition severely limits the security of the cylinder and the maximum expansion of the system, and often leads to key interchange.

**unidirectional cylinder** A cylinder whose key can turn in only one direction from the key pull position, often not making a complete rotation.

**Vac** ac volts.

**VATS** Vehicle Anti-Theft System.

**VATS decoder** A device designed for determining which VATS key blank to use to duplicate a VATS key or to make a VATS first key.

**VATS key** A key designed to operate a vehicle equipped with VATS.

**Vdc** dc volts.

**vehicle anti-theft system** An electromechanical system used in many General Motors vehicles to deter theft. Sometimes referred to as the PASS-Key system.

**vertical group master key (VGM)** The two pin master key for all combinations listed in all blocks in a line down a page in the standard progression format.

**visual key control (VKC)** A specification that all keys and the visible portion of the front of all lock cylinders be stamped with standard keying symbols.

**volt** A unit of measure for voltage.

**voltage** The force that pushes electrical current. Voltage (or electromotive force) is measured in volts.

**voltage drop** The change in voltage across an electrical component (such as a resistor) in a circuit.

**vom** Volt-ohm-milliammeter. A type of multitester.

**ward** A usually stationary obstruction in a lock or cylinder that prevents the entry and/or operation of an incorrect key.

**ward cut** A modification of a key that allows it to bypass a ward.

**watt** A unit of measure of electrical power.

**X** Symbol used in hardware schedules to indicate a cross keyed condition for a particular cylinder—for example, XAA2, X1X (but not AX7).

**zero bitted** Of or pertaining to a cylinder that is combinated to keys cut to the manufacturer's reference number "0" bitting.

# Index